Fuzzy Logic and its Application in Technology and Management

Fuzzy Logic and its Application in Technology and Management

Editors

Debjani Chakraborty

S. Nanda

D. Dutta Majumder

Narosa Publishing House

New Delhi Chennai Mumbai Kolkata

Editors
Debjani Chakraborty
Department of Mathematics, IIT Kharagpur, India

S. Nanda
Department of Mathematics, IIT Kharagpur
Presently, Vice Chancellor, North Orissa University, Orissa, India

D. Dutta Majumder
Professor Emeritus, Indian Statistical Institute (ISI) &
Director-Secretary, Institute of Cybernetics Systems & Information Technology (ICSIT), Kolkata, India

Reprint 2008

NAROSA PUBLISHING HOUSE PVT. LTD.

22 Delhi Medical Association Road, Daryaganj, New Delhi 110 002
35-36 Greams Road, Thousand Lights, Chennai 600 006
306 Shiv Centre, Sector 17, Vashi, Navi Mumbai 400 703
2F-2G Shivam Chambers, 53 Syed Amir Ali Avenue, Kolkata 700 019

www.narosa.com

Printed from the camera-ready copy provided by the Editors.

ISBN 978-81-7319-733-8

Published by N.K. Mehra for Narosa Publishing House Pvt. Ltd.,
22 Delhi Medical Association Road, Daryaganj, New Delhi 110 002

Printed in India

Preface

During last few decades, adoption of fuzzy logic in computation and realizing a large complex system has created tremendous sensation among the engineers / technologists. There are a few books on application of fuzzy logic in technology and fewer books on fuzzy logic in management separately. However, the application of fuzzy logic, bringing technology and management together, is still a dark continent to be discovered yet. Using the state-of-the-art and especially in the language of technologists and management researchers, the proposed book is an excellent blend of the fuzzy mathematics, information technology and operations management.

The edited volume contains the selected papers from the conference *Fuzzy Logic and its Application in Technology and Management* (FLATeM 04) held in Indian Institute of Technology Kharagpur, organized by jointly Department of Mathematics of IIT Kharagpur, Vinod Gupta School of Management of IIT Kharagpur and Indian Society of Fuzzy Mathematics and Information Processing (ISFUMIP), an 'Institutional member' of *International Fuzzy Systems Association* (IFSA).

We tried to keep all contributions in a specific format and tried our best to keep the volume free of typographical errors. The book is divided into five parts covering different aspects. First part includes twelve contributions on basic fuzzy / soft / rough set theory and analysis. Four papers are on fuzzy application of data mining and information processing and put in the next part. Part three contains six papers on fuzzy modeling and decision making. Part four includes different areas of fuzzy applications on technology and management consists of fourteen papers. Last part has four papers on fuzzy application in medical technology.

We take this opportunity to express our gratitude to all the persons / organizations who have given the financial supports. We cordially thank Mr. Vinod Gupta, Chairman & CEO, InfoUSA and the distinguished alumni of IIT Kharagpur. We give our sincere thanks to All India Council of Technical Education (AICTE, New Delhi, India) and Council of Scientific & Industrial Research (CSIR, New Delhi, India) for their generous financial support. We are grateful to Prof. S.K. Dube, the Director, IIT Kharagpur for his kind support and encouragement. We also thank all functionaries, the faculty members, employees and research scholars of IIT Kharagpur, especially Chandan Chakraborty. We must thank all the contributors and reviewers. We would like to convey our sincere thanks to all the members of ISFUMIP with special gratitude to Prof. Mihir Chakraborty, University of Calcutta, Kolkata,

Prof. N.R. Pal, ISI Kolkata and Prof. A.K. Srivastava, BHU, Varanasi. Lastly we are very much thankful to Atanu Sengupta for his generous help in all respect.

Debjani Chakraborty
S. Nanda
D. Dutta Majumder

Contents

Part 2: Fuzzy Data Mining and Information Processing

Part 3: Fuzzy Modeling and Decision Making

Part 4: Application of Fuzzy Logic in Technology & Management

Part 1

Fuzzy, Rough, Soft Set Theory and Analysis

Fuzzy Logic and its Application in Technology and Management
Editors: Debjani Chakraborty, S. Nanda and D. Dutta Majumder

A Partition of a Set Through a GIFS

M. Panigrahi and S. Nanda

Department of Mathematics, Indian Institute of Technology Kharagpur-721302
e-mail: {motilal, snanda}@maths.iitkgp.ernet.in

Abstract: The purpose of this paper is to give a partition of a non-empty crisp set X through a Generalized Intuitionistic Fuzzy Set (GIFS) A defined over X. We will introduce some new numbers through the partition and generate some GIFSs through these numbers and study their properties.

Key words: Intuitionistic fuzzy set (IFS), Generalized intuitionistic fuzzy set (GIFS).

1. Introduction

In 1965, in the classic paper [17], Zadeh introduced the concept of fuzzy subsets. Later many authors developed the subject and many researchers generalized the idea of fuzzy subsets in many different directions, like vague sets, rough sets, soft sets etc. After two decades, Atanassov [1, 2, 3] introduced the concept of intuitionistic fuzzy sets, which is a generalization of fuzzy subsets.

In fuzzy subset, only membership function is relevant. Let X be a nonempty set. Then a fuzzy subset A of X is nothing but a function $\mu_A : X \to [0,1]$, called a membership function. Whereas an Intuitionistic Fuzzy Set (IFS) A on X is an object of the from

$$A = \{\langle x, \mu_A(x), \nu_A(x)\rangle | \, x \in X\} \tag{1}$$

where the functions $\mu_A : X \to [0,1]$, and $\nu_A : X \to [0,1]$, satisfy the rule

$$\mu_A(x) + \nu_A(x) \leq 1 \qquad \forall x \in X \tag{2}$$

called the intuitionistic condition (IC).

After the introduction of IFS, many authors like Burillo & Bustince [7-11], Coker [12], Demicri [13], including Atanassov [1-6] have developed and applied the theory of IFSs.

Recently in 2002, Mondal & Samanta [15], further generalized the concept of IFSs to Generalized Intuitionistic Fuzzy Sets (GIFSs). Here instead of IC, if the functions μ_A and ν_A satisfy the following rule

$$\mu_A(x) \wedge \nu(x) \leq 0.5 \qquad \forall x \in X \tag{3}$$

called the Generalized Intuitionistic Condition(GIC), then A is said to be a GIFS on X.

For each $x \in X$, $\mu_A(x)$ and $\nu_A(x)$ are called the degree of membership and degree of non-membership, respectively of x to A.

Here we may note that an IFS is a GIFS but the converse is not true. We denote, the set of all GIFSs on X by $G(X)$ and the set of all IFSs on X by $F(X)$.

For the shake of compieteness here we write the following definitions.

Definition 1.1. For every two GIFSs (or IFSs [3]) A and B on X, we define

1. $A \subseteq B$ *iff* $(\forall x \in X)(\mu_A(x) \le \mu_B(x)$ *and* $\nu_A(x) \ge \nu_B(x))$,
2. $A = B$ *iff* $A \subseteq B$ *and* $B \subseteq A$,
3. $\overline{A} = \{\langle x, \nu_A(x), \mu_A(x)\rangle | x \in X\}$
4. $A \cap B = \{\langle x, \min(\mu_A(x), \mu_B(x)), \max(\nu_A(x), \nu_B(x))\rangle | x \in X\}$,
5. $A \cup B = \{\langle x, \max(\mu_A(x), \mu_B(x)), \min(\nu_A(x), \nu_B(x))\rangle | x \in X\}$.

We organize the paper as follows: Section 1 is Introduction and provides some preliminary concepts. In section 2 we have introduced the partition of a non-empty crisp set X through a GIFS A. We have established some results on the subsets, which are obtained through partition. Then we have defined some new numbers through a given partition. Through these numbers we have generated some sets and shown that they are indeed GIFSs. Here some relationships among these new sets are obtained.

2. A Partition on a Set

Definition 2.1. Let X be a non-empty set. Let A be a GIFS (or IFS) on X. Then we partition X into the following three subsets of X, defined as

Strong subset of $X = \breve{X}_A = \{x \in X : \mu_A(x) > 0.5\}$ (4)

Weak subset of $X = \hat{X}_A = \{x \in X : \nu_A(x) > 0.5\}$ (5)

Intermediate subset of $X = \overset{o}{X}_A = \{x \in X : \mu_A(x) \le 0.5, \nu_A(x) \le 0.5\}$ (6)

Clearly, $X = \breve{X}_A \dot{\cup} \hat{X}_A \dot{\cup} \overset{o}{X}_A$ (7)

where, $\dot{\cup}$ represents disjoint union in crisp set theory.

Here we will use the same symbols $\subseteq, \supseteq, =, \cap, \cup$ throughout with their usual meaning context wise, that is, for crisp sets, IFS, GIFS etc.

Theorem 2.2. For $A, B \in G(X)$ if $A \subseteq B$ then

(i) $\breve{X}_A \subseteq \breve{X}_B$ (ii) $\hat{X}_A \supseteq \hat{X}_B$ (iii) $\overset{o}{X}_A \subseteq \overset{o}{X}_B \dot{\cup} \breve{X}_B$

Proof: We have $A \subseteq B \Rightarrow \forall x \in X, \mu_A(x) \le \mu_B(x)$ and $\nu_A(x) \ge \nu_B(x)$.

(i) Now $x \in \breve{X}_A \Rightarrow \mu_A(x) > 0.5 \Rightarrow \mu_B(x) \geq \mu_A(x) > 0.5 \Rightarrow x \in \breve{X}_B$. Hence $\breve{X}_A \subseteq \breve{X}_B$.

(ii) Again $x \in \hat{X}_B \Rightarrow \nu_B(x) > 0.5 \Rightarrow \nu_A(x) \geq \nu_B(x) > 0.5 \Rightarrow x \in \hat{X}_A$. Hence $\hat{X}_A \supseteq \hat{X}_B$.

(iii) Again $x \in \overset{o}{X}_A \Rightarrow \mu_A(x) \leq 0.5$ *and* $\nu_A(x) \leq 0.5$. As $A \subseteq B$,

we have, $\nu_B(x) \leq \nu_A(x) \leq 0.5$. Now if $\mu_B(x) \leq 0.5$ then $x \in \overset{o}{X}_B$, otherwise $\mu_B(x) > 0.5 \Rightarrow x \in \breve{X}_B$. Hence $\overset{o}{X}_A \subseteq \overset{o}{X}_B \dot{\cup} \breve{X}_B$.

Theorem 2.3. For $A, B \in G(X)$, we have

(i) $\breve{X}_{A \cup B} = \breve{X}_A \cup \breve{X}_B$, (ii) $\hat{X}_{A \cup B} = \hat{X}_A \cap \hat{X}_B$, (iii) $\overset{o}{X}_{A \cup B} \subseteq \overset{o}{X}_A \cup \overset{o}{X}_B$.

Proof: (i) $x \in \breve{X}_{A \cup B} \Leftrightarrow \mu_{A \cup B}(x) = \max(\mu_A(x), \mu_B(x)) > 0.5$

$\Leftrightarrow$ either $\mu_A(x) > 0.5$ or $\mu_B(x) > 0.5$ $\Leftrightarrow$ either $x \in \breve{X}_A$ or $x \in \breve{X}_B$.

(ii) The proof is similar to (i).

(iii) $x \in \overset{o}{X}_{A \cup B} \Leftrightarrow \mu_{A \cup B}(x) \leq 0.5, \nu_{A \cup B}(x) \leq 0.5$

$\Rightarrow \max(\mu_A(x), \mu_B(x)) \leq 0.5$ and $\min(\nu_A(x), \nu_B(x)) \leq 0.5$

$\Rightarrow \{\mu_A(x) \leq 0.5$ and $\mu_B(x) \leq 0.5\}$ and $\{$either $\nu_A(x) \leq 0.5$ or $\nu_B(x) \leq 0.5\}$

$\Rightarrow$ either $x \in \overset{o}{X}_A$ *or* $x \in \overset{o}{X}_B \Rightarrow x \in \overset{o}{X}_A \cup \overset{o}{X}_B$.

Theorem 2.4. For $A, B \in G(X)$, we have

(i) $\breve{X}_{A \cap B} = \breve{X}_A \cap \breve{X}_B$, (ii) $\hat{X}_{A \cap B} = \hat{X}_A \cup \hat{X}_B$, (iii) $\overset{o}{X}_{A \cap B} \subseteq \overset{o}{X}_A \cup \overset{o}{X}_B$.

Proof: The proof is similar to the Theorem 2.3.

Theorem 2.5. For $A, B \in G(X)$, we have $\overset{o}{X}_{A \cap B} \cup \overset{o}{X}_{A \cup B} = \overset{o}{X}_A \cup \overset{o}{X}_B$.

Proof: By Theorem 2.3, and Theorem 2.4, we have $\overset{o}{X}_{A \cap B} \cup \overset{o}{X}_{A \cup B} \subseteq \overset{o}{X}_A \cup \overset{o}{X}_B$.

Conversely, let $x \in \overset{o}{X}_A \cup \overset{o}{X}_B$. Then without loss of generality, let $x \in \overset{o}{X}_A$ that is, $\{\mu_A(x) \leq 0.5$ *and* $\nu_A(x) \leq 0.5\}$. But as $B \in G(X)$, we have $\min(\mu_B(x), \nu_B(x)) \leq 0.5 \Rightarrow$ either $\mu_B(x) \leq 0.5$ or $\nu_B(x) \leq 0.5$

Therefore,

either$\{\mu_A(x) \le 0.5$ and $\nu_A(x) \le 0.5$ *and* $\mu_B(x) \le 0.5\}$

or $\{\mu_A(x) \le 0.5$ and $\nu_A(x) \le 0.5$ *and* $\nu_B(x) \le 0.5\}$

$\Rightarrow$ either $\{\max(\mu_A(x), \mu_B(x)) \le 0.5$ *and* $\min(\nu_A(x), \nu_B(x)) \le 0.5\}$

or $\{\min(\mu_A(x), \mu_B(x)) \le 0.5$ and $\max(\nu_A(x), \nu_B(x)) \le 0.5\}$

$\Rightarrow$ either $x \in \overset{o}{X}_{A\cup B}$ or $x \in \overset{o}{X}_{A\cap B}$

$\Rightarrow x \in \overset{o}{X}_{A\cup B} \cup \overset{o}{X}_{A\cap B}$

Hence, $\overset{o}{X}_A \subseteq \overset{o}{X}_{A\cap B} \cup \overset{o}{X}_{A\cup B}$. Similarly, $\overset{o}{X}_B \subseteq \overset{o}{X}_{A\cap B} \cup \overset{o}{X}_{A\cup B}$. Hence the proof.

Definition 2.6. For non-empty $\breve{X}_A$, $\hat{X}_A$ and $\overset{o}{X}_A$, we define the following numbers in [0, 1]:

$$\breve{K} = \max_{x\in\breve{X}_A} \mu_A(x),\quad \breve{L} = \min_{x\in\breve{X}_A} \nu_A(x),\quad \breve{k} = \min_{x\in\breve{X}_A} \mu_A(x),\quad \breve{l} = \max_{x\in\breve{X}_A} \nu_A(x),$$

$$\hat{K} = \max_{x\in\hat{X}_A} \mu_A(x),\quad \hat{L} = \min_{x\in\hat{X}_A} \nu_A(x),\quad \hat{k} = \min_{x\in\hat{X}_A} \mu_A(x),\quad \hat{l} = \max_{x\in\hat{X}_A} \nu_A(x),$$

$$\overset{o}{K} = \max_{x\in\overset{o}{X}_A} \mu_A(x),\quad \overset{o}{L} = \min_{x\in\overset{o}{X}_A} \nu_A(x),\quad \overset{o}{k} = \min_{x\in\overset{o}{X}_A} \mu_A(x),\quad \overset{o}{l} = \max_{x\in\overset{o}{X}_A} \nu_A(x).$$

From the definition of the above numbers we have the following observation.

Observation 2.7. (i) $\breve{K} \ge \hat{K},\ \breve{K} \ge \overset{o}{K},\ \breve{K} \ge \breve{k},\ \breve{K} \ge \overset{o}{k}$, (ii) $\hat{L} \ge \breve{L},\ \hat{L} \ge \overset{o}{L}$,

(iii) $\breve{k} \ge \hat{k},\ \breve{k} \ge \overset{o}{k}$, (iv) $\hat{l} \ge \breve{l},\ \hat{l} \ge \overset{o}{l},\ \hat{l} \ge \breve{L},\ \hat{l} \ge \hat{L},\ \hat{l} \ge \overset{o}{L}$.

Definition 2.8. For $A \in G(X)$ and non-empty $\breve{X}_A$, $\hat{X}_A$ and $\overset{o}{X}_A$, we define the following operators:

(i) $\breve{C}(A)=\{\langle x, \breve{K}, \breve{L}\rangle : x \in X\}$, (ii) $\hat{C}(A)=\{\langle x, \hat{K}, \hat{L}\rangle : x \in X\}$,

(iii) $\overset{o}{C}(A)=\{\langle x, \overset{o}{K}, \overset{o}{L}\rangle : x \in X\}$, (iv) $\breve{I}(A)=\{\langle x, \breve{k}, \breve{l}\rangle : x \in X\}$,

(v) $\hat{I}(A)=\{\langle x, \hat{k}, \hat{l}\rangle : x \in X\}$, (vi) $\overset{o}{I}(A)=\{\langle x, \overset{o}{k}, \overset{o}{l}\rangle : x \in X\}$.

Theorem 2.9. The operators defined in Definition 2.8 are all GIFSs.

Proof: We will only prove (i), all the remaining can be proved similarly.

Proof of (i): Since $\breve{K} = \max_{x \in \breve{X}_A} \mu_A(x) > 0.5$ (as $\mu_A(x) > 0.5 \ \forall x \in \breve{X}_A$),

and $\breve{L} = \min_{x \in \breve{X}_A} \nu_A(x) \leq 0.5$ (as $\nu_A(x) \leq 0.5 \ \forall x \in \breve{X}_A$). So GIC is satisfied for $\breve{C}(A)$. Thus $\breve{C}(A) \in G(X)$.

Theorem 2.10. For $A \in G(X)$ and non-empty $\breve{X}_A$, $\hat{X}_A$ and $\overset{o}{X}_A$, we have

(i) $\breve{C}(A) \supseteq \hat{C}(A)$, (ii) $\breve{I}(A) \supseteq \hat{I}(A)$, (iii) $\breve{C}(A) \supseteq \breve{I}(A)$,

(iv) $\hat{C}(A) \supseteq \hat{I}(A)$, (v) $\overset{o}{C}(A) \supseteq \overset{o}{I}(A)$.

Proof: We will prove (i), rest all can be proved similarly. From Observation 2.7, we have $\breve{K} \geq \hat{K}$ and $\hat{L} \geq \breve{L}$ therefore, $\breve{C}(A) \supseteq \hat{C}(A)$.

Definition 2.11. For $A \in G(X)$ and non-empty $\breve{X}_A$, $\hat{X}_A$ and $\overset{o}{X}_A$, we define the following operators:

(i) $\overset{\vee o}{C}(A) = \{\langle x, \breve{K}, \overset{o}{L} \rangle : x \in X\}$, (ii) $\overset{o \vee}{C}(A) = \{\langle x, \overset{o}{K}, \breve{L} \rangle : x \in X\}$,

(iii) $\overset{\wedge o}{C}(A) = \{\langle x, \hat{K}, \overset{o}{L} \rangle : x \in X\}$, (iv) $\overset{o \wedge}{C}(A) = \{\langle x, \overset{o}{K}, \hat{L} \rangle : x \in X\}$,

(v) $\overset{\vee o}{I}(A) = \{\langle x, \breve{k}, \overset{o}{l} \rangle : x \in X\}$, (vi) $\overset{o \vee}{I}(A) = \{\langle x, \overset{o}{k}, \breve{l} \rangle : x \in X\}$,

(vii) $\overset{\wedge o}{I}(A) = \{\langle x, \hat{k}, \overset{o}{l} \rangle : x \in X\}$, (viii) $\overset{o \wedge}{I}(A) = \{\langle x, \overset{o}{k}, \hat{l} \rangle : x \in X\}$.

Theorem 2.12. The operators defined in Definition 2.11 are all GIFSs.
Proof: The proof is straightforward.

Theorem 2.13. For $A \in G(X)$ and non-empty $\breve{X}_A$, $\hat{X}_A$ and $\overset{o}{X}_A$, we have the following relations:

(i) $\overset{o \vee}{C}(A) \supseteq \overset{o \wedge}{C}(A)$, (ii) $\overset{o \wedge}{C}(A) \subseteq \overset{\vee o}{C}(A)$, (iii) $\overset{\vee o}{I}(A) \supseteq \overset{o \wedge}{I}(A)$, (iv) $\overset{\vee o}{I}(A) \supseteq \overset{\wedge o}{I}(A)$,

(v) $\overset{\wedge o}{C}(A) \subseteq \overset{\vee o}{C}(A)$, (vi) $\overset{o \vee}{I}(A) \supseteq \overset{o \wedge}{I}(A)$, (vii) $\overset{o \wedge}{I}(A) \subseteq \overset{o \wedge}{C}(A)$, (viii) $\overset{\wedge o}{I}(A) \subseteq \overset{\wedge o}{C}(A)$,

(ix) $\overset{\vee o}{I}(A) \subseteq \overset{\vee o}{C}(A)$, (x) $\overset{o \vee}{I}(A) \subseteq \overset{o \vee}{C}(A)$.

Proof: We will only prove (i), rest all can be proved similarly. From observation 2.8, we have $\overset{o}{K} = \overset{o}{K}\ and\ \breve{L} \leq \hat{L}$, therefore, $\overset{o\vee}{C}(A) \supseteq \overset{o\wedge}{C}(A)$.

Acknowledgement

The first author thanks G. H. Patel College of Engg. and Tech., Vallabh Vidyanagar for the grant of leave, the QIP for the financial assistance to carry out the research Programme and to IIT Kharagpur for providing the research facilities.

Reference

1. K.T. Atanassov, & S. Stoeva (1983), Intuitionistic Fuzzy Sets, in: *Polish Symp. on Interval & Fuzzy Mathematics*, Poznan (Aug.): 23-26.
2. K.T. Atanassov (1984), Intuitionistic Fuzzy Sets, in: V. Sgurev, Ed., *VII ITKR's Session*, Sofia, June1984 (Central S&T Library, Bulgeria Academy of Sciences).
3. K.T. Atanassov(1986), Intuitionistic FuzzySets, *Fuzzy Sets and Systems* 20: 87-96.
4. K.T. Atanassov (1989), More on Intuitionistic Fuzzy Sets, *Fuzzy Sets and Systems* 33(1): 37-46.
5. K.T. Atanassov (1994), New Operations Defined over The Intuitionistic Fuzzy Sets, *Fuzzy Sets and Systems* 61: 137-142.
6. K.T. Atanassov (2000), Two Theorems for Intuitionistic Fuzzy Sets, *Fuzzy Sets and Systems* 110: 267-269.
7. P. Burillo & H. Bustince (1995), Two Operators on Interval-valued Intuitionistic Fuzzy Sets: Part-II, *C. R. Acad. Bulg. Sci.* 48(1): 17-20.
8. P. Burillo & H. Bustince (1996), Entropy on Intuitionistic Fuzzy Sets and Interval-valued Fuzzy Sets Part-II, *Fuzzy Sets and Systems* 78: 305-316.
9. H. Bustince (1994), *Conjuntos Intuicionistas e Intervalo valorados Difusos: Propiedades y Construccion, Relaciones Intuicionistas Fuzzy*, Thesis, Universidad Publica de Vavarra.
10. H. Bustince & P. Burillo (1995), Correlation of Interval Valued Intuitionistic Fuzzy Sets, *Fuzzy Sets and Systems* 74: 237-244.
11. H. Bustince & P. Burillo (1995), Intuitionistic Fuzzy Relations (Part 1), *Mathware and Soft Computing* 2: 5-38.
12. D. Coker (2003), An Introduction to Intuitionistic Fuzzy Topological Spaces, *Fuzzy Sets and Systems* 131:193-209.
13. D. Coker & M. Demirci (1996), An Introduction to Intuitionistic Fuzzy Topological Spaces in Sostak's Sense, *Busefal* 67: 67-76.
14. Tapas Kumar Mondal & S. K. Samanta (2002), On Intuitionistic Gradation of Openness, *Fuzzy Sets and Systems* 131: 323-336.
15. Tapas Kumar Mondal & S. K. Samanta (2002), Generalized Intuitionistic Fuzzy Sets, *The Journal of Fuzzy Mathematics* 10 (4): 839-861.
16. Guo-jun Wang & Ying-Yu He (2000), Intuitionistic Fuzzy Sets and L-fuzzy Sets, *Fuzzy Sets and Systems* 110: 271-274.
17. L. A. Zadeh (1965), Fuzzy Sets, *Inform. and Control* 8: 338-353.

Fuzzy Logic and its Application in Technology and Management
Editors: Debjani Chakraborty, S. Nanda and D. Dutta Majumder

Fuzzy Logic on Affine CL-Algebra

Jayanta Sen[1] *and Mihir Kumar Chakraborty*[2]

[1]Department of Mathematics, Haldia Government College, Debhog, Midnapore
E-mail: jsen_01@yahoo.co.in
[2]Department of Pure Mathematics, University of Calcutta, 35 Ballygaunge Circular Road, Calcutta 700 019
E-mail: mihirc99@cal3.vsnl.net.in

Abstract: A fuzzy set theory on affine classical linear algebra is developed which is a generalisation of the existing bold fuzzy set theory. A logic (sequent calculus) for this fuzzy set theory is presented.

Keywords: Fuzzy logic, MV-algebra, Linear logic.

1. Introduction

Since 1965, after the publication of Zadeh's paper [12], fuzzy set theory has established itself as a good mathematical framework to deal with vagueness or non-probabilistic uncertainty. During the past four decades, the theory has undergone enormous development in various directions both in theory and applications. In spite of all these achievements, the foundational aspects of the theory are not yet fully settled. In this paper we shall carry out some investigations in this regard and shall bring into focus a special type of fuzzy sets that have so far escaped proper attention. We shall also present a logic for this kind of fuzzy sets.

A fuzzy set is a function from the universe of discourse to some "suitable" algebraic structure. In Zadeh's original paper the structure was taken to be the set [0,1] with '*min*', '*max*' and '*1-x*' as the basic operations. These are the algebraic counterparts of 'and', 'or' and 'not' of the corresponding logic. Subsequently, more general sets of operations have been proposed from various 'practical' necessities. These operations were later formalised and given the names: '*t*-norm', '*s*-norm' and 'generalized negation'. A '*t*-norm' $*$ is a mapping from $[0,1] \times [0,1] \rightarrow [0,1]$ satisfying

(i) commutativity and associativity
(ii) $x \le y \Rightarrow x * z \le y * z$
(iii) $1 * x = x$, $0 * x = 0$

An *s*-norm $\oplus$ is a similar mapping satisfying the conditions

(i′) commutativity and associativity
(ii′) $x \le y \Rightarrow x \oplus z \le y \oplus z$

(iii′) $1 \oplus x = 1,\ 0 \oplus x = x$.

A generalized negation ~ is a mapping $[0,1] \rightarrow [0,1]$ such that

(iv) $x \leq y \Rightarrow \sim y \leq x$

(v) $\sim\sim x = x$

(vi) $\sim 0 = 1,\ \sim 1 = 0$.

For a discussion of *t-norm*, *s-norm* and generalized negations one may be referred to [6,8]. '*min*' and '*max*' are the boundaries of the whole set of t-norms and s-norms respectively in the sense that

$$x * y \leq min(x,y) \leq max(x,y) \leq x \oplus y.$$

A major part of the (*t*-norm, negation) pairs fall in a class of algebraic structures called many-valued algebras (MV-al∩gebras), originally proposed by Chang in [3] and brought to the limelight in the fuzzy set context by Sessa, Di Nola, Mundici, Belluce, Hajek, Novak and others [1,5,7,9,]. These authors have studied fuzzy sets where the membership values are taken from an arbitrary MV-algebra instead of the particular ones on [0,1]. It is almost claimed that MV-algebras are 'the' structures suitable for a 'reasonable' fuzzy set theory (the so called Bold fuzzy sets).

We show in this paper that the above claim is not true. There are other, more general algebraic structures viz., affine classical linear algebra (ACL-algebra) that can be another structure on which an interesting fuzzy set theory may be developed. This class of structures includes MV-algebras as a proper subclass. Quite a number of *t*-norms and negation operations on [0,1] in the existing literature constitute ACL-algebras but they are not MV-algebras.

***Definition 1*:** $(X, \cap, \cup, \sim, *, 1)$ is an *affine classical linear algebra* (ACL-algebra) [10,11] if and only if

(i) $(X, \cap, \cup, 1)$ is a lattice with topmost 1 (let '$\leq$' be the lattice-ordering).

(ii) $(X, *, 1)$ is a commutative monoid with unit 1.

(iii) if $x \leq y$ then $\sim y \leq \sim x$ and $x * z \leq y * z$, for all $x, y, z \in X$.

(iv) $x * y \leq z$ if and only if $x \leq \sim(y * \sim z)$, for all $x, y, z \in X$.

(v) $\sim\sim x = x$.

An MV-algebra [4] is an ACL-algebra with the property

i) $\sim(\sim(x * \sim y) * \sim y) = \sim(\sim(y * \sim x) * \sim x)$

Because of this property, the lattice structure in the MV-algebra is definable in terms of negation (~) and the monoidal operator. In fact, $x \cup y = \sim(\sim(x * \sim y) * \sim y)$ in an MV-algebra.

Now we shall give examples of ACL-algebras which are not MV-algebras.

***Example 1*:** Let $L = \{0, a, b, 1\}$. Lattice-ordering, $*$ and ~ tables are the following

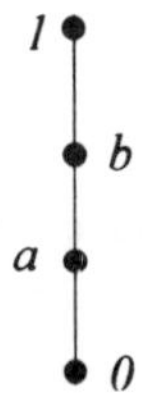

*	0	A	b	1
0	0	0	0	0
a	0	0	0	a
b	0	0	b	b
1	0	A	b	1

	0	a	b	1
~	1	b	a	0

Clearly, L is an ACL-algebra

But $\sim(\sim(b{*}\sim a) {*}\sim a) = \sim(\sim(b{*}b){*}b) = \sim(\sim b{*}b) = \sim(a{*}b) = \sim 0 = 1$

and $\sim(\sim(a{*}\sim b){*}\sim b) = \sim(\sim(a{*}a){*}a) = \sim(\sim 0{*}a) = \sim(1{*}a) = \sim a = b$.

So, this is not an MV-algebra.

Example 2: In [0,1], let $a * b = 0$, if $a + b \leq 1$

$= min\ \{a,b\}$, otherwise

and $\sim a = 1 - a$.

Here $*$ is a t-norm [6]. It can be shown that $([0,1], \cap, \cup, \sim, *, 1)$ is an ACL-algebra but $\left(\frac{1}{2} \to \frac{5}{8}\right) \to \frac{5}{8} = 1 \to \frac{5}{8} = \frac{5}{8}$ and $\left(\frac{5}{8} \to \frac{1}{2}\right) \to \frac{1}{2} = \frac{1}{2} \to \frac{1}{2} = 1$, where $a \to b = \sim(a{*}\sim b)$. So this is not an MV-algebra.

A few properties of ACL-algebra are the following

(i) $x*(y \cup z) = (x* y) \cup (x* z)$

(ii) $x*y \leq x \wedge y \leq x \cap y \leq x \leq x \cup y \leq x \vee y \leq x + y$, where $x \vee y = \sim(\sim(x{*}\sim y){*}\sim y)$, $x \wedge y = \sim(\sim(x \vee \sim y)$ and $x + y = \sim(\sim x{*}\sim y)$.

The first property is also shared by an MV-algebra. However, the following properties of MV-algebra is generally not shared by an ACL-algebra.

i) $x{*}y = x$ and $y \cup \sim x = 1$ are equivalent

ii) if $x + z = y + z$, $x \leq \sim z$ and $y \leq \sim z$ then $x = y$.

Note that, in general, $x \vee y \neq y \vee x$ in an ACL-algebra. It is shown in the above examples.

An ACL-algebra and an MV-algebra have many similarities. Both possess two binary operations $*$ and 6, the first giving the monoid and the second the lattice, both possess the negation ~. But whereas in the first, the lattice is independent of the monoid only being linked through some axioms, in the second, the ordering is totally defined by the monoidal product and the negation. In what follows we shall characterize all the MV-algebra embedded in an ACL-algebra in the following sense.

Definition 2: A subset of an ACL-algebra is an *embedded MV-algebra* if and only if it is an MV-algebra relative to the same operations $*$, ~ (and hence +) and has the same ordering as the ACL-algebra.

For a discussion on similar embedding, see [2].

Theorem 1: $(A, +, *, \sim, 0, 1)$ is an MV-algebra embedded in an ACL-algebra $(X, \cap, \cup, \sim, *, 1)$ if and only if

i) $A \subseteq X$.
ii) *A is* closed relative to $*$ *and* $\sim$.
iii) $x \vee y = y \vee x$ for all $x, y \in A$.

Now, we shall present a sequent calculus AMALL for ACL-algebra.

The language consists of the alphabet: variables p_i; logical symbols $\sim$, $*$, $\cup$. The axioms and the rules are the following

Ax $\quad A \Rightarrow A$ $\qquad$ **Cut** $\quad \dfrac{\Gamma, A \Rightarrow \Delta \quad \Gamma' \Rightarrow A, \Delta}{\Gamma, \Gamma' \Rightarrow \Delta, \Delta}$

LW $\quad \dfrac{\Gamma \Rightarrow \Delta}{\Gamma, A \Rightarrow \Delta}$

L~ $\quad \dfrac{\Gamma \Rightarrow A, \Delta}{\Gamma, \sim A \Rightarrow \Delta}$ $\qquad$ **R~** $\quad \dfrac{\Gamma, A \Rightarrow \Delta}{\Gamma \Rightarrow \sim A, \Delta}$

L$*$ $\quad \dfrac{\Gamma, A, B \Rightarrow \Delta}{\Gamma, A * B \Rightarrow \Delta}$ $\qquad$ **R$*$** $\quad \dfrac{\Gamma \Rightarrow A, \Delta \quad \Gamma' \Rightarrow B, \Delta'}{\Gamma, \Gamma' \Rightarrow A * B, \Delta, \Delta'}$

L$\cup$ $\quad \dfrac{\Gamma, A \Rightarrow \Delta \quad \Gamma, B \Rightarrow \Delta}{\Gamma, A \cup B \Rightarrow \Delta}$ $\qquad$ **R$\cup$** $\quad \dfrac{\Gamma \Rightarrow A, \Delta}{\Gamma \Rightarrow A \cup B, \Delta} \quad \dfrac{\Gamma \Rightarrow B, \Delta}{\Gamma \Rightarrow A \cup B, \Delta}$

In this calculus, Γ, Δ, Γ' and Δ' are finite multisets of formulae (possible empty) and A, B are formulae.

Definition 3: An *ACL-model* $((X, \cap, \cup, \sim, *, 1), [\![\ ]\!])$ is an ACL-algebra with a valuation $[\![\ ]\!]$ assigning values in X to each propositional variable p_i.

The valuation $[\![\ ]\!]$ is extended by $\sim[\![A]\!] = [\![\sim A]\!]$, $[\![A]\!] * [\![B]\!] = [\![A * B]\!]$, $[\![A]\!] \cap [\![B]\!] = [\![A \cap B]\!]$ and $[\![A]\!] \cup [\![B]\!] = [\![A \cup B]\!]$. Also $[\![1]\!] = 1$.

A sequent $A_1, \ldots, A_m \Rightarrow B_1, \ldots, B_n$ is valid if and only if $[\![A_1]\!] * \ldots * [\![A_m]\!] \le [\![B_1]\!] + \ldots + [\![B_n]\!]$ in an ACL-model, where + is defined as $x + y = \sim(\sim x * \sim y)$.

$\Rightarrow B$ is valid if and only if $[\![1]\!] \le [\![B]\!]$.

$A \Rightarrow$ is valid if and only if $[\![A]\!] \le [\![\sim 1]\!]$.

Theorem 2 (Soundness): If $\Gamma \Rightarrow \Delta$ is derivable in the AMALL, $\Gamma \Rightarrow \Delta$ is valid in every ACL-model.

Theorem 3 (Completeness): If $\Gamma \Rightarrow \Delta$ is valid in every ACL-model, it is derivable in AMALL.

Just like fuzzy set theories on MV-algebras, it is now possible to develop a similar theory on ACL-algebras. We shall here develop a fuzzy system on the examples of ACL-algebra given above.

An L-fuzzy subset $\tilde{F}$ is a mapping from some universe X to L. We define two kinds of intersection $\cap$, $\otimes$, two kinds of union $\cup$, $\oplus$, and a complement $(^c)$ by

$\tilde{A}\theta\tilde{B}(x) = \tilde{A}(x)\eta\tilde{B}(x)$, where $\theta \in \{\cap, \otimes, \cup, \oplus\}$ and $\eta \in \{\cap, *, \cup, +\}$ $(\tilde{A}^c)(x) = \sim \tilde{A}(x)$. $\tilde{A} \subseteq \tilde{B}$ if and only if $\tilde{A}(x) \leq \tilde{B}(x)$ for all $x \in X$.

Two designated fuzzy subsets are $\tilde{\phi}, \tilde{U}$ by $\tilde{\phi}(x) = 0$ and $\tilde{U}(x) = 1$ for all $x \in X$.
Some properties of L-fuzzy subsets are

1. $\tilde{A} \otimes \tilde{A} \subseteq \tilde{A} \subseteq \tilde{A} \oplus \tilde{A}$
2. $\tilde{A} \otimes \tilde{A}^c = \tilde{\phi}, \quad \tilde{A} \oplus \tilde{A}^c = \tilde{U}$
3. $(\tilde{A} \otimes \tilde{B})^c = \tilde{A}^c \oplus \tilde{B}^c, \quad (\tilde{A} \cap \tilde{B})^c = \tilde{A}^c \cup \tilde{B}^c$
4. $((\tilde{A})^c)^c = \tilde{A}$
5. $\tilde{A} \subseteq \tilde{B}$ if and only if $\tilde{B}^c \subseteq \tilde{A}^c$
6. commutativity, associativity with respect to all binary operators hold
7. $\oplus, \cup$ are distributive over $\otimes$
8. $\tilde{A} \oplus \tilde{\phi} = \tilde{A} = \tilde{A} \otimes \tilde{U} = \tilde{A} \cup \tilde{\phi} = \tilde{A} \cap \tilde{U}$
9. $\tilde{\phi}^c = \tilde{U}, \quad \tilde{U}^c = \tilde{\phi}$
10. $\tilde{A} \otimes \tilde{B} \subseteq \tilde{A} \cap \tilde{B} \subseteq \tilde{A} \subseteq \tilde{A} \cup \tilde{B} \subseteq \tilde{A} \oplus \tilde{B}$
11. $\tilde{A} \cup \tilde{A} = \tilde{A} = \tilde{A} \cap \tilde{A}$.

This fuzzy set theory does not have the following properties which hold in fuzzy set theory developed on MV-algebra.

1. $\tilde{A} \otimes \tilde{B} = \tilde{A}$ and $\tilde{B} \cup \tilde{A}^c = \tilde{U}$ are equivalent.
2. If $\tilde{A} \oplus \tilde{C} = \tilde{B} \oplus \tilde{C}$, $\tilde{A} \subseteq \tilde{C}^c$ and $\tilde{B} \subseteq \tilde{C}^c$ then $\tilde{A} = \tilde{B}$.
3. $\left((\tilde{A} \otimes \tilde{B}^c)^c \otimes \tilde{B}^c\right)^c = \left((\tilde{B} \otimes \tilde{A}^c)^c \otimes \tilde{A}^c\right)^c$

Conclusion

Further algebraic studies of ACL-algebras is pending. Besides, some specific applications that endorse incorporation of this algebra as the truth set should also be investigated.

References

1. L. P. Belluce, A. Di Nola & S. Sessa (1991), Triangular norms, MV-algebras and bold fuzzy set theory, *Math Japonica*, 36: 481-487.
2. M. K. Chakraborty & J. Sen (1991), MV-algebras embedded in a CL-algebra, *Int J Approximate Reasoning*, 18: 217-229.
3. C. C. Chang (1958), Algebraic analysis of many-valued logics, *Trans Am Math Soc*, 88: 467-490.
4. R. L. O. Cignoli, I. M. L. D'Ottaviano & D. Mundici (2000), *Algebraic foundations of many-valued reasoning* : Kluwer Academic Publishers.

5. Di Nola & S. Sessa (1995), On MV-algebras of continuous functions, in : *Non-classical logics and their applications to fuzzy subsets*, U. Höhle and E. P. Klement, eds., Kluwer Academic Publishers, 1995, pp. 23-32.
6. J. Fodor, R. R. Yager (2000), Fuzzy set-theoretic operators and quantifiers, in: *Fundamentals of fuzzy sets*. D. Dubois and H. Prade, eds., Kluwer Academic Publishers,.
7. P. Hájek (1998), *Metamathematics of fuzzy logic*, Kluwer Academic Publishers, (Trends in Logic, Studia Logica Library).
8. G. J. Klir & B. Yuan (1997), *Fuzzy sets and fuzzy logic: Theory and applications*, Prentice-Hall of India Pvt. Ltd.
9. V. Novák (1986), *Fuzzy sets and their applications*, Adam Hilger.
10. J. Sen (2000), *Some embeddings in linear logic and related issues*: Ph.D. dissertation, University of Calcutta.
11. S. Troelstra (1992), *Lectures on linear logic*, 29, CSLI, Stanford.
12. L. A. Zadeh (1965), Fuzzy sets, *Information and Control*, 8: 338-353.

Fuzzy Logic and its Application in Technology and Management
Editors: Debjani Chakraborty, S. Nanda and D. Dutta Majumder

Vagueness, Sorites and Graded Consequence

Sanjukta Basu[1] *and Mihir K. Chakraborty*[2]

[1]City College, Raja Rammohan Sarani, Kolkata 700 009
[2]Department of Pure Mathematics, University College of Sciences
University of Calcutta, 35 Ballygaunge Circular Road, Kolkata 700 019

Abstract: At the heart of the problems arising from the use of vague predicates like 'heap', 'bald', 'red' etc., is the problem about Sorites paradox. Sorites paradox is the name given to a type of paradoxical arguments, which arise due to lack of sharp boundary between applicability and non-applicability of the predicates involved in these arguments. This paradox brings to the surface the untenability of classical logic in the context of vagueness. It'll be argued here that the notion of graded consequence [2] is best suited to solve the paradox.

1. Introduction

In crisp context we never call a set consisting of a formula 'α' and its negation 'not-α' to be consistent. Such sets are absolutely inconsistent from the bivalent standpoint. However within multivalence, for some 'α', both 'α' and 'not-α' can be assigned some values, none of which cancels the other. Take, for example the sentence "this patch of colour is red" where "this" refers to an instance of the predicate 'red', of which neither can it be determinately said that it is red, nor can it be determinately said that it is not red. To assign any of the two classical truth-values to such a sentence and its reverse value to the negation, would be counter-intuitive. In a multiple truth-value set-up, such sentences and their negations can be assigned some value in between 0 and 1. In that case a set whose elements include both a sentence like the above one and its negation may not be absolutely inconsistent, rather may have a degree of consistency, which, though not 1, is not 0 also. It is well known that in crisp logic there is a close connection between the notion of consistency and consequence. If an argument is logically valid, then the set consisting of all the premises and the negation of the conclusion of the argument is inconsistent; and conversely if a set X of well-formed formulas together with the negation of a well-formed formula α is inconsistent, then α is a logical consequence of X. A similar link can be expected between these two notions in multivalent context also. In fact, equivalence between the notion of graded consequence and the notion of graded inconsistency has been established in [3]. Thus, when the set $X \cup \{\text{not-}\alpha\}$ is inconsistent to some degree q, $0 < q < 1$, then instead of calling the argument $X \setminus \therefore \alpha$ invalid, one could assign it some degree of validity which though less than 1, may still be significantly high, in proportion as q is high. In this way degrees allowed to consistency can be further extended to the notion of consequence also, giving rise to the notion of graded consequence.

It will be argued here that the notion of graded consequence is best suited for modelling such human reasoning as are involved in the context of vagueness, especially in Sorites type arguments.

2. Graded consequence

Formally speaking graded consequence is a fuzzy relation from the power set $P(F)$ of a set of well-formed formulas F to a non-empty set L with suitable structure, when this relation satisfies certain conditions *GI, GII, GIII* to be given later [2,3]. We take $L \equiv < L, \leq, *, \Rightarrow, 0, 1>$ to be a *residuated* lattice, where '0' and '1' are the least and the greatest elements of the lattice $< L, \leq >$ respectively; '*' and '⇒' are binary operations on L, satisfying the properties:

L-(i)	$a*b = b*a$
L-(ii)	$a*(b*c) = (a*b)*c$
L-(iii)	$1*a = a$
L-(iv)	$a*b \leq c$ if and only if $a \leq b \Rightarrow c$, for $a, b, c \in L$.

A special case of L is the set [0, 1] with the usual ordering and suitably defined binary operations in it.

The conditions to be satisfied by a graded consequence relation as given below are generalisations of the conditions defining a consequence relation in crisp context [7]:

G-(I):	If $\alpha \in X$, $gr(X \vdash \alpha) = 1$, (Overlap),
G-(II):	If $X \subseteq Y$, $gr(X \vdash \alpha) \leq gr(Y \vdash \alpha)$, (Dilution or Monotonicity),
G-(III):	$gr(X \vdash \alpha) \geq gr(X \cup Y \vdash \alpha) * \inf_{\beta \in Y} gr(X \vdash \beta)$, (Cut)

Where $X, Y \subseteq F$ and $\alpha, \beta \in F$.

As such, grade of a consequence, notationally $gr(X \vdash \alpha)$ which is an element of L, is the grade of *derivability* measuring the strength of logical relatedness of a conclusion α to a given set X of premises. This grade is not any value assigned to the conclusion indicating the degree to which the *conclusion* as derived from the premises *can be true*.

A justification for introducing grade of derivability can be obtained from the semantic consideration in the situation of multiple-valuedness. A situation or a state-of-affairs in multivalent context is a function T_i from the set of well-formed formulas F to the truth set L. Given a collection of situations, namely $\{T_i\}_{i \in I}$, the notion of graded semantic consequence ($\models$) can be formulated as:

$$gr(X \models \alpha) = \inf_{i \in I} [\inf_{\beta \in X} T_i(\beta) \Rightarrow T_i(\alpha)].$$

This equation reduces in two-valued situation to the following formulation of Shoesmith and Smiley [10]: given a set of states-of-affairs $\{T_i\}_{i \in I}$, α is a semantic consequence from the premise set X, if and only if, for any state-of-affairs T_i, every member of X is true in T_i implies α is true in T_i. Thus the above formulation of graded semantic consequence comes naturally as an extension of the classical notion of semantic consequence in multivalent context.

3. Vagueness and Sorites Paradox

At the heart of the problems arising out of vagueness is the problem regarding Sorites paradox. The Sorites is a type of arguments which proceed in a chain of deductions, involving predicates like 'heap', 'bald', 'tall', 'old', 'red', 'orange' etc. which lack sharp boundaries between their application and non-application. Centuries back Eubulides, the Megarian thinker, observed that a chain of apparently valid deductions setting off with evidently true premises ultimately leads to patently false conclusion. The paradox as originally formulated by Eubulides is known as the paradox of heap ('Soros' is the Greek word for heap). Suppose from a heap of wheats one grain of wheat is taken away. Obviously it remains a heap. If another grain is taken away, there still is a heap, and so on. In this way at the end one would arrive at an absurd consequence that zero number of grain makes a heap. Extension of any such predicate can be arranged in a series, each member of which is indiscriminable from its immediate successor and the latter involves insignificantly small change from its immediate predecessor. Because these predicates tolerate insignificantly small changes, one cannot pin point exactly at which point in the series, the predicate ceases to apply. Such predicates being essentially vague it is in fact meaningless to point out the exact location of the cross over. Nevertheless, indiscriminability is a relation which becomes non-transitive in a long chain. This actually is the source of the paradox. One popular form in which Sorites argument is presented in the literatures is conditional form, where the argument proceeds by a series of conditionals and repeated application of Modus Ponens:

F_{a_n}

If F_{a_n} then $F_{a_{n-1}}$

If $F_{a_{n-1}}$ then $F_{a_{n-2}}$

$\rightarrow$

If $F_{a_{n-9999}}$ then $F_{a_{n-10,000}}$ $(= F_{a_m})$

$\therefore F_{a_m}$

Here 'F' is a vague predicate, say "... is a heap"; n is an arbitrarily large number $\geq$ 10,000 and m is either 0 or is a small number. For each i, $m \leq i \leq n$, a_i is the set containing i-number of grains. Then 'F_{a_i}' means the set containing i grains is a heap.

Literatures on vagueness abound in discussions regarding Sorites paradox and its solutions [4,5,6,8,9,11,12,13,14,15]. No logic of vagueness can earn any reckoning without addressing the Sorites paradox. Some of the approaches to Sorites paradox consider the classical reasoning process behind the Sorites to be valid [4,5,11,12,13,14], although considering some of the conditional premises to be either false [11,13,14], truth-value less [5] or indeterminate [12]. But thus considering one of the premises to be not true amounts to admitting the possibility of drawing a sharp line between the positive and the negative instances of the vague predicate involved at the object level or of the truth predicate at the meta level.

A radically different approach is taken by fuzzy logicians Zadeh [15], Goguen [8], Gains [6] and some others within a semantics that admits a continuous and infinite truth value set. According to this approach repeated application of Modus Ponens on conditional premises, truth values of which are not necessarily 1, causes a gradual decrease in the truth value of the conclusion derived at each successive stage in a chain of deductions; and the conclusion receives a very a very low value at the end of the computation.

Approaching Sorites with the notion of graded consequence harps on showing that as the chain of deductions become longer and longer, strength of derivability at each successive stage of the argument, diminishes bit by bit, turning the whole argument process less and less reliable. To see how this is so, let us first define the notion of derivation in a theory of graded consequence.

Derivation proceeds from given premises and/or axioms to new formulas with the help of certain rules of inference. The underlying consequence relation of a rule of inference which is a one-step deduction, must have a strength which may not be 1, depending on the degrees of vagueness of the premises and the conclusion involved in it. This grade of an inference rule R_j is given by:

$$\inf_{X,\alpha} [\mathrm{gr}(X \models \alpha)] \text{ such that } X\, R_j \alpha \text{ holds.}$$

A derivation $XD\alpha$ of a well-formed formula α from a set X of well-formed formulas is a pair $<\{\alpha_1, \ldots, \alpha_n\};\{|\alpha_1|, \ldots, |\alpha_n|\}>$ of sequences $\{\alpha_i\}$ of well-formed formulas and values $|\alpha_i| \in L$, $i = 1, \ldots, n$, satisfying the following conditions:

(i) $\alpha_n = \alpha$ and for each i, $1 \leq i \leq n$, either $\alpha_i \in X$ or $\alpha_i \in A_L$ (set of logical axioms) or $\{\alpha_{i_1}, \ldots, \alpha_{i_n}\} R_j \alpha_i$ holds, and R_j is a rule of inference.

(ii) $|\alpha_i| = 1$ if $\alpha_i \in X$

$|\alpha_i| = \mathrm{Taut}(\alpha_i)$ if $\alpha_i \in A_L \backslash X$, where $\mathrm{Taut}(\alpha_i) = \inf_{i \in I} T_i(\alpha_i)$ is the degree of tautologihood of α_i.

$|\alpha_i| = |R_j|$ if $\{\alpha_{i_1}, \ldots, \alpha_{i_n}\} R_j \alpha_i$ holds. $|R_j|$ is the value attached to a rule of inference R_j.

It can be shown that in a deductive system with graded consequence, all the rules of classical logic will receive the grade 1 in crisp context. In vague situations this grade is not necessarily 1. The value of the derivation $XD\alpha$, denoted $\mathrm{val}(XD\alpha) = \pi\{|\alpha_1|, \ldots, |\alpha_n|\}$, where $\pi\{a_1, \ldots, a_n\} = a_1 * \ldots * a_n$, for $a_i \in L$, $i = 1, \ldots, n$.

With the above method of computing grades of derivations, we can recast any conditional Sorites argument in the form of the following chain of graded derivations

F_{a_n} …	… 1		(a_n being the given premise)
$F_{a_n} \rightarrow F_{a_{n-1}}$ …	1		(being the given premise
$F_{a_{n-1}}$	… …	$\lvert MP \rvert =$	$\inf_{\beta, \beta\rightarrow\gamma} \mathrm{gr}[(\beta \wedge \beta \rightarrow \gamma) \models \gamma]$
$F_{a_{n-1}} \rightarrow F_{a_{n-2}}$	…	1	(given premise)
$F_{a_{n-2}}$	… …	$\lvert MP \rvert$	
$\rightarrow$			
$F_{a_{n-9999}} \rightarrow F_{a_{n-10,000}}$ …		1	
$F_{a_{n-10,000}}$ … …	$\lvert MP \rvert$		

In the above derivation the rule MP is applied 10,000 times. Value of the above derivation is computable as:

$$1*1*|MP|*\uparrow*1*|MP|*\uparrow*|MP| = \overbrace{|MP|*\cdots\cdots*|MP|}^{10{,}000\text{ times}}$$

In the present vague context, the value of MP as given by $\inf_{\beta,\beta\to\gamma} gr[(\beta\wedge\beta\to\gamma)\approx\gamma]$ is not necessarily 1 [1]. Thus the value of the whole Sorites argument as given by the above '*' operation (which in some case may be the operation of product) is not 1. As the number of steps in the derivation increases, value of the derivation decreasingly approaches zero, because $a_1 *\uparrow* a_n < a_1 *\uparrow* a_{n-1}$ when $a \neq 1, a \neq 0$.

4. Conclusion

The above is a syntactic way of showing that there is nothing paradoxical about the 'Sorites'. Without requiring to know the actual values of the premises and the conclusion, it can be said that derivation at each successive stage gradually becomes too weak to establish its conclusion. At the ultimate stage, value of the derivation as a whole lowers so much that the conclusion arrived at the end of this derivation loses all its credibility. Other approaches within a framework of infinite degrees of truth, seek to solve the paradox by showing how the value of the conclusion derived at each successive stage of the argument, as expected, becomes lesser than the value of the conclusion at the preceding stage.

In the traditional fuzzy logical approaches, we have no answer why despite each derivation in the chain being sound in absolute sense, i.e., to degree 1, value of the conclusion at each successive stage of the Sorites gradually decreases, so much so that it loses all its credibility. The notion of graded consequence which is a notion of some degree of the strength of the derivation as a process, serves to explain why in case of Sorites the longer the derivation becomes, the more it becomes precarious for establishing a conclusion as true. In this connection we would like to mention the following comment of Kenton Machina, known through a communication: "I don't see how a proper solution to Sorites paradoxes can be obtained without just such a notion of graded consequence".

References

1. S. Basu (2003), *A Study in Logical and Philosophical Implications of Graded Consequence*, Ph.D. thesis, Jadavpur University, Kolkata.
2. M.K. Chakraborty (1988), *Use of Fuzzy Sets in Introducing Graded Consequence in Multiple-valued Logic* in *Fuzzy Logic in Knowledge-Based Systems, Decision and Control*, M. M. Gupta and T. Yamakawa (eds.), Elsevier Science Publishers B.V. North-Holland, pp. 247-257.
3. M.K. Chakraborty and S. Basu (1997), Graded Consequence and Some Metalogical Notions Generalised, *Fundamenta Informaticae*, 32, IOS Press, Amsterdam, pp. 299-311.
4. M. Dummett (1970), Wang's Paradox, *The Journal of Symbolic Logic*, 1970; reprinted in *Truth and Other Enigmas*, Michael Dummett, Duckworth, London, 1978, pp. 248-268.
5. Kit Fine (1975), Vagueness, Truth and Logic, *Synthese*, 30: 265-300.

6. B.R. Gaines (1977), *Foundations of Fuzzy Reasoning* in *Fuzzy Automata and Decision Processes*, M. M. Gupta, G. N. Saridis, B. R. Gaines (eds.), Elsevier North-Holland Inc., New York, pp. 19-75.
7. G. Gentzen (1969), *Investigations into Logical Deduction* in *The Collected Papers of Gerhard Gentzen*, M. E. Szabo (ed.), North-Holland Publishing Company, Amsterdam, London, pp. 68-131.
8. J. A. Goguen (1969), The Logic of Inexact Concepts, *Synthese*, 19: 325-373.
9. Rohit Parikh (1983), *The Problem of Vague Predicates* in *Language, Logic and Method*, R. C. Cohen and M. W. Wartofsky (eds.), D. Reidel Publishing Company, Dordrecht, Holland, pp. 241-261.
10. D. J. Shoesmith and T. J. Smiley (1978), *Multiple Conclusion Logic*, Cambridge University Press.
11. R. Sorensen (1988), *Blindspots*, Clarendon Press, Oxford.
12. M. Tye (1994), *Sorites paradoxes and the semantics of vagueness* in *Philosophical Perspectives: Logic and Language*, J. Tomberlin (ed.), Ridgeview: Atascadero, California.
13. T. Williamson (1994), *Vagueness*, Routledge, London.
14. C.C.J. Wright (1975), On the Coherence of Vague Predicates, *Synthese*, 30: 325-365.
15. L.A. Zadeh (1975), Fuzzy Logic and Approximate Reasoning, *Synthese*, 30: 407-428.

Fuzzy Logic and its Application in Technology and Management
Editors: Debjani Chakraborty, S. Nanda and D. Dutta Majumder

On Fuzzy Real-valued Sequence Space $\ell(p)^F$

Binod Chandra Tripathy[1] and Amar Jyoti Dutta

Mathematical Science Division, IASST, Khanapara, Guwahati-781022
[1]E-mail : tripathybc@yahoo.com; tripathybc@rediffmail.com

Abstract: In this article we study the different properties of fuzzy real-valued sequence based $\ell(p)^F$ like solidity, symmetricity, convergence free, sequence algebra and monotonicity. We obtain some inclusion results and equality of the space is also established.

Keywords: Fuzzy real numbers, solid space, symmetric space, convergence free, sequence-algebra and monotone.

1. Introduction

The concept of fuzzy set was introduced by Zedah [5]. Applying the notion of fuzzy real numbers different types of fuzzy real valued sequences was introduced and studied by Nanda[1], Tripathy and Nanda[4], Subrahmanyam[3], Nuray and Savas[2] and many others. The $\ell(p)$ spaces of sequences of fuzzy numbers was introduced by Nuray and Savas [2]. They studied only the completeness property.

Following known definitions, results and operations on fuzzy real numbers are reproduced as a preliminary concept of fuzzy real-valued sequences.

A fuzzy set A in X is a function with domain X and values in [0,1] (Zedah [5]). A fuzzy real number is fuzzy set on R is a mapping X: $R \to I$ (= [0,1]), which is bounded, convex and normal.

The class of all upper semi-continuous normal convex fuzzy real number is denoted by $R(I)$.

If $X \in R(I)$, then for any $\alpha \in [0,1]$, α -level set is denoted by $[X]^\alpha$ is defined as

$$[X]^\alpha = \{t \in R : X(t) \geq \alpha\}, \text{ for } \alpha > 0 ;$$
$$= \{t \in R : X(t) > 0\}, \text{ for } \alpha = 0 .$$

Let X, $Y \in R(I)$ and the α-level sets be $[X]^\alpha = \left[a_1^\alpha, b_1^\alpha\right]$, $[Y]^\alpha = \left[a_2^\alpha, b_2^\alpha\right]$, $\alpha \in [0,1]$. Then the arithmetic operations on $R(I) \times R(I)$ are defined as follows :

$$(X \oplus Y)(t) = \sup_{s \in R}\{X(s) \wedge Y(t-s)\}, t \in R$$
$$(X - Y)(t) = \sup_{s \in R}\{X(s) \wedge Y(s-t)\}, t \in R$$

$$(X \otimes Y)(t) = \sup_{s \neq 0, s \in R} \{X(s) \wedge Y(t/s)\}, t \in R$$

$$(X \div Y)(t) = \sup_{s \in R} \{X(st) \wedge Y(s)\}, t \in R .$$

The arithmetic operations for α- level sets are defined as follows:

$$[X \oplus Y]^\alpha = [a_1^\alpha + a_2^\alpha, \ b_1^\alpha + b_2^\alpha]$$
$$[X - Y]^\alpha = [a_1^\alpha - b_2^\alpha, b_1^\alpha - a_2^\alpha]$$
$$[X \otimes Y]^\alpha = \left[\min_{i \in \{1,2\}} a_i^\alpha b_i^\alpha, \max_{i \in \{1,2\}} a_i^\alpha b_i^\alpha \right]$$
$$[Y^{-1}]^\alpha = \left[\frac{1}{b_2^\alpha}, \frac{1}{a_2^\alpha} \right], \ 0 \notin Y.$$

The additive identity and the multiplicative identity in $R(I)$ are denoted by $\bar{0}$ and $\bar{1}$ respectively and defined as

$$\bar{0}(t) = \begin{cases} 1, & for \ t = 0 \\ 0, & for \ t \neq 0. \end{cases} \quad \text{and} \quad \bar{1}(t) = \begin{cases} 1, & for \ t = 1 \\ 0, & for \ t \neq 1. \end{cases}$$

The absolute value of $X \in R(I)$ is defined as

$$|X|(t) = max \{X(t), X(-t)\}, \text{ for } t \geq 0,$$
$$= \bar{0}, \text{ otherwise}$$

Let D be the set of all closed bounded intervals $X = [X^L, X^R]$ on the real line R. Then we write

$X \leq Y$ if and only if $X^L \leq Y^L$ and $X^R \leq Y^R$

and

$d(X, Y) = \max \{ |X^L - Y^L|, |X^R - Y^R| \}.$

Then clearly (D, d) is a complete metric space.

Now define a mapping $\bar{d} : I(R) \times I(R) \rightarrow R$ by

$$\bar{d}(X, Y) = \sup_{0 \leq \alpha \leq 1} d(X^\alpha, Y^\alpha), \text{ for } X, Y \in I(R).$$

Then $\bar{d}$ defines a metric on $I(R)$.

2. Definition and background

A fuzzy real-valued sequence is denoted by $\{X_n\}$,where $X_n \in R(I)$, for all $n \in N$.

***Definition*:** A sequence $X = \{X_k\}$ of fuzzy real numbers is said to be *convergent* to the fuzzy number X_0 if for every $\varepsilon > 0$, there exist $k_0 \in N$ such that

$\bar{d}(X_k, X_0) < \varepsilon$, for all $k > k_0$.

Definition: A fuzzy real-valued sequence $\{X_n\}$ is said to be level convergent to the fuzzy real number X, if for each $\alpha \in (0,1]$ $\lim a_n^\alpha = a^\alpha$ and $\lim b_n^\alpha = b^\alpha$, where $X_n^\alpha = [a_n^\alpha, b_n^\alpha]$, $X^\alpha = [a^\alpha, b^\alpha]$.

The set E^F of sequence taken from $R(I)$ is said to be a sequence of fuzzy real number if for $\{X_n\}, \{Y_n\} \in E^F$, $r \in R$ $X_n, Y_n \in R(I)$

$\{X_n\} + \{Y_n\} = \{X_n + Y_n\} \in E^F$

and

$r\{X_n\} = \{rX_n\} \in E^F$,

where $rX_n(t) = \begin{cases} X_n(r^{-1}t) & \text{if } r \neq 0 \\ \bar{0} & \text{if } r = 0 \end{cases}$

In [2], Savas and Nuray introduced $\ell(p)^F$ – spaces of sequences of fuzzy numbers as follows :

$$\ell(p)^F = \left\{ X = \{X_k\} : \sum_{k=1}^{\infty} [\bar{d}(X_k, \bar{0})]^{p_k} < \infty \right\},$$

where, (p_k) is a bounded sequence of strictly positive real numbers .

Further it is proved by them that $\ell(p)$ is a complete metric space with the metric defined by

$$\rho(X,Y) = \left(\sum_k [\bar{d}(X_k, Y_k)]^{p_k} \right)^{\frac{1}{M}}$$

where, $X = \{X_k\}$ and $Y = \{Y_k\}$ are sequences of fuzzy numbers which are in $\ell(p)$ and $M = \max(1, \sup_k p_k)$.

A sequence space E^F is said to be *normal* (or *solid*) if $\{Y_i\} \in E^F$ whenever $|Y_i| \le |X_i|$ for all $i \in N$ and $\{X_i\} \in E^F$.

A sequence space E^F in R(I) is said to be *monotone* if E^F contains the canonical pre-images of all its step spaces.

A sequence space E^F is said to be *symmetric* if $(X_{\pi(n)}) \in E^F$, whenever $(X_n) \in E^F$, $\pi(n)$ is a permutation of N.

A sequence space E^F is said to be *sequence algebra* if $(X_n \odot Y_n) \in E^F$, whenever $(X_n), (Y_n) \in E^F$.

A sequence space is E^F is said to be *convergence free* if $(Y_n) \in E^F$, whenever $(X_n) \in E^F$ and $X_n = \bar{0}$ implies $Y_n = \bar{0}$.

3. Main Results

In this section we verify different properties like solidity, sequence algebra, convergence free, symmetricity and inclusion results for the above mentioned space.

Theorem 1: *$\ell(p)^F$ is solid.*

Proof: Let $\{X_k\}$ and $\{Y_k\}$ be two fuzzy real-valued sequences. Let $|Y_k| \le |X_k|$, for all $k \in N$. Let $\{X_k\} \in \ell(p)^F$.

Then $\sum_{k=1}^{\infty} \bar{d}\left(X_k,\bar{0}\right)^{p_k} < \infty$ and $\sum_{k=1}^{\infty} \bar{d}(Y_k,\bar{0})^{p_k} \le \sum_{k=1}^{\infty} \bar{d}(X_k,\bar{0})^{p_k} < \infty$.

Thus $\{Y_k\} \in \ell(p)^F$. Hence $\ell(p)^F$ is solid.

Theorem 2: *$\ell(p)^F$ is a sequence algebra.*

Proof: Let $\{X_k\}$ and $\{Y_k\}$ be two fuzzy real-valued sequences in $\ell(p)^F$. Then

$$\sum_{k=1}^{\infty} \bar{d}(X_k,\bar{0})^{p_k} < \infty \text{ and } \sum_{k=1}^{\infty} \bar{d}(Y_k,\bar{0})^{p_k} < \infty, \text{ for all } k \in N.$$

Then,

$$\sum_{k=1}^{\infty} \bar{d}(X_k Y_k,\bar{0})^{p_k} \le \sum_{k=1}^{\infty} [\bar{d}(X_k,\bar{0})\bar{d}(Y_k,\bar{\bar{0}})]^{p_k}$$

$$\le \sum_{k=1}^{\infty} \bar{d}(X_k,\bar{0})^{p_k} \sum_{k=1}^{\infty} \bar{d}(Y_k,\bar{0})^{p_k}.$$

Hence, $(X_k Y_k) \in \ell(p)^F$ and thus $\ell(p)^F$ is a sequence algebra.

Theorem 3: *$\ell(p)^F$ is not convergence free.*

Proof: The proof follows from the following example.

Example 1: Consider the fuzzy sequence $\{X_k\}$ defined by $X_k = \overline{k^{-2}}$, for each $k \in N$ and p_k is defined by $p_k = 1$ for all $k \in N$. Then $\{X_k\} \in \ell(p)^F$. Consider $\{Y_k\}$ defined by $Y_k = \bar{1}$. Then $\{Y_k\} \notin \ell(p)^F$. Hence $\ell(p)^F$ is not convergence free.

Theorem 4: *$\ell(p)^F$ is not symmetric.*

Proof: The proof follows from the following example.

Example 2: Consider the fuzzy sequence $\{X_k\}$ in $\ell(p)^F$, defined by $X_k = \overline{k^{-2}}$, for k odd and $X_k = \overline{k^{-1}}$, for k even and (p_k) be defined by $p_k = 1$, for k odd and $p_k = 2$, for k even Then $\sum_{k=1}^{\infty} \bar{d}(X_k,\bar{0})^{p_k} = \sum_{k=1}^{\infty} k^{-2}$, which is convergent. Let $\{Y_k\}$ be an arrangement of $\{X_k\}$ given by

$$\{Y_k\} = \{X_2, X_1, X_4, X_3, X_6, \ldots\}.$$

Then we have $\sum_{k=1}^{\infty} \bar{d}(Y_k,\bar{0})^{p_k} > \sum_{k,even} k^{-1} = \infty$.

Hence $\{Y_k\} = \left\{X_{\pi(k)}\right\} \notin \ell(p)^F$. Thus $\ell(p)^F$ is not symmetric.

For $p_k = p$, for all $k \in N$, we denote the space $\ell(p)^F$ by ℓ_p^F. The proof of the following result is a routine verification.

Theorem 5: *Let $0 < p < q$. Then $\ell_p^F \subset \ell_q^F$ and the inclusion is proper.*

Reference

1. S. Nanda (1989), On sequences of fuzzy numbers; *Fuzzy sets and Systems*; 33: 123-126.
2. F. Nuray & E. Savas (1995), Statistical convergence of sequences of fuzzy real numbers; *Math. Slovaca*; 45(3): 269-273.
3. P. V. Subrahmanyam (1999), Cesaro summability of fuzzy real numbers; *J. Analysis*; 7: 159-168.
4. B. K. Tripathy & S. Nanda (2000), Absolute value of fuzzy real numbers and fuzzy sequence spaces, *J. Fuzzy Mathathematics*, 8(4): 883-892.
5. L. A. Zadeh (1965), Fuzzy sets, *Information and Control*, 8: 338-353.

Fuzzy Logic and its Application in Technology and Management
Editors: Debjani Chakraborty, S. Nanda and D. Dutta Majumder

Topological Concepts in Image Analysis: The Fuzzy Angle

Debajit Hazarika

Department of Mathematical Sciences, School of Science and Technology
Tezpur University, Napaam - 784028
Email: debajit@tezu.ernet.in

Abstract: A growing number of computer application areas have problems with some kind of topological character. Research efforts have been directed to tackle these problems via formulation of various topological notions in the digital framework. Fuzzy sets play an indispensable role in this regard on various counts. In this paper, we have explored of the role of fuzzy sets and logic in the use of topological concepts in application areas such as in image analysis. We have also examined the possible utility of a more topological approach in this direction.

1. Introduction

With more computer application areas arising with in built problems of topological character, research efforts are growing to tackle these problems in different directions. Image processing is one such significant area (computer graphics, modeling being two other such areas) where topological concepts appear to have considerable influence. Topological notions such as connectedness have great utility in image analysis and have been utilized by researchers in different forms. A newly emerging branch of study called digital topology tries to form a mathematical framework for the study of digital pictures in terms of topological properties [3, 6].

An important aspect of this development is that fuzzy sets play an indispensable role in this field on various counts. Images produced by any imaging device are inherently fuzzy which arises from reasons ranging from resolution limitation to material heterogeneity of objects [16]. To tackle such uncertainty elements as well as intensity gradations, a fuzzy approach is undoubtedly more suitable. Classical digital topology primarily concerns itself in the study of black-white images in the digital plane, it invariably needs to take care of the gray level images that can be represented by fuzzy sets as noted by Rosenfeld [10]. A strong theoretical background of the theory of fuzzy sets and fuzzy logic that has emerged over the years can provide the necessary groundwork for this development. Another associated aspect is that various features, properties and relationship in digital images that are not crisply defined can be regarded only as fuzzy subsets of images. Of the various topological relationships among parts of a digital picture, notions such as connectedness and adjacency play an important role in image analysis.

Conventionally, properties and relationships are defined for subsets of the picture extracted from it by various segmentation processes of different types. The strong commitment of segmentation process can be weakened by extracting fuzzy subsets rather than ordinary subsets from the image [10]. This necessitates extension of concepts digital topology (which is to a good extent a well established branch of theory) to fuzzy subsets. So the first approach constitutes of carrying out a fuzzy segmentation of the given image first so that a fuzzy subset of the image domain is obtained, with every image element having fuzzy membership. The topological concepts are developed thereon. Rosenfeld initiated work with this approach [10, 11, 12]. The works of Dellepiane and Fontana [2] suggested methods for utilising these developments for image segmentation, which is an integral part in image analysis.

Alternatively, these concepts can be defined directly on the given image and integrate them with the process of segmentation. Utilising this methodology, topological and geometric notions, particularly that of fuzzy connectedness has been developed to treat the analysis of images [16] in a comprehensive manner, particularly for analysis of medical and related images. This notion is found to be useful for image segmentation.

It may be noted that despite the name 'digital topology', graph theoretic approaches have been applied in preference to topological methods [5, 15].

In this paper, we have noted the essentiality of the role of fuzzy sets in the use of topological concepts in application areas such as in image analysis. We have also examined the possible utility of a more topological approach in this direction.

2. Digital Topology

We provide a brief background of digital topology in this section to indicate how the topological concept, particularly the notion of connectedness is developed into the digital images. Connectedness is indeed one of the very important and intuitive topological notions. Due to its geometric nature, it has been found to be very useful in the study of properties of images.

In essence, digital topology is the study of the topological properties of image arrays. Such questions arise in a wide range of diverse applications including computer graphics, computer tomography, pattern analysis, and robotic design. Its purpose is to provide a sound mathematical basis for various image processing operations. They may include operations such as *image segmentation*, *image thinning*, *border following*, *contour filling*, *object contouring* and so on.

Traditionally image arrays are considered to be binary arrays, all of whose elements have value 0 or 1. But many parts of digital topology needs to be generalised to fuzzy digital topology, which deals with unsegmented gray scale image arrays whose elements lie in the range $0 \le x \le 1$ [10, 11, 12].

A binary image is typically obtained by thresholding a gray scale image array, but then at many situations it means loss of vital information.

Our discussions are mostly confined to 2-D and 3-D images. Higher framework for higher dimensions has also obtained by several authors.

Elements of 2-D image array (considered unbounded in all directions for convenience) are called pixels and elements of 3-D image array are called voxels. Elements of higher (n)-dimensions have been called n-spels (space elements) by some authors.

With each pixel and voxel, a lattice point with integer coordinates is associated in the plane or in the 3-D space. In 2-D, two lattice points are 8-adjacent if they are distinct and each coordinate of one differs for the corresponding coordinate of the other by at most 1; 4-adjacent if they are 8-adjacent and differ in at most one of their coordinates. In 3-D, two points are 26-adjacent if they are distinct and each coordinate of one differs from the corresponding coordinate by at most 1; 18-adjacent if 26-adjacent and differ in at most two of their coordinates; 6-adjacent if 26-adjacent and differ at most in one coordinate. For n = 4, 6, 8, 18 or 26 an n-neighbour of a point is an n-adjacent point. A set S of lattice points is said to be *n-connected* if S cannot be partitioned into two subsets that are not n-adjacent to each other. An *n-component* of a set of lattice points S is a non-empty n-connected subset of S that is not n-adjacent to any other point in S. A lattice point associated with a pixel or voxel that has value 1 is called a black point and one with value 0 is called a white point. The study of digital topology started with a common practice of using different adjacency relations for black and white points to avoid certain paradoxes (known as connectivity paradoxes). For instance, if 4-adjancey is used for all pairs of points then the black points are totally disconnected but still separate the set of white points into two components. If 8-adjacency is used for all pairs of points then the black point do not separate the white points but black points form two separate components. This paradox can be settled by using different adjacency relations for black and white points. There are other methods of avoiding this paradox by other methods, usually by selection of images.

3. Digital Fuzzy Concepts

In this section, we highlight how the existing topological concepts are sought to be developed when gray scale images are taken in to consideration. Inevitably fuzzy sets play the lead role in this direction.

Let Σ be a digital array of lattice points, in either 2-D or 3-D. A *fuzzy subset* of Σ is a function $\sigma:\Sigma \rightarrow [0,1]$. For any $P \in \Sigma$, the value $\sigma(P)$ is called the *degree of membership* of P in σ. An ordinary subset S of Σ can be regarded as fuzzy subset for which the function takes on only the values 0 and 1, where points of S map into 1 and points of $\overline{S}$ into 0. The function $1-\sigma$, which takes each $P \in \Sigma$ into $1-\sigma(P)$, is called the *complement* of σ. For convenience, it is assumed that for all points P on the border of Σ, $\sigma(P)=0$.

In the sequel, we briefly describe the formulation of the notion of fuzzy connectedness in digital topology for the study of images.

3.1. Fuzzy connectedness and related notions

We describe below how the fuzzy sets have been used to describe the concept of conncectedness in digital spaces, with particular reference to the 2-D case [10].

Let σ be a fuzzy subset of Σ, and let $\rho: P = P_1, P_2, \ldots, P_n = Q$ be any path between two points of Σ. The strength $s_\sigma(\rho)$ of ρ (w.r.t. σ) is defined as $\min_{0 \le i \le n} \sigma(P_i)$.

The *degree of connectedness* of P and Q (w.r.t. σ) is defined as $C_\sigma(P,Q) = \max_\rho s_\sigma(\rho)$, where max is over all paths from P to Q. Thus, $0 \le s_\sigma(\rho) \le 1$ for all ρ, and $0 \le C_\sigma(P,Q) \le 1$ for all P,Q.

For all P and Q, we have $C_\sigma(P,P) = \sigma(P)$ and $C_\sigma(P,Q) = C_\sigma(Q,P)$.

If σ is into {0,1}, let $S = \sigma^{-1}(1) \equiv \{P \mid P \in \Sigma \text{ and } \sigma(P) = 1\}$; then $s_\sigma(P,Q) = 1$, iff ρ consists entirely of points of S and $C_\sigma(P,Q) = 1$ iff P and Q are connected in S. The degree of connectedness generalises the ordinary concept of connectedness. In general, $C_\sigma(P,Q) = 1$ iff there exist a path from P to Q all of whose points are mapped into 1 by σ. For any set $T \subseteq \Sigma$, we can define the *degree of connectedness* of T (w.r.t. σ) as $C_\sigma(T) = \min_{p,q \in T} C_\sigma(P,Q)$. Though C_σ is reflexive and symmetric, it is not transitive.

Even though C_σ is not an equivalence relation, Rosenfeld used it to define a notion of connected component with respect to σ. A *plateau* in σ is a maximal connected subset of Σ on which σ has a constant value. Any $P \in \Sigma$ belongs to exactly one plateau. Plateau Π is called a *top* if its σ value is a local maximum, i.e., if $\sigma(P) > \sigma(Q)$ for all pairs of neighbouring points $P \in \Pi, Q \notin \Pi$. If its value is a local minimum, Π is called a *bottom*. Obviously, Π is a plateau in σ iff it is a plateau in $1-\sigma$. It is a bottom in σ iff a top in $1-\sigma$ and vice-versa. Tops and bottoms can be regarded as generalisations of connected components. In the non-fuzzy case, the plataeus are just the connected components of S and $\bar{S}$. Rosenfeld has further shown that Σ is connected w.r.t. σ iff there exists a unique top in σ. The number of components of σ has simply been defined as the number of its tops. σ is said to be connected if, for all P, Q in Σ, there exists a path $P = P_0, P_1, \ldots, P_n = Q$ such that each $\sigma(P_i) \ge \min(\sigma(P), \sigma(Q))$ and it was proved that σ is connected iff σ has a unique top.

Rosenfeld further developed the topological concepts of *fuzzy surroundedness* and also the concepts of *holes* and *genus* [11].

A notable aspect of the above development is that the concepts are more graph theory based despite the name digital topology. Though easily handled by computers, this has approach is disadvantageous while handling complicated objects.

4. A More Topological Approach

Kong et al. [5] suggested another approach showing that a more topological approach is possible without completely avoiding the graph theoretic notions. This involves the application of specially devised topology for the digital spaces. A connected topology on Z (regarded as an ordered set) is used for this purpose, which was originally defined by Khalimsky [18], several variations of this topology nevertheless exist in the literature. This induces a product topology on $Z \times Z$ on 2-D and on $Z \times Z \times Z$ in 3-D. A *digital path* and *digital arc* are defined analogously to the real case by replacing the real interval by an interval subset of Z (i.e., a set of the form $\{n, n+1, ..., n+k\}$ with this topology). The Jordan curve theorem in 2-D digital plane has been proved by in this platform by Khalimsky et al. [17]. The Jordan surface theorem in $Z \times Z \times Z$ can also be developed with this approach invoking the Classical Jordan-Brower theorem, which underlines the promise of this approach. One advantage of the topological approach is backed by strong background of classical topological theory. Several other authors have introduced different topologies on the digital spaces suitable for different purposes [4, 7].

As noted earlier, the classical theory is unable to handle the grayscale images and other occurrences of fuzzy images or fuzzy subsets which is a definite drawback. This gives rise to the need for the development of a framework based on suitable fuzzy topology. Fuzzy topological spaces were introduced by Chang [1] in the late sixties as collection of fuzzy subsets preserved under arbitrary sups and finite infs. Fuzzy images can typically be considered as fuzzy spaces. One major advantage here is that the theory of fuzzy topology has advanced considerably during the late eighties and thereafter. Most of the theories have now been generalized to lattice valued topological spaces. Well-developed theories exist in connectedness and related notions [8, 9, 13, 14].

Two major bottlenecks for the applications of the concepts of fuzzy spaces can hardly escape observation. The first is that normally several generalizations of the same classical topological notion occur in the fuzzy framework occur due to the existence of the stratum structures. Secondly, the existing theories, though well developed theoretically, do not indicate or facilitate much about their possible applications. It therefore requires choosing a suitable framework and also to choose or adapt a suitable notion for the application purpose. This indeed appears to be highly challenging area of research.

References

1. C. L. Chang (1968), Fuzzy topological spaces, *J. Math. Anal. Appl.* 24: 182-190.
2. S. Dellepiane and F. Fontana (1995), Extraction of intensity connectedness for image processing, *Pattern Recognition Letters*, 16: 313-324.
3. U. Eckhardt and L. Latecki, *Digital Topology*, Electronic reprint.
4. C. O. Kiselman (2001), *Digital Jordan Curve Theorem*, Electronic preprint, Department of mathematics, Uppsala University (2001) 1-11.
5. T. Y. Kong, R. Kooperman and P. R. Meyer (1991), A topological approach to digital topology, *American Mathematical Monthly*, 98 (10): 901-917.
6. T. Y. Kong, A. Rosenfeld (1989), Digital topology, Introduction and Survey, *Computer Vision, Grpahics, and Image Processing*, 48: 357-393.
7. L. Latecki, U. Eckhardt, and R. Rosenfeld (1995), Well-composed sets, *Computer Vision and Image understanding*, 61 (1): 70-83.
8. R. Lowen (1981), Connectedness in fuzzy topological spaces, *Rocky Mountain J. Math.* 11 (3): 427-433.
9. S. E. Rodabaugh (1982), Connectivity and the L-fuzzy unit interval, *Rocky Mountain J. Math.* 12, 113-121.

10. A. Rosenfeld (1979), Fuzzy Digital Topology, *Information and Control*, 40(1) 76-87.
11. A. Rosenfeld (1983), On connectivity properties of grayscale pictures, *Pattern Recognition*, 16: 47-50.
12. A. Rosenfeld (1984), The fuzzy geometry of image subsets, *Pattern Recognition Letters*, 2: 311-317.
13. R. D. Sarma, D. Hazarika, and A. Sharfuddin (2003), Extremal disconnectedness in L-topological spaces, *J. Fuzzy mathematics*, 11(3).
14. R. D. Sarma, D. Hazarika, and A. Sharfuddin, Some aspects of C_i-connectedness in L-topological spaces, *Communicated*.
15. J. Slapal (2003), Closure operations for digital topology, *Theoretical Computer Science*, 305: 457-471.
16. J. K. Udupa and P. K. Saha (2003), Fuzzy Connectedness and Image Segmentation, *Proceedings of the IEEE*, 91 (10): 1649-1669.
17. E. D. Khalimsky, R. Kooperman and P. R. Meyer (1990), Computer Graphics and connected topologies on finite ordered sets, *Topology and its Applications*, 36: 1-17.
18. E. D. Khalimsky (1969), On topologies of generalised segments, *Soviet Math. Dokalady*, 10: 1508-1511.

Fuzzy Logic and its Application in Technology and Management
Editors: Debjani Chakraborty, S. Nanda and D. Dutta Majumder

Finitisticness of L-topological Spaces Under Sum and Lowen Functor Omega

Dalip Singh Jamwal[1] *and Shakeel Ahmed*[2]

[1] Department of Mathematics, University of Jammu, Jammu, J&K
[2] Department of Mathematics, Govt. Degree College, Rajouri, J&K
Email: dalipsj@yahoo.com, shakeelar2002@yahoo.co.in

Abstract: In this paper we have studied finitisticness of L-topological spaces under sum and also checked whether the concept of finitisticness is preserved by Lowen Functor Omega or not.

Keywords: L-topology, Finitistic Space, Sum space.

1. Introduction

It was in 1968, that Chang [1] introduced the concept of fuzzy topology and made an attempt to develop the basic topological notions for such spaces. Goguen [4] generalized the notion of fuzzy topology and introduced the concept of L-fuzzy topology. Hohle-Rodabaugh [6] have used the terms L-topology and L-topological space for the terms L-fuzzy topology and L-fuzzy topological space respectively. We are also using the terms L-topology and L-topological space for the terms L-fuzzy topology and L-fuzzy topological space respectively.

Let $\{(X_t, \delta_t) : t \in T\}$ be a family of L-topological spaces, different X_t's be disjoint, $X = \bigcup_{t\in T} X_t$. Then *Sum Topology* of $\{\delta_t : t \in T\}$ on X is denoted by $\bigoplus_{t\in T} \delta_t$ and it is defined as follows:

$$\forall U \in L^X,\ U \in \bigoplus_{t\in T} \delta_t \Leftrightarrow \forall t \in T,\ U|_{X_t} \in \delta_t .$$

The L-topological space $(X, \bigoplus_{t\in T} \delta_t)$ is called the *Sum Space* of $\{(X_t, \delta_t) : t \in T\}$, [7].

Let $L=[0,1]$. For every $(X,T) \in \text{Ob}(Top)$, let $\omega_L(T)$ denotes the family of all lower semicontinuous mappings from (X,T) to L. Note that $\omega_L(T)$ is an L-topology on X. It means $(X, \omega_L(T)) \in \text{Ob}(L\text{-}Top)$. Here *L-Top* means category of all L-topological spaces and their continuous maps and *Top* means category of all ordinary topological spaces and their continuous functions. $\omega_L : Top \rightarrow L\text{-}Top$ is called the *Lowen Functor*. It assigns an object (X,T) of *Top* to an object $(X, \omega_L(T))$ of *L-Top*, [7].

Let (X, δ) be an L-topological space. A subfamily $\mathcal{A} \subset \delta$ is said to be an *open cover* of (X, δ) if $\vee \mathcal{A} = \underline{1}$, [7].

Let $\Lambda \neq \varnothing$ and $\mathcal{A} = \{A_\lambda : \lambda \in \Lambda\}$ be a family of L-fuzzy subsets of a nonempty set *X*. The *order* of $\mathcal{A}$ is defined as under:

Case-I: When $A_\lambda \neq \underline{0}$ for atleast one value of $\lambda \in \Lambda$. Then order of $\mathcal{A}$ is the largest nonnegative integer n for which there exists a subset M of Λ with n+1 elements such that $\underset{\lambda \in \Lambda}{\wedge} A_\lambda \neq \underline{0}$, or is ∞ if there is no such largest integer n.

Case-II: When $A_\lambda = \underline{0}$ for all $\lambda \in \Lambda$. Then the order of $\mathcal{A}$ is -1.

An L-topological space (X, δ) is said to be *finitistic* if each open cover of (X, δ) has a finite order open refinement ([2], [3]).

Throughout this paper *L* is a complete lattice having atleast two elements with an order reversing involution $' : L \rightarrow L$. When we write $(L, ')$ is a complete lattice, it means *L* is a complete lattice with atleast two elements and $' : L \rightarrow L$ is an order reversing involution.

2. Main Results

Theorem 1: Let $\mathcal{A}$ and $\mathcal{B}$ be two families of L-fuzzy subsets of a set *X* such that order of $\mathcal{A}$= *m* and order of $\mathcal{B}$= *n*. Then order of $(\mathcal{A} \cup \mathcal{B}) \leq m+n+1$.

Proof: Let M be any subfamily of $\mathcal{A} \cup \mathcal{B}$ having $m+n+3$ members. We show that $\wedge M = \underline{0}$.

Case-I: When $M \subset \mathcal{A}$. Then clearly $\wedge M = \underline{0}$ because total number of members in *M* is greater than $m+2$.

Case-II: When $M \subset \mathcal{B}$. Then also $\wedge M = \underline{0}$ because order of $\mathcal{B}$= *n* and total number of members in *M* is greater than $n+2$.

Case-III: When *M* contains more than $m+2$ members of $\mathcal{A}$. Then also $\wedge M = \underline{0}$ because order of $\mathcal{A}$ is *m*.

Case-IV: When *M* contains exactly $m+2$ members of $\mathcal{A}$. Then also $\wedge M = \underline{0}$ because order of $\mathcal{A}$ is *m*.

Case-V: When *M* contains less than $m+2$ members of $\mathcal{A}$. It means *M* contains more than or equal to $n+2$ members of $\mathcal{B}$. This also implies that $\wedge M = \underline{0}$ because order of $\mathcal{B}$ is n. Hence in all the possible cases, we have shown that $\wedge M = \underline{0}$. It means order of $(\mathcal{A} \cup \mathcal{B}) \leq m+n+1$.

Theorem 2: Let (X, δ_1) and (Y, δ_2) be two L-topological spaces. Then $(X \cup Y, \delta_1 \oplus \delta_2)$ is finitistic if and only if (X, δ_1) and (Y, δ_2) are finitistic.

Proof: We know by definition of sum L-topological spaces that $X \cap Y = \varnothing$.

Suppose (X, δ_1) and (Y, δ_2) are finitistic. Let $\mathcal{U} = \{U_\lambda : \lambda \in \Lambda\}$ be any open cover of $(X \cup Y, \delta_1 \oplus \delta_2)$. Then clearly $\mathcal{U}_X = \{U_\lambda |_X : U_\lambda \in \mathcal{U}\}$ and $\mathcal{U}_Y = \{U_\lambda |_Y : U_\lambda \in \mathcal{U}\}$ are open covers of (X, δ_1) and (Y, δ_2). Since (X, δ_1) and (Y, δ_2) are finitistic, therefore $\mathcal{U}_X$ and $\mathcal{U}_Y$ have finite order open refinement, say $\mathcal{V}_X = \{V_\alpha : \alpha \in \Delta_1\}$ and $\mathcal{V}_Y = \{W_\beta : \beta \in \Delta_2\}$. $\forall V_\alpha \in \mathcal{V}_X$, define,

$$R_\alpha = \begin{cases} V_\alpha & on\ X \\ \underline{0} & on\ Y\ . \end{cases}$$

Then clearly $R_\alpha \in L^{X \cup Y}$. Also, $\forall W_\beta \in$ V_Y, define

$$S_\beta = \begin{cases} W_\beta & on\ X \\ \underline{0} & on\ Y \end{cases}.$$

Then clearly $S_\beta \in L^{X \cup Y}$.

Let $A = \{ R_\alpha : R_\alpha = V_\alpha$ on X and $R_\alpha = \underline{0}$ on Y, $\forall V_\alpha \in$ $V_X\}$ and $B = \{ S_\beta : S_\beta = W_\beta$ on Y and $S_\beta = \underline{0}$ on X, $\forall W_\beta \in$ $V_Y\}$.

Let $U_l = A \cup B$. Then clearly U_l is an open cover of $(X \cup Y,\ \delta_1 \oplus \delta_2)$ and U_l is also a refinement of U. Now we show that order of U_l is finite. Since order of V_X and V_Y is finite, therefore order of A and B must be finite. Let order of $A = m$ and order of $B = n$. Then by Theorem 1, order of U_l is $\leq m + n + 1$. It means order of U_l is finite. Hence $(X \cup Y,\ \delta_1 \oplus \delta_2)$ is finitistic.

Converse: Suppose $(X \cup Y,\ \delta_1 . \oplus \delta_2)$ is finitistic. We have to show that $(X,\ \delta_1)$ and $(Y,\ \delta_2)$ are finitistic. Let $U = \{ V_\lambda : \lambda \in \Lambda\}$ be any open cover of $(X,\ \delta_1)$. $\forall V_\lambda \in$ U, define

$$R_\lambda = \begin{cases} V_\lambda & on\ X \\ \underline{1} & on\ Y \end{cases}.$$

Then clearly $R_\lambda \in L^{X \cup Y}$.

Let $A = \{ R_\lambda : R_\lambda = V_\lambda$ on X and $R_\lambda = \underline{1}$ on Y, $\forall V_\lambda \in U\}$. Then clearly A is an open cover of $(X \cup Y,\ \delta_1 \oplus \delta_2)$. Since $(X \cup Y,\ \delta_1 \oplus \delta_2)$ is finitistic, therefore A has a finite order open refinement, say $A_1 = \{ U_\alpha : \alpha \in \Delta \}$. Then clearly $U_l = \{ U_\alpha |_X : U_\alpha \in A_l\}$ is a finite order open refinement of U. Hence $(X,\ \delta_1)$ is finitistic. Similarly we can show that $(Y,\ \delta_2)$ is finitistic.

Theorem 3: Let the set {order(W_t) : W_t is a finite order open refinement of some open cover of (X_t, δ_t), $\forall t \in T$ } be finite. Then the sum space $(X,\ \underset{t \in T}{\oplus} \delta_t)$ of a family $\{(X_t, \delta_t) : t \in T\}$ of L-topological spaces is finitistic if and only if each L-topological space (X_t, δ_t) is finitistic.

Proof: Here all X_t's are disjoint.

Suppose each (X_t, δ_t) is finitistic. We have to show that $(X,\ \underset{t \in T}{\oplus} \delta_t)$ is finitistic. (Here $X = \underset{t \in T}{\cup} X_t$). Let $U = \{U_\lambda : \lambda \in \Lambda\}$ be any open cover of $(X,\ \underset{t \in T}{\oplus} \delta_t)$. Then clearly $U_t = \{ U_\lambda |_{X_t} : U_\lambda \in U\}$ is an open cover of $(X_t,\ \delta_t)$. Since $(X_t,\ \delta_t)$ is finitistic, therefore U_t has a finite order open refinement, say $V_t = \{ V_\alpha^t : \alpha \in \Delta \}$. For every $V_\alpha^t \in V_t$, define

$$R_\alpha^t = \begin{cases} V_\alpha^t & on\ \ X_t \\ \underline{0} & on\ \ X - X_t \end{cases}.$$

Let $W_t = \{ R^t_\alpha = V^t_\alpha$ on X_t and $R^t_\alpha = \underline{0}$ on X-X_t, $\forall V^t_\alpha \in V_t \}$. Let $V = \underset{t\in T}{\cup} W_t$. Then clearly V is an open cover of $(X, \underset{t\in T}{\oplus} \delta_t)$ and V is also a refinement of U. Now we show that order of V is finite. Since order of each V_t is finite, therefore order of each W_t is finite. By given condition, the set {order(W_t) : $t \in T$} is finite. Let $m = max$ {order(W_t) : $t \in T$}. Here order (W_t) means order of W_t. Let m = order (W_{to}). Let A be any subfamily of V having $m + 2$ distinct members. We show that $\wedge A = \underline{0}$. Let $A = \{S_1, S_2, S_3 ... S_{m+2}\}$. Two cases arise, either $A \subset W_{to}$ or there exists at least two distinct members S_i, $S_j \in A$ and there exists two families W_{ti} and W_{tj} among W_t's such that $S_i \in W_{ti}$ and $S_j \in W_{tj}$.

***Case-I*:** If $A \subset W_{to}$, then clearly $\wedge A = \underline{0}$ because order of W_{to} is m.

***Case-II*:** There exists two distinct members S_i, $S_j \in A$ and there exists two families W_{tj}, W_{tj} among W_t's such that $S_i \in W_{ti}$ and $S_j \in W_{tj}$. Let $x \in X$. It means $x \in \underset{t\in T}{\cup} X_t$. Then we have three possible cases:

(i) $x \in X_{t_i}$ and $x \notin X_{t_j}$

(ii) $x \notin X_{t_i}$ and $x \in X_{t_j}$.

(iii) $x \notin X_{t_i}$ and $x \notin X_{t_j}$ (x cannot belong to both X_{t_i} and X_{t_j} as $X_{t_i} \cap X_{t_j} = \varnothing$)

(i) $x \notin X_{t_j} \Rightarrow S_j(x) = 0 \Rightarrow (\overset{m+2}{\underset{k=1}{\wedge}} S_k)(x) = 0 \Rightarrow \wedge A = \underline{0}$ (because x is arbitrary).

(ii) $x \notin X_{t_i} \Rightarrow S_i(x) = 0 \Rightarrow (\overset{m+2}{\underset{k=1}{\wedge}} S_k)(x) = 0 \Rightarrow \wedge A = \underline{0}$.

(iii) $x \notin X_{t_i}$ and $x \notin X_{t_j} \Rightarrow S_i(x) = 0$ and $\Rightarrow S_j(x) = 0 \Rightarrow (\overset{m+2}{\underset{k=1}{\wedge}} S_k)(x) = 0 \Rightarrow \wedge A = \underline{0}$.

It means order of V is not exceeding m. Hence $(X, \underset{t\in T}{\oplus} \delta_t)$ is finitistic.

***Converse*:** Suppose $(X, \underset{t\in T}{\oplus} \delta_t)$ is finitistic. We have to show that each L-topological space (X_t, δ_t) is finitistic. Let $U_t = \{ U_\lambda : \lambda \in \Lambda\}$ be an open cover of (X_t, δ_t). $\forall U_\lambda \in U_t$, define

$$R_\lambda = \begin{cases} U_\lambda & on \quad X_t \\ \underline{1} & on \quad X - X_t \end{cases} \quad where \ X = \underset{t\in T}{\cup} X_t$$

Then clearly $U = \{ R_\lambda : R_\lambda = U_\lambda$ on X_t and $R_\lambda = \underline{1}$ on X - X_t, $\forall U_\lambda \in U_t \}$ is an open cover of $(X, \underset{t\in T}{\oplus} \delta_t)$. Since $(X, \underset{t\in T}{\oplus} \delta_t)$ is finitistic, therefore U has a finite order open refinement, say $V = \{ V_\alpha : \alpha \in \Delta \}$. Then clearly $V_t = \{ V_\alpha |_{X_t} : V_\alpha \in V\}$ is a finite order open refinement of U_t. Hence (X_t, δ_t) is finitistic $\forall t \in T$.

Note that Theorem 2 is a particular case of Theorem 3; We have intentionally added Theorem 2 here because it's direct part has been proved by using Theorem 1.

***Theorem 4*:** Let $I = [0,1]$ and (X,T) be any ordinary topological space. Then $(X, \omega_L(T))$ is finitistic implies (X,T) is finitistic, where $\omega_L : Top \rightarrow L\text{-}Top$ is the Lowen functor.

***Proof*:** We know $\omega_L(T)$ is the set of all lower semicontinuous functions from (X, T) into L. Also $\omega_L(T)$ is an L-topology on X, where $L= [0,1]$. Suppose $(X, \omega_L(T))$ is finitistic. Let $\mathcal{U} = \{U_\lambda : \lambda \in \Lambda\}$ be any open cover of (X, T). It can be easily shown that $\mathcal{V} = \{\chi_{U_\lambda} : U_\lambda \in \mathcal{U}\}$ is an open cover of $(X, \omega_L(T))$. Since $(X, \omega_L(T))$ is finitistic, therefore $\mathcal{V}$ has a finite order open refinement, say $\mathcal{V}_l = \{V_\alpha : \alpha \in \Delta\}$. Let order of $\mathcal{V}_l = n$. We can show that $\mathcal{U}_l = \{((V_\alpha)_{(0)} : V_\alpha \in \mathcal{V}_l\}$ is an open refinement of $\mathcal{U}$ and order of $\mathcal{U}_l$ is not exceeding n. Hence (X, T) is finitistic.

***Example 5*:** Converse of Theorem 4 is not true.

Let $X = \{x\}$. Then $T = \{\varnothing, X\}$ is the unique ordinary topology on X. Let $L = [0,1]$. Then it can be easily shown that $\omega_L(T) = \{\underline{\alpha} : \alpha \in L\}$. Here clearly (X, T) is finitistic but $(X, \omega_L(T))$ is not finitistic because $\mathcal{U} = \{\underline{a_n} : a_n = 1 - 1/n, \forall n \in N\}$ is an open cover of $(X, \omega_L(T))$ which has no finite order open refinement.

***Note 6*:** Theorem 4 is also true for any complete lattice $(L, ')$, where L is a chain or meet of any finite number of nonzero elements of L is nonzero.

***Theorem 7*:** :Let (X, T) be an ordinary topological space and $L= P(X)$. Let $\delta = \{\underline{U} : U \in T\}$. Then (X, T) is finitistic if and only if (X, δ) is finitistic.

***Proof*:** Here $(L, ')$ is a complete lattice, where $\leq$ on L is defined as $A \leq B$ if $A \subset B$, $\forall A, B \in L$ and $' : L \rightarrow L$ is defined as A'= complement of A in X. Here $0 = \varnothing$ and $1 = X$. Infact $(L, ')$ is a Boolean algebra. Here $\underline{U}$ means U-valued constant function from X to L. It can be easily checked that (X, δ) is an L-topological space.

Suppose (X, T) is finitistic. Let $\mathcal{U} = \{\underline{U_\alpha} : \alpha \in \Delta\}$ be any open cover of (X, δ). We can easily show that $\mathcal{V} = \{U_\alpha : \underline{U_\alpha} \in \mathcal{U}\}$ is an open cover of (X, T). Since (X, T) is finitistic, therefore $\mathcal{V}$ has a finite order open refinement say $\mathcal{V}_l = \{V_\beta : \beta \in \Delta_1\}$. Again it can be easily shown that $\mathcal{U}_l = \{\underline{V_\beta} : V_\beta \in \mathcal{V}_l\}$ is a finite order open refinement of $\mathcal{U}$. Hence (X, δ) is finitistic.

***Converse*:** Suppose (X, δ) is finitistic. Let $\mathcal{U} = \{U_\alpha : \alpha \in \Delta\}$ be any open cover of (X,T). Then clearly $\mathcal{V} = \{\underline{U_\alpha} : U_\alpha \in \mathcal{U}\}$ is an open cover of (X, δ). Since (X, δ) is finitistic, therefore $\mathcal{V}$ has a finite order open refinement say $\mathcal{V}_l = \{\underline{V_\beta} : \beta \in \Delta_1\}$. We can show that $\mathcal{U}_l = \{V_\beta : \underline{V_\beta} \in \mathcal{V}_l\}$ is a finite order open refinement of $\mathcal{U}$. Hence (X,T) is finitistic.

***Note 8*:** Theorem 4 and Theorem 7 give a relationship between the concepts of finitisticness in ordinary topology and finitisticness in L-topology.

References

1. C. Chang (1968), Fuzzy topological spaces, *J. Math. Anal.*, 24: 182-190.
2. D.S. Jamwal and Shakeel Ahmed (2002), *Covering Dimension and Finitistic Spaces in L-topology*, Conf. Proc., Fuzzy Set Theory, BHU, Allied Pub. : 117-122.
3. D.S. Jamwal and Shakeel Ahmed (2004), On Covering Dimension of L-fuzzy subsets and L-topological spaces, *J. Fuzzy Math*, 12(1).
4. J. Goguen (1973), The Fuzzy Tychnoff Theorem, *J. Math. Anal. Appl.*, 43: 734-742.
5. S. Deo (1987), Topology of finitistic spaces and related topics, *Bull. Allahabad Math. Soc.* : 231-61.
6. U. Hohle and S.E. Rodabaugh (1999), *Mathematics of Fuzzy Sets, Logic, Topology and Measure Theory*, Kluwer Academic Pub.
7. Y.M. Liu and M.K. Luo (1997), *Fuzzy Topology*, World Scientific Pub.

Fuzzy Logic and its Application in Technology and Management
Editors: Debjani Chakraborty, S. Nanda and D. Dutta Majumder

Complex Fuzzy Matrices

Susanta Kumar Khan and Madhumangal Pal[1]

Department of Applied Mathematics with Oceanology and Computer Programming,
Vidyasagar University, Midnapore-721102
[1]Email: madhumangal@lycos.com

Abstract: In this paper, the concept of complex fuzzy matrices (CFMs) is introduced. Some definitions regarding this CFM are also presented. Various properties of CFMs are presented here. The complex fuzzy determinant and adjoint of CFM are defined and presented some properties of them.

Keywords: Complex fuzzy set, Complex fuzzy matrix, Complex fuzzy determinant.

1. Introduction

Ample uses of matrices are in science and engineering but there remaining uncertainty in those spheres, the old matrix theory can not solve them. To solve those problems researchers found out fuzzy matrix. Fuzzy matrix deals with membership value of an object. The merits of utilizing fuzzy matrix for modeling uncertainty, representing street join, honesty or modesty of human being etc. As fuzzy matrix was found out by extending crisp matrix, complex fuzzy matrix is found out likewise by further extension of the fuzzy matrix.

The membership value of fuzzy matrix is of one dimension. But when the dimension of membership value will be two, the problems of them could not be solved with fuzzy matrix. Hence we introduced complex fuzzy matrix. No change in the basic concept of fuzzy matrix has been required for that. All the more complex fuzzy matrix was profited with gaining the amplitude and phase of membership value of element of matrix.

Definition 1. Complex Fuzzy Set: [3] A complex fuzzy set *S*, defined on a universe of discourse *U*, is characterized by a membership function $\mu_s(x)$ that assigns any element $x \in U$ a complex valued grade of membership in *S*. By definition, the values of $\mu_s(x)$ may receive all values within the unit circle in the complex plane, and of the form $r_s(x)\cdot e^{j.w_s(x)}$, where $j=\sqrt{-1}$, $\mathrm{r_s}(x)$ and $w_s(x)$ are both real-valued, with $r_s(x)\in[0,1]$.

The complex fuzzy set S may be represented as the set of ordered pairs

$$S=\{(x,\mu_s(x))|x\varepsilon U\}.$$

Definition 2. Complex Fuzzy Matrix (CFM): A complex fuzzy matrix A of order $m \times n$ is defined as $A = [\langle x_{pq}, a_{pq\mu} \rangle]$ where $a_{pq\mu} = a_{pqr} \cdot e^{j.a_{pqw}}$ is called complex valued membership function of x_{pq} in A, where the range of $a_{pq\mu}$ is not limited to [0, 1] but may receive all values within the unit circle in the complex plane.

Definition 3. Complement of Complex Fuzzy Matrix: Let $A = [\langle x_{pq}, a_{pq\mu} \rangle]$ be a complex fuzzy matrix and $a_{pq\mu}$ is the degree of membership of x_{pq}. Let $\overline{A}$ denote the fuzzy complement of A, defined by the function $\overline{a}_{pq\mu} : [0,1] \rightarrow [0, 1]$, which assigns a value $\overline{a}_{pq\mu}.\overline{a}_{pq\mu}$ may be interpreted both as the degree to which x_{pq} is a member of $\overline{A}$, and as the degree to which x_{pq} is not a member of A. The fuzzy complement function $\overline{a}_{pq\mu}$ must satisfy at least the following two axiomatic requirements:

Axiom 1. $\overline{0} = 1$ and $\overline{1} = 0$;

Axiom 2. For all $x_{pq\mu}, y_{pq\mu} \in [0,1]$, if $x_{pq\mu} \leq y_{pq\mu}$, then $\overline{x}_{pq\mu} \geq \overline{y}_{pq\mu}$.

In some cases, the complement may satisfy the following requirements:

Axiom 3. $\overline{a}_{pq\mu}$ is a continuous function;

Axiom 4. $\overline{a}_{pq\mu}$ is involutive, i.e. for all $x_{pq\mu} \in [0,1], \overline{(\overline{x}_{pq\mu})} = x_{pq\mu}$.

1.1. Operations Between Two Cmfs

Let A and B are two CFMs, where

$a_{pq\mu} = a_{pqr}.e^{j;a_{pqw}}$ *and* $b_{pq\mu} = b_{pqr}.e^{j.b_{pqw}}$.

Then $A \cup B = [(a_{pqr} \oplus b_{pqr})\ e^{j(apqw \cup bpqw)}]$. The operators $\oplus$ and $\cup$ are defined in many different ways. Some of them are given below.

Maximum (Standard Union): $a_{pqr} \oplus b_{pqr} = \max[a_{pqr}, b_{pqr}]$,

Algebraic Sum: $a_{pqr} \oplus b_{pqr} = a_{pqr} + b_{pq\,r} - a_{pqr} \cdot b_{pqr}$,

Bounded Sum: $a_{pqr} \oplus b_{pqr} = \min[1, a_{pqr} + b_{pqr}]$.

The operator $\bigcup$ is defined as

Sum: $a_{pqw} \bigcup b_{pqw} = a_{pqw} + b_{pqw}$,

Max: $a_{pqw} \bigcup b_{pqw} = \max[a_{pqw}, b_{pqw}]$,

Min: $a_{pqw} \bigcup b_{pqw} = \min[a_{pqw}, b_{pqw}]$,

"Winner Talk All" : $a_{pqw} \bigcup b_{pqw} = a_{pqw}$, if $a_{pqr} > b_{pqr}$

$= b_{pqw}$, if $a_{pqr} < b_{pqr}$.

The appropriate function must be selected according to the specific applications. The intersection between two CFMs is defined as $A \bigcap B = [(a_{pqr} * b_{pqr})\ e^{j(apqw \cap bpqw)}]$.

The operators * and $\bigcap$ are defined as follows:

Minimum (Standard Intersection): $a_{pqr} * b_{pqr} = \min[a_{pqr}, b_{pqr}]$,

Algebraic Product: $a_{pqr} * b_{pqr} = a_{pqr} \cdot b_{pqr}$,

Bounded Difference: $a_{pqr} * b_{pqr} = \max[0, a_{pqr} + b_{pqr} - 1]$.

The operator $\bigcap$ is defined as

Sum : $a_{pqw} \bigcap b_{pqw} = a_{pqw} + b_{pqw}$,
Max : $a_{pqw} \bigcap b_{pqw} = \max[a_{pqw}, b_{pqw}]$,
Min : $a_{pqw} \bigcap b_{pqw} = \min[a_{pqw}, b_{pqw}]$,
"Winner Talk All" : $a_{pqw} \bigcap b_{pqw} = a_{pqw}$, if $a_{pqr} > b_{pqr}$
$= b_{pqw}$, if $a_{pqr} < b_{pqr}$.

Definition 4.Complex Fuzzy Determinant (CFD): A complex fuzzy determinant (CFD) function f_1:$M \rightarrow F$ is a function on the set M of CFMs of order $n \times n$ over F (the set of complex numbers within a unit circle) such that if $A \in M$ then f_1(A) or |A| or det(A) belongs to F and is given by $|A| = \bigcup_{\sigma \in S_n} \bigcap_{i=1}^{n} a_{i\sigma(i)\mu}$, where $a_{i\sigma(i)\mu} = a_{i\sigma(i)\mu}.e^{ja_{i\sigma(i)w}}$ and S_n denotes the symmetric group of all permutations of the symbols {1, 2, …, n}.

Definition 5. Transpose of a CFM: Let A=[$a_{pq\mu}$] be a CFM order A. The transpose A^T of A is defined as A^T=[$a_{qp\mu}$].

Definition 6. Adjoint of Complex Fuzzy Matrix: The adjoint of CFM of order $n \times n$, which is denoted by Adj A, is defined by [A_{qp}], where A_{qp} is the determinant of the CFM A of order $(n-1) \times (n-1)$ formed by suppressing row q and column p of the CFM A. In other word [A_{qp}] can be written in the form $\bigcup_{\sigma \in S_{n_p n_q}} \bigcap_{t \in n_q} a_{t\sigma(t)r}.e^{j.a_{t\sigma(t)w}}$ when $n_q = \{1,2,...n\} \setminus \{q\}$ and $S_{n_p n_q}$ is the set of all permutation of set n_q over the set n_p.

Definition 7. Symmetric CFM: A square CFM A = [$a_{pq\mu}$] where $a_{pq\mu} = a_{pqr}.e^{j.apqw}$ is said to be symmetric CFM if $A = A^T$, i.e. $a_{pq\mu} = a_{qp\mu}$, for all p, q.

Definition 8. Conjugate CFM: Let A is a CFM of order $m \times n$. Then the elements of a matrix A are replaced by their conjugate elements, then that type of matrix is called conjugate CFM and is denoted by A^*.

Definition 9. Tranjugate CFM: The transpose of A^* denoted by $(A^*)^T$ is called tranjugate CFM.

Definition 10. Let A = [$a_{pq\mu}$] where $a_{pq\mu} = a_{pqr}.e^{j.apqw}$ be a CFM of order $m \times n$. The rotation of A by θ radians is defined as

$$A_{ROT} = \left[a_{pqr}.e^{j(a_{pqw}+\theta)}\right]$$

and the reflection of A is defined as

$$A_{REF} = \left[a_{pqr}.e^{-j.a_{pqw}}\right].$$

Property 1. If any two rows (or columns) of a square CFM are interchanged then determinant of that CFM remain unchanged.

Proof. Let $A = \left[a_{pqr}.e^{j.a_{pqw}}\right] and\ B = \left[b_{pqr}.e^{j.b_{pqw}}\right]$ is obtained from A by interchanging the rth row and sth row of A. Without loss of generality, we assume $r < s$.
Then

$$b_{pqr}.e^{j.b_{pqw}} = a_{pqr}.e^{j.a_{pqw}}, p \neq r, p \neq s.$$

$$b_{mqr}.e^{b_{mqw}} = a_{sqr}.e^{j.a_{sqw}},$$

$$b_{sqr}.e^{jb_{sqw}} = a_{mqr}.e^{j.a_{mqw}}.$$

Now,

$$|B| = \bigcup_{\sigma \in s_n} b_{1\sigma(1)r}.e^{j.b_{1\sigma(1)w}} \cap b_{2\sigma(2)r}.e^{j.b_{2\sigma(2)w}} \ldots \cap b_{m\sigma(m)r}.e^{j.b_{m\sigma(m)w}} \ldots \cap b_{sr(s)r}.e^{j.b_{s\sigma(s)w}} \cap \ldots$$

$$\cap b_{n\sigma(nr)}.e^{j.b_{n\sigma(n)w}}$$

$$= \bigcup_{\sigma \in s_n} a_{1\sigma(1)r}.e^{j.a_{1\sigma(1)w}} \cap a_{2\sigma(2)r}.e^{j.a_{2\sigma(2)w}} \cap \ldots \cap a_{s\sigma(m)r}.e^{j.a_{s\sigma(m)w}} \ldots \cap a_{m\sigma(s)r}.e^{j.a_{m\sigma(s)w}} \ldots$$

$$\cap a_{n\sigma(n)r}.e^{j.a_{n\sigma(n)w}}.$$

Let $\lambda = \begin{pmatrix} 1 & 2 \ldots m \ldots s \ldots n \\ 1 & 2 \ldots s \ldots m \ldots n \end{pmatrix}$

Then λ is a transposition interchanging *m* and *s*.

Let $\sigma\lambda = \psi$, as σ runs over the set of all permutations of *{1, 2,..., n}*, ψ does so, because $\sigma_1\lambda = \sigma_2\lambda \Rightarrow \sigma_1 = \sigma_2$.

Now $\psi = \sigma\lambda$

$$= \begin{pmatrix} 1 & 2 & \ldots & m & \ldots & s & \ldots & n \\ \sigma(1) & \sigma(2) & \ldots & \sigma(m) & \ldots & \sigma(s) & \ldots & \sigma(n) \end{pmatrix} \begin{pmatrix} 1 & 2 \ldots m \ldots s \ldots n \\ 1 & 2 \ldots s \ldots m \ldots n \end{pmatrix} = \begin{pmatrix} 1 & 2 & \ldots & m & \ldots & s & \ldots & n \\ \sigma(1) & \sigma(2) & \sigma(s) & \ldots & \sigma(m) & \ldots & \sigma(n) \end{pmatrix}$$

Therefore, $\psi(p) = \sigma(p), p \neq m, p \neq s$.

$\psi(m) = \sigma(s), \psi(s) = \sigma(m).$

$$|B| = \bigcup_{\sigma \in s_n} a_{1\sigma(1)r}.e^{j.a_{1\sigma(1)w}} \cap a_{2\sigma(2)r}.e^{ja_{2\sigma(2)w}} \ldots \cap a_{s\sigma(m)r}.e^{j.a_{s\sigma(m)w}} \ldots \cap a_{m\sigma(s)r}.e^{j.a_{m\sigma(s)w}} \ldots$$

$$\cap a_{n\sigma(n)r}.e^{j.a_{n\sigma(n)w}}$$

$$= \bigcup_{\psi \in s_n} a_{1\psi(1)r}.e^{j.a_{1\psi(1)w}} \cap a_{2\psi(2)r}.e^{j_{a2\psi2(w)}} \cap \ldots \cap a_{m\psi(m)r}.e^{j.a_{m\psi(m)w}} \ldots \cap a_{s\psi(s)r}.e^{j.a_{s\psi(s)w}} \ldots$$

$$\cap a_{n\psi(n)r}.e^{j.a_{n\psi(n)w}} = |A|.$$

Property 2. For a square CFM $(AdjA^T) = (AdjA)^T$.

Proof. Let $B = AdjA$, $C = Adj\ A^T$.

Therefore,

$$b_{pq\mu} = \bigcup_{\sigma \in S_{n_q n_p}} \bigcap_{t \in n_p} a_{t\sigma(t)r}.e^{j.a_{t\sigma(t)w}} \text{ and } C_{pq\mu} = \bigcup_{\sigma \in S_{n_p n_q}} \bigcap_{t \in n_p} a_{t\sigma(t)r}.e^{j.a_{t\sigma(t).w}} = b_{pq\mu}$$

Therefore, $AdjA^T = (AdjA)^T$.

Property 3. For a CFM *A*, we have $|A| = |AdjA|$.

Proof. $AdjA = [A_{qpr}.e^{j.A_{qpw}}]$

Where, $A_{qpr}.e^{j.A_{qpw}}$ is the co-factor of the element $a_{pq\mu}$ in *A*.

Therefore, $|AdjA| = \bigcup_{\sigma \in S_n} A_{1\sigma(1)r}.e^{j.A_{1\sigma(1)w}} \cap A_{2\sigma(2)r} \cdot e^{j.A_{2\sigma(2)W}} \ldots \cap A_{n\sigma(n)r} \cdot e^{j.A_{n\sigma(n)w}}$

$$= \bigcup_{\sigma \in S_n} \bigcap_{p=1}^{n} A_{p\sigma(p)r}.e^{j.A_{p\sigma(p)w}}$$

$$= \bigcup_{\sigma \in S_n} \left[\bigcap_{p=1}^{n} \left(\bigcup_{\theta \in S_{n_p n_{\sigma(p)}}} \bigcap_{t \in n_p} a_{t\theta(t)r} \cdot e^{j.a_{t\theta(t)w}} \right) \right]$$

$$= \bigcup_{\sigma \in S_n} \left[\left(\bigcap_{t \in n_1} a_{t\theta_1(t)r}.e^{j.a_{t\theta_1(t)w}} \right) \left(\bigcap_{t \in n_2} a_{t\theta_2(t)r}.e^{j.a_{t\theta_2(t)w}} \right) \left(\bigcap_{t \in n_n} a_{t\theta_n(t)r}.e^{j.a_{t\theta_n(t)w}} \right) \right]$$

for some $\theta_1 \in S_{n_1 n_{\sigma(1)}}, \theta_2 \in S_{n_2 n_{\sigma(2)}}, \ldots \theta_n \in S_{n_n n_{\sigma(n)}}$.

$$= \bigcup_{\sigma \in S_n} \left[\left(a_{2\theta_1(2)r}.e^{j.a_{2\theta_1(2)w}} \bigcap a_{3\theta_1(3)r}.e^{ja_{3\theta_1(3)w}} \bigcap \ldots \right. \right.$$

$$\left. \bigcap a_{n\theta_1(n)r}.e^{j.a_{n\theta_1(n)w}} \right) \left(a_{1\theta_2(1)r}.e^{j.a_{1\theta_2(1)w}} \bigcap a_{3\theta_2(3)r}.e^{j.a_{3\theta_2(3)w}} \ldots \bigcap a_{n\theta_2(n)r}.e^{j.a_{n\theta_2(n)w}} \right)$$

....

$$\left(a_{1\theta_n(1)r}.e^{ja_{1\theta_n(1)w}} \bigcap a_{2\theta_n(2)r}.e^{j.a_{2\theta_n(2)w}} \bigcap \ldots \bigcap a_{n-1}\theta_{n(n-1)r}e^{ja_{n-1\theta_n(n-1)w}} \right)$$

$$= \bigcup_{\sigma \in S_n} \left[\left\{ \left(a_{1\theta_2(1)r}.e^{j.a_{1\theta_2(1)w}} \right) \bigcap \left(a_{1\theta 3(1)r}.e^{j.a_{1\theta_3(1)w}} \right) .. \bigcap \left(a_{1\theta_n(1)r}e^{j.a_{1\theta_n(1)w}} \right) \right\} \right.$$

$$\left\{ \left(a_{2_{\theta_1(2)r}}.e^{j.a_{2\theta_1(2)w}} \right) \bigcap a_{2\theta_3(2)r}.e^{j.a_{2\theta_3(2)w}} \ldots \bigcap \left(a_{2\theta_n(2)r}.e^{j.a_{2\theta_n(2)w}} \right) \right\} .. \left\{ \left(a_{n\theta_1(n)r}.e^{j.a_{n\theta_1(n)w}} \right) \bigcap \right.$$

$$\left. \left. \left(a_{n\theta_2(n)r}.e^{j.a_{n\theta_2(n)w}} \right) \bigcap \ldots \left(a_{n\theta_{n-1}(n)r}.e^{j.a_{n\theta_{n-1}(n)w}} \right) \right\} \right]$$

$$= \bigcup_{\sigma \in S_n} \left[\left(a_{1\theta_{f_1}(1)r}.e^{j.a_{1\theta_{f_1}(1)w}} \right) \bigcap \left(a_{2\theta_{f_2}(2)r}.e^{j.a_{2\theta_{f_2}(2)w}} \right) \bigcap \ldots \bigcap \left(a_{n\theta_{fn}(n)r}.e^{j.a_{n\theta_{f_n}(n)w}} \right) \right]$$

Where $f_{\hat{\theta}} \in \{1,2,\ldots,n\} \setminus \{\hat{\theta}\}, \hat{\theta} = 1,2,\ldots,n$.

But since, $a_{\hat{\theta}\theta_{f_\theta}(\hat{\theta})r}.e^{j.a_{\theta\theta_{f_\theta}(\theta)w}} = a_{n\sigma(n)r}.e^{j.a_n\sigma(n)w}$

Therefore, $|AdjA| = \bigcup_{\sigma \in S_n} a_{1\sigma(1)r}.e^{j.a_{1\sigma(1)w}} \bigcap a_{2\sigma(2)r}.e^{ja_{2\sigma(2)r}} \bigcap \ldots a_{n\sigma(n)r}.e^{ja_{n\sigma(n)w}}$

$= |A|$.

Property 4. If A be square CFM then $|A| = |A^T|$.

Proof. Let $A = [a_{pq\mu}]$ where $a_{pq\mu} = a_{pqr}.e^{j.a_{pqw}}$

and $A^T = B = [b_{pq\mu}]$ where $b_{pq\mu} = b_{pqr}.e^{j.b_{pqw}}$.

$= [a_{pq\mu}]$

Now, $|B| = \bigcup_{\sigma \in S_n} b_{1\sigma(1)r}.e^{j.b_{1\sigma(1)w}} \bigcap b_{2\sigma(2)r}.e^{j.b_{2\sigma(2)w}} \ldots \bigcap b_{n\sigma(n)r}.e^{j.b_{n\sigma(n)w}}$

$= \bigcup_{\sigma \in S_n} a_{\sigma(1)1r}.e^{j.a_{\sigma 1(1)w}} \bigcap a_{\sigma(2)2r}.e^{j.a_{\sigma 2(2)w}} \ldots \bigcap a_{\sigma(n)nr}.e^{j.a_{\sigma n(n)w}}$.

Let ψ be the permutation of $\{1,2,\ldots,n\}$ such that $\psi\sigma = I$, the identity permutation. Then $\psi = \sigma^{-1}$. As σ runs over the whole set of permutation so does ψ.

Let $\sigma(p) = q, p = \sigma^{-1}(q) = \psi(q)$.

Therefore, $a_{\sigma(p)pr}.e^{j.a_{\sigma(p)pw}} = a_{q\psi(q)r}.e^{j.a_{q\psi(q)pw}}$ for all p, q.

As p runs over the set {1, 2, …, n}, q does so.

Now $a_{\sigma(1)1r}.e^{j.a_{\sigma(1)1w}} \cap a_{\sigma(2)2r}.e^{j.a_{\sigma(2)2w}} \cap \ldots \cap a_{\sigma(n)nr}.e^{j.a_{\sigma(n)nw}} = a_{1\psi(1)r}.e^{ja_{1\psi(1)w}} \cap a_{2\psi(2)r}.e^{ja_{2\psi(2)w}} \ldots \cap a_{n\psi(n)r}.e^{ja_{n\psi(n)w}}$.

Therefore, $|B| = \bigcup_{\sigma \in S_n} a_{\sigma(1)1r}e^{ja\sigma(1)1w} \cap a_{\sigma(2)2r}e^{ja_{\sigma(2)2w}} \ldots \cap a_{\sigma(n)nr}.e^{j.a_{\sigma(n)nw}}$

$$= \bigcup_{\psi \in S_n} \left(a_{1\psi(1)r}.e^{ja_{1\psi(1)w}} \cap a_{2\psi(2)r}e^{ja_{2\psi(2)w}} \cap a_{n\psi(n)r}.e^{ja_{n\psi(n)w}}\right)$$

$$= |A|.$$

Property 5. For any three CFMs A, B and C of same order then

i) $A \cup B = B \cup A$

ii) $(A \cup B) \cup C = A \cup (B \cup C)$

iii) $A \cap B = B \cap A$

iv) $(A \cap B) \cap C = A \cap (B \cap C)$

v) $A \cap (B \cup C) = (A \cap B) \cup (A \cap C)$

vi) $A \cup (B \cap C) = (A \cup B) \cap (A \cup C)$.

References

1. J.J. Buckley (1987), Fuzzy complex numbers, in *Proc. ISFK*, Guangzhou, China, pp 597-700.
2. G.J. Klir & B. Yuan (1995), *Fuzzy Sets and Fuzzy Logic: Theory and Applications*, NJ, Prentice Hall.
3. D. Ramot, R. Milo, M. Friedman & A. Kandel (2002), Complex fuzzy sets, *IEEE Transactions on Fuzzy Systems*, 10 (2): 171-186.

Fuzzy Logic and its Application in Technology and Management
Editors: Debjani Chakraborty, S. Nanda and D. Dutta Majumder

Distance Between Fuzzy Matrices and Its Application

Amiya K. Shyamal and Madhumangal Pal[1]

Department of Applied Mathematics with Oceanology and Computer Programming
Vidyasagar University, Midnapore-721 102
[1]Email: palmadhumangal@rediff.com

Abstract: Two basic distances, Hamming distance and Euclidean distance between two fuzzy matrices are defined in this paper and a lot of algebraic properties are investigated using these two distances. By measuring the Hamming distance between two Boolean fuzzy matrices a method is developed to recognize a character.

Keywords: Fuzzy matrix, Fuzzy operators, Hamming and Euclidean distances.

1. Introduction

In real life situation, matrices play an important role in different branches of science and technology. But, due to presence of various types of uncertainties, the traditional classical matrices may not be sufficient for the precise description of the characteristics of any system, pattern, etc. Fuzzy matrices have attracted the attention of many authors and they have presented a number of results on the convergence of power sequence of fuzzy matrices [3].

In this paper, we define two basic distances Hamming distance, Euclidean distance and we have investigated a lot of properties between Hamming distance and Euclidean distance. One application using distance is also cited here.

In the age of automation, men having the natural propensity of performing all types of work automatically instead of human interference, the character recognition is one of the most successful and epoch-making applications of automatic pattern recognition. The objective of character recognition system is to recognise a character automatically by machine without typing or human interaction. There are various types of optical character recognition system in different languages. Several works on optical character recognition system are available in literature [1,4,5]. Here we discuss and analyze a new system of character recognition. In this process recognition of character is possible by measuring the Hamming distance.

2. Definitions

In this section, fuzzy matrix and boolean fuzzy matrix are defined and some operations on fuzzy matrices are also introduced.

***Fuzzy Matrix*:** A fuzzy matrix (*FM*) A of order $m \times n$ is defined as $A=[<a_{ij}, a_{ij\mu}>]_{m\times n}$ where $a_{ij\mu}$ is the membership value of the element a_{ij} in A. For simplicity, we write A as $A = [a_{ij}]_{m\times n}$.

***Boolean Fuzzy Matrix*:** A fuzzy matrix $A = [a_{ij}]_{m\times n}$ is said to be a boolean fuzzy matrix of order $m\times n$ if all elements of A are either 0 or 1.

Let $A = [a_{ij}]$ and $B = [b_{ij}]$ be any two fuzzy matrices of order $m\times n$. The following operations are defined for any two elements x and y of a fuzzy matrix.

(i) $x \vee y = max\,(x,y)$
(ii) $x \wedge y = min\,(x,y)$
(iii) $x \oplus y = x+y-x.y$
(iv) $x \odot y = x.y$
(v) $x \ominus y = \begin{cases} x, & \text{if } x>y \\ 0, & \text{if } x\leq y \end{cases}$
(vi) $x^{(\alpha)} = x$, if $x>\alpha$, otherwise 0.

The following operations are defined for any two fuzzy matrices $A = [a_{ij}]$ and $B = [b_{ij}]$ of order $m\times n$ as follows:

(i) $A\oplus B = [a_{ij}+b_{ij} - a_{ij}.b_{ij}]$
(ii) $A \odot B = [a_{ij}.\, b_{ij}]$
(iii) $A\vee B = [a_{ij} \vee b_{ij}]$
(iv) $A\wedge B = [a_{ij} \wedge b_{ij}]$
(v) $A \ominus B = [a_{ij} \ominus b_{ij}]$
(vi) $A^{[k+1]} = A^{[k]} \odot A, A^{[1]}=A, k=1, 2, \ldots$
(vii) $[k+1]A = [k]A \oplus A,\ [1]A=A,\ k=1, 2, \ldots$
(viii) $A@B = [\frac{a_{ij}+b_{ij}}{2}]$
(ix) $A\$B = [\sqrt{a_{ij}.b_{ij}}\,]$
(x) $A^{(\alpha)} = [a_{ij}^{(\alpha)}]$
(xi) $A' = [a_{ji}]$ (the transpose of A)
(xii) $A^c = [1-a_{ij}]$ (the complement of A)
(xiii) $A \leq B$ if and only if $a_{ij} \leq b_{ij}$ for all i, j
(xiv) For any two fuzzy matrices A and B, $min\,(A, B) = A$, if $A \leq B$ and B if $B \leq A$.

The operators '+', '−' and '.' are ordinary addition, subtraction and multiplication.

For an $n\times n$ fuzzy matrix $A = [a_{ij}]$ we have,

(i) A is reflexive if and only if $a_{ii} = 1$ for all $i=1, 2, \ldots, n$.
(ii) A is irreflexive if and only if $a_{ii} = 0$ for all $i=1, 2,\ldots, n$.
(iii) A is nearly irreflexive if and only if $a_{ii} \leq a_{ij}$ for all $i, j = 1, 2, \ldots, n$.
(iv) A is symmetric if and only if $A' = A$.
(v) A is identity if and only if $a_{ii} = 1$ and $a_{ij} = 0$ for all $i\neq j$, for all i, j.
The identity matrix of order $n\times n$ is generally denoted by I_n.
(vi) A is weakly reflexive if $a_{ii} \geq a_{ij}$ for all i, j.
(vii) A is diagonal if $a_{ii} \geq 0$ and $a_{ij} = 0,\ i\neq j$ for all i, j.

3. Distances Between FMs

In this section, we define two basic distances between FMs. The distance δ between FMs A and B is a mapping from the set of FMs (M) to the set of real numbers (R) i.e., $\delta : M \times M \rightarrow R$.

3.1. Hamming Distance

The Hamming distance between two FMs A and B of order $m \times n$ is

$$H(A,B) = \sum_{i=1}^{m} \sum_{j=1}^{n} |a_{ij} - b_{ij}|$$

It is obvious that $0 \leq H(A,B) \leq m.n.$

If $a_{ij} = 1$ and $b_{ij} = 0$ for all i, j or $a_{ij} = 0$ and $b_{ij} = 1$ for all i, j then $H(A,B) = m.n.$

(1)Ifa_{ij}=b_{ij} for all $i\,j$ then $H(A,B) = 0$.

(2)The Hamming distance $H : M \times M \rightarrow R$ satisfies the following conditions:

(i) $H(A,B) \geq 0$ for all $A, B \in M$,

(ii) $H(A,B) = 0$ iff $A = B$ for all $A, B \in M$,

(iii) $H(A,B) = H(B,A)$ for all $A, B \in M$ (symmetry),

(iv) $H(A,B) \leq H(A,C) + H(C,B)$ for all $A, B, C \in M$ (triangular inequality).

Thus the distance function H is a metric on M.

3.2. Normalized Hamming Distance

The normalized Hamming distance between two FMs A and B is denoted by $H^*(A,B)$ and is defined as $H^*(A,B) = H(A,B)/m.n$. It may be noted that $0 \leq H^*(A,B) \leq 1$.

3.3. Euclidean Distance

The Euclidean distance between two FMs A and B of order $m \times n$ is defined as

$$E(A, B) = \sqrt{\sum_{i=1}^{m} \sum_{j=1}^{n} (a_{ij} - b_{ij})^2}.$$

It is also obvious that $0 \leq E(A, B) \leq \surd\{m.n\}$. The sign of equality holds when (1) and (2) are satisfied. The Euclidean distance E is also a metric on M.

3.4. Normalized Euclidean Distance

The Normalized Euclidean distance between two FMs A and B is denoted by $E(A,B)$ and is defined as $E^*(A, B) = E(A, B) / \surd\{m.n\}$. Here also $0 \leq E^*(A, B) \leq 1$. The maximum Hamming distance between two boolean fuzzy matrices A and B of order $m \times n$ is $m.n$. It is possible only when one of the following conditions hold: $a_{ij} = 1$ and $b_{ij} = 0$ or $a_{ij} = 0$ and $b_{ij} = 1$. Again if $a_{ij} = b_{ij}$ then distance will be zero, it is the minimum distance between two matrices.

Depending on the values of $H^*(A,B)$ or $E^*(A,B)$ we can define the following terms.

(i) If $H^*(A,B) = 1$ or $E^*(A,B) = 1$ then A and B are *totally opposite*.

(ii) If $H^*(A,B) = 0.5$ or $E^*(A,B) = 0.5$ then A and B are *nearly equal*.

(iii) If $H^*(A,B) = 0.25$ or $E^*(A,B) = 0.25$ then A and B are *almost equal*.

(iv) If $H^*(A,B) = 0$ or $E^*(A,B) = 0$ then A and B are *exactly equal*.

4. Properties of Distances

A large number of properties on Hamming distance between two fuzzy matrices are investigated in this section. The similar results are also valid for Euclidean distance.

***Property 1*:** For any fuzzy matrix A,

(i) $H(A, A^{[2]}) \leq H(A, A^{[k]})$; $k = 2,3,...$
(ii) $H(A,[2]A) \leq H(A,[k]A)$; $k = 2,3,...$
(iii) $H(A, A^{[2]}) = H(A,[2]A)$

***Proof*:** (i) Let $A = [a_{ij}]$.
Then, $A^{[2]} = A \circ A = [a_{ij}^2]$. For any positive integer k, $A^{[k]} = [a_{ij}^k]$.
Since $0 \leq a_{ij} \leq 1$ then, $a_{ij}^2 \geq a_{ij}^3 \geq ... \geq a_{ij}^k$, for any positive integer k.
Hence, $H(A, A^{[2]}) \leq H(A, A^{[k]})$, $k=2,3,...$

(ii) Let $A = [a_{ij}]$. Then, $[2]A = A \oplus A = [2a_{ij} - a_{ij}^2] = [1-(1-a_{ij})^2]$.
For any positive integer k, we have
$[k]A = [1-(1-a_{ij})^k]$.

Now, $H(A,[2]A) = \sum_{i=1}^{n}\sum_{j=1}^{n}\left| a_{ij} - 1 + (1-a_{ij})^2 \right|$ and $H(A,[k]A) = \sum_{i=1}^{n}\sum_{j=1}^{n}\left| a_{ij} - 1 + (1-a_{ij})^k \right|$.

Since, $1 - a - (1-a)^2 \leq 1 - a - (1-a)^k$ for $k \geq 2$.
Therefore, $| a - 1 + (1-a)^2 | \geq | a - 1 + (1-a)^k |$.
Since, $0 \leq a_{ij} \leq 1$ we have, $(1-a_{ij})^2 \geq (1-a_{ij})^k$ for any positive integer k.
Hence, $H(A, [2]A) \leq H(A, [k]A)$.

(iii) $H(A, A^{[2]}) = \sum_{i=1}^{n}\sum_{j=1}^{n}\left| a_{ij} - a_{ij}^2 \right|$ and $H(A, [2]A) = \sum_{i=1}^{n}\sum_{j=1}^{n}\left| a_{ij} - 1 + (1-a_{ij})^2 \right|$.

We have, $(a-1) + (1-a)^2 = a(a-1)$ and $a - a^2 = a(1-a)$.
Therefore, $| a_{ij} - a_{ij}^2 | = | a_{ij} - 1 + (1-a_{ij})^2 |$.
Hence, $H(A, A^{[2]}) = H(A, [2]A)$.

***Property 2*:** Let I_n be the identity fuzzy matrix of order $n \times n$ and $A = [a_{ij}]$ be any FM of the same order then,

(i) I_n^c and $A \ominus I_n$ are irreflexive FMs.
(ii) $I_n \vee A$ and $I_n \oplus A$ are reflexive FMs.
(iii) $I_n \wedge A$ and $I_n \circ A$ are diagonal FMs.
(iv) $I_n @ A$ is weakly reflexive FM.
(v) $I_n \oplus (A \oplus A')$ is reflexive and symmetric FM.

***Property 3*:** Let I_n be the identity FM of order $n \times n$ and $A = [a_{ij}]$ be any FM of the same order then,

(i) $H(I_n \circ A, I_n \wedge A) = 0$, (ii) $H(I_n \oplus A, I_n \vee A) = 0$,
(iii) $H(I_n \circ A, I_n \$ A) \leq n$, (iv) $H(I_n, I_n \circ A) \leq n$,
(v) $H(I_n, I_n \wedge A) \leq n$, (vi) $H(I_n, I_n \$ A) \leq n$,
(vii) $H(I_n, I_n^c) = n^2$, (viii) $H(I_n \oplus (A \oplus A'), I_n \oplus A) \leq n^2 - n$,
(ix) $H(I_n^c, A \ominus I_n) \leq n^2 - n$, (x) $H(I_n, I_n \oplus A) \leq i^2 - n$, (xi) $H(I_n, I_n \vee A) \leq n^2 - i$.

***Property 4*:** For any two fuzzy matrices $A = [a_{ij}]$ and $B = [b_{ij}]$ of the same order,

(i) $H(A, A \wedge B) = H(B, A \vee B)$, (viii) $H(A, A @ B) = H(B, A @ B)$,
(ii) $H(A, A \vee B) = H(B, A \wedge B)$, (ix) $H(A @ B, A \vee B) = H(A @ B, A \wedge B)$,
(iii) $H(A, A \oplus B) = H(B, A \circ B)$, (x) $H(A \oplus B, A @ B) = H(A \circ B, A @ B)$,
(iv) $H(A, A \circ B) = H(B, A \oplus B)$, (xi) $H(A, A @ B) = H(A @ B, A \wedge B)$,
(v) $H(A \oplus B, A \vee B) = H(A \circ B, A \wedge B)$, (xii) $H(A^c \oplus B^c, (A \oplus B)^c) = H(A^c \circ B^c, (A \circ B)^c)$,
(vi) $H(A, B) = H(A \vee B, A \wedge B)$, (xiii) $H(A^c \oplus B^c, (A \circ B)^c) = H((A \oplus B)^c, A^c \circ B^c)$.
(vii) $H(A \circ B, A \vee B) = H(A \oplus B, A \wedge B)$,

Proof: (vii) We have, $H(A \odot B, A \vee B) = \sum_{i=1}^{n}\sum_{j=1}^{n} |a_{ij}.\, b_{ij} - \max\{a_{ij}, b_{ij}\}|$ and

$$H(A \oplus B, A \wedge B) = \sum_{i=1}^{n}\sum_{j=1}^{n} |a_{ij} + b_{ij} - a_{ij}.b_{ij} - \max\{a_{ij}, b_{ij}\}|.$$

Two cases will arise:

Case-I. When $a_{ij} \geq b_{ij}$ then

$$H(A \odot B, A \vee B) = \sum_{i=1}^{n}\sum_{j=1}^{n} |a_{ij}.\, b_{ij} - a_{ij}| \text{ and also } H(A \oplus B, A \wedge B) = \sum_{i=1}^{n}\sum_{j=1}^{n} |a_{ij} - a_{ij}.\, b_{ij}|.$$

Case-II. When $b_{ij} \geq a_{ij}$ then $H(A \odot B, A \vee B) = \sum_{i=1}^{n}\sum_{j=1}^{n} |a_{ij}.\, b_{ij} - b_{ij}|$ and

$$H(A \oplus B, A \wedge B) = \sum_{i=1}^{n}\sum_{j=1}^{n} |b_{ij} - a_{ij}.\, b_{ij}|.$$

Hence, from both the cases we have, $H(A \odot B, A \vee B) = H(A \oplus B, A \wedge B)$.
Other proofs are similar.

Property 5: For any two fuzzy matrices $A = [a_{ij}]$ and $B = [b_{ij}]$ of same order,

(i) $H(A, A \odot B) \geq H(A, A \wedge B)$,
(ii) $H(A, A \odot B) \geq H(A \oplus B, A \vee B)$,
(iii) $H(A, A \oplus B) \geq H(A, A \vee B)$,
(iv) $H(A, A \oplus B) \geq H(A \oplus B, A \vee B)$,
(v) $H(A \odot B, A \vee B) \geq H(A \vee B, A \wedge B)$,
(vi) $H(A \vee B, A \wedge B) \geq H(A @ B, A \vee B)$
(vii) $H(A \oplus B, A @ B) \geq H(A @ B, A \vee B)$,
(viii) $H(A \odot B, A \vee B) \geq H(A \oplus B, A @ B)$,
(ix) $H(A \oplus B, A \odot B) \geq H(A \oplus B, A @ B)$,
(x) $H(A \oplus B, A \wedge B) \geq$ i$(A \oplus B, A \vee B)$,
(xi) $H(A \oplus B, A \ominus B) \geq H(A \oplus B, A \vee B)$,
(xii) $H(A \odot B, A \vee B) \geq H(A \oplus B, A \vee B)$,
(xiii) $H(A \oplus B, A \wedge B) \geq H(A \vee B, A @ B)$,
(xiv) $H(A \vee B, A \$ B) \geq H(A \vee B, A \wedge B)$,
(xv) $H(A \vee B, A \$ B) \geq H(A \wedge B, A \$ B)$,
(xvi) $H(A \oplus B, A \$ B) \geq H(A \vee B, A \$ B)$,
(xvii) $H(A \oplus B, A \wedge B) \geq H(A \vee B, A \wedge B)$,
(xviii) $H(A \ominus B, A \oplus B) \geq H(A \ominus B, A \vee B)$.

5. Application to Character Recognition

For a long time human being tried to perform all types of work automatically instead of manual procedures. The character (alphabet) recognition is one of them. Several conventional methods are available to recognize the characters. But recognition of a character gets easier by measuring the Hamming distance. Here we discuss about the recognition of a character (printed only) by the help of Hamming distance. The recognition of hand written characters much difficult problem arises due to the large variability of data. Specially, in the case of hand written character variation of shape occurs due to writing habit, style, mood, etc. Which will differ from man to man and also some other factors such as the writing instruments, writing surface, etc. Now we shall illustrate how character recognition is possible by this process. There are 52 fixed printed English alphabets in binary matrix form including upper case and lower case alphabets. Now a printed manuscript of same font, same size and shape (length, width etc.) is taken to serve the purpose. A scanner, which is placed on the top of the printed manuscript and scanner, scanned each of the character. As a scanning process the scanner will draw a horizontal scan line when it will find black pixels of the character at the time of its top-down movement from the top of the manuscript. It will draw another horizontal scan line i.e., parallel to the first horizontal line at the fag end of that character. After completion of downwards movement the scanner will starts its horizontal movement and it will draw two parallel vertical scan lines touching at least one black pixel in the left and right side of the same character. Thus we formed a minimum sized rectangular box called *bounding box* which is needed to enclose the character. The black pixels of the character within the bounding box are filled up by 1 and white pixels are filled up by 0. It is nothing but a binary matrix of the character is formed. Now we measure the Hamming distance between the worked out binary matrix and the former binary matrix representation of 52 given characters. For which specific binary matrix of 52 former given characters the Hamming distance is zero, the worked out character will be the same specific character. In this case, there are some limitations. The characters have same font, same size and shape. Here for example, we consider two printed characters, say, 'h' and 'g'. Every pixel of both the characters (tracing path of the character) is filled up by 1 and the other pixels (back ground of the character) is filled up by 0. The boolean matrices of these two characters are shown in Figure 1. The boolean matrices for the characters 'h' and 'g' are respectively denoted by X and Y.

11110000000000000	00000111111000111
01110000000000000	00011111111100111
01110000000000000	00111111111110111
01110000000000000	01111110000111111
01110000000000000	01111000000001111
01110000000000000	11111000000001111
01110000000000000	11110000000001111
01110000000000000	11110000000000111
01110000000000000	11110000000000111
01110000000000000	11110000000000111
01110001111000000	11110000000000111
01110111111111000	11110000000001111
01111111111111100	01111000000001111
01111100000111100	01111000000011111
01111000000011110	01111100000111111
01110000000001110	00111111111110111
01110000000001110	00001111111100111
01110000000001110	00000011110000111
01110000000001110	00000000000000111
01110000000001110	00000000000000111
01110000000001110	01100000000000111
01110000000001110	01110000000000111
01110000000001110	01110000000001110
01110000000001110	00111000000011110
01110000000001110	00111100000111110
01110000000001110	00011111111111100
01110000000001110	00001111111111000
11111000000011111	00000011111100000
(a)	(b)

Figure 1: Binary matrices (a) for 'h' and (b) for 'g'.

The Hamming distance between the matrices X and Y is, $H(X,Y)= 215$.

This implies that the characters 'h' and 'g' for the matrices X and Y are not same, i.e., they are different from each other. But $H(X,X)=0$ and $H(Y,Y)=0$. Here Hamming distance in both the cases are zero. These imply that both the characters are same in each of the cases. Because same character has the same font, same size and shape, therefore the boolean matrix representation of a particular character will be exactly same. Hence all the time Hamming distance will be zero, if we measure the distance between the same characters.

References

1. Mantas (1986), An overview of character recognition methodologies, *Pattern Recognition*, 19: 425-430.
2. M.Z. Ragab and E.G. Emam (1994), The determinant and adjoint of a square fuzzy matrix, *Fuzzy Sets and Systems*, 61: 297-307.
3. M.G. Thomason (1977), Convergence of powers of a fuzzy matrix, *J.Math Anal. Appl.*, 57: 476-480.
4. K.Y. Wang, R.G. Casey and F.M. Wahl (1982), Document analysis system, *IBJM. Res. Dev.* 26: 647-656.
5. P.S.P. Wang (1991), *Character and Handwritten Recognition*, World Scientific, Singapore.

Fuzzy Logic and its Application in Technology and Management
Editors: Debjani Chakraborty, S. Nanda and D. Dutta Majumder

Partial and Multiple Correlations and Information Energy on Interval-Valued Fuzzy Sets

Nishi Kanta Jana and Madhumangal Pal[1]

Department of Applied Mathematics with Oceanology and Computer Programming
Vidyasagar University, Midnapore-721 102
[1]Email: mmpal@vidyasagar.ac.in

Abstract: In this paper, we define correlation of interval-valued fuzzy sets and studied some results on it. Also we studied the concept of partial and multiple correlation coefficients of interval-valued fuzzy sets and some properties. Finally the idea to measure the fuzzy ness of the IVFSs by the concept of information energy is introduced.

Keywords: Interval-valued fuzzy sets, Correlation, Partial and multiple correlations, Information energy.

1. Introduction

The correlation of fuzzy sets has been discussed using the concept of conventional statistics. The value of the correlation coefficient lies in the interval [-1,1]. Similar work have been done for different sets in [2, 4, 6]. The notation of interval-valued fuzzy sets (IVFSs) were introduced for the first time by Gorzalczany [5]. They had been applied in the field of approximate inference, signal transmission and control, etc. IVFS is a general case of fuzzy set. It is always not possible to settle up the exact membership value for an element. But, it is better to set up the membership value of an element by an interval, say, $[a, b]$. This means that the membership value of that element lies between a and b including a and b. This interval must be the sub interval of [0,1]. In this paper, we define the correlation between two IVFSs. Also we define the partial and multiple correlation coefficients of IVFS. The fuzziness of an IVFS is measure by calculating the information energy of an IVFS.

2. Definition and Preliminaries

Let $D[0,1]$ be a set of closed sub-intervals of the interval [0,1]. Let $M=[M_L, M_U]$ be an interval, where M_L and M_U are the lower and upper extreme respectively. Two intervals M and N are equal if $M_L=N_L$ and $M_U=N_U$. Similarly, $M\leq N$ iff $M_L\leq N_L$ and $M_U\leq N_U$ respectively. Let M^c be the complement interval of M. Then $M^c=I-M=[1-M_U, 1-M_L]$ with I the interval [1, 1] and O the interval [0, 0].

The amplitude of the interval M is denoted by W_M, which is equal to M_U-M_L.

Definition 1. Let X be the universal set. An IVFS in X is an expression given by $A=\{<x, M_A(x)> / x\in X\}$,where $M_A : X\rightarrow D[0,1]$ be the interval of degree of membership of the element x to the set A.

For all $A, B \in$IVFSs the following expressions are defined in [1, 7].

1. $A=B \Leftrightarrow M_A(x) = M_B(x)$ for all $x \in X$.
2. $A\leq B \Leftrightarrow M_{AL}(x) \leq M_{BL}(x)$ and $M_{AU}(x) \geq M_{BU}(x)$ for all $x\in X$.
3. $A^c = \{<x, M^c_A(x) > / x\in X\}$.
4. $A+B=\{<x, [M_{AL}(x)+ M_{BL}(x)-M_{AL}(x)M_{BL}(x), M_{AU}(x)+M_{BU}(x)-M_{AU}(x)M_{BU}(x)]>/x \in X\}$.
5. $A.B=\{<x, [M_{AL}(x).M_{BL}(x), M_{AU}(x).M_{BU}(x)] > /x \in X\}$.
6. $A@B=\{<x, [\{M_{AL}(x)+M_{BL}(x)\}/2,\{M_{AU}(x)+M_{BU}(x)\}/2]>/x \in X\}$.
7. $A\$B=\{<x, [\{M_{AL}(x).M_{BL}(x)\}^{1/2},\{M_{AU}(x).M_{BU}(x)\}^{1/2}]>/x \in X\}$.

Let us consider another IVFS D such that $D = kA$, $0<k\leq 1$. Then the degree of membership of D is $M_D(x)= kM_A(x) = [kM_{AL}(x),kM_{AU}(x)]$.

Definition 2. Let $X=\{x_1, x_2, x_3, ..., x_n\}$ be the universal set containing n members. Let $A=\{<x,M_A(x)> / x\in X\}$ be an IVFS. Then the mean (m_A) and variance ($var(A)$) of membership function of A defined on X can be written as $m_A=1/n\Sigma\{M_{AL}(x_i)+M_{AU}(x_i)\}/2$ and $Var(A)=1/(n-1)\Sigma\{(M_{AL}(x_i)-m_A)^2+(M_{AU}(x_i)-m_A)^2\}/2$. Then the standard deviation will be $S_A=\{Var(A)\}^{1/2}$.

Except the mean and variance of IVFSs, our interest very often is on the information about the relationship among the IVFSs. The correlation of two IVFSs is a relative measure of how they related to one another. Which gives a rough indication of any association that may exist between two IVFSs. The correlation coefficient on the other hand is a measure to describe the type and strength of the relationship of two IVFSs.

3. Simple Correlation of IVFSs

Definition 3. Let X be the universal set. Let A and B be any two IVFSs, where $A = \{<x, [M_{AL}(x), M_{AU}(x)] > / x\in X\}$and $B = \{<x, [M_{BL}(x), M_{BU}(x)] > / x\in X\}$. Then the correlation between A and B is denoted by $C_{IVFS}(A,B)$ and defined as

$C_{IVFS}(A, B) = \Sigma\{M_{AL}(x_i).M_{BL}(x_i)+M_{AU}(x_i).M_{BU}(x_i)\}/2$.

The correlation coefficient of the IVFSs A and B is defined as

$K_{IVFS}(A, B) = C_{IVFS}(A, B)/\{(C_{IVFS}(A, A))^{1/2}.(C_{IVFS}(B, B))^{1/2}\}$.

For all $A, B \in$ IVFSs, the following results are obviously follows from definition.

1. If $A=B$ then $K_{IVFS}(A, B)=1$.
2. $K_{IVFS}(A, B) = K_{IVFS}(B, A)$.

4. Some Results on Correlation

Property 1. (i) If A be an IVFS, then $C_{IVFS}(A, kA)= C_{IVFS}(kA, A)=k\, C_{IVFS}(A, A)$, where $0<k\leq 1$.

(ii) $K_{IVFS}(A, kA)=1$, $0<k\leq 1$.

Property 2. For any two IVFSs A and B, $0\leq K_{IVFS}(A, B)\leq 1$.

Proof. Since $C_{IVFS}(A, B)\geq 0$, we need only to prove that

$C_{IVFS}(A, B) \leq C_{IVFS}(A, A))^{1/2}.((C_{IVFS}(B, B))^{1/2}$.

$C_{IVFS}(A, A). C_{IVFS}(B, B)=1/(2^2)\Sigma[M_{AL}^2(xi)+M_{AU}^2(x_i)].\Sigma[M_{BL}^2(x_i)+M_{BU}^2(x_i)]$

$\geq 1/(2^2)\{\Sigma[\ M_{AL}^2(xi) + M_{AU}^2(x_i)]^{1/2}\ [M_{BL}^2(x_i)+M_{BU}^2(x_i)]^{1/2}\}^2$
$=1/(2^2)\{\Sigma[\ (M_{AL}^2(xi) + M_{AU}^2(x_i))(\ M_{BL}^2(x_i)+ M_{BU}^2(x_i))]^{1/2}\}^2$
$=1/(2^2)\{\Sigma[\ M_{AL}^2(xi).\ M_{BL}^2(x_i)+M_{AL}^2(x_i).\ M_{BU}^2(x_i) + M_{AU}^2(x_i).\ M_{BL}^2(x_i)$
$+ M_{AU}^2(x_i).\ M_{BU}^2(x_i)]^{1/2}\}^2$
$\geq \{1/2\Sigma[\{\ M_{AL}(xi).\ M_{BL}^2(x_i)+ M_{AU}(x_i).\ M_{BU}^2(x_i)\}^2]^{1/2}\}^2=[C(A,B)]^2.$

That is, $C_{IVFS}(A, A\ ).\ C_{IVFS}(B, B\ )\geq [C_{IVFS}(A, B\ )]^2$.

Therefore, $C_{IVFS}(A, B\ )/\{\ C_{IVFS}(A, A\ ).\ C_{IVFS}(B, B\ )\}^{1/2} \leq 1$.

Hence, $0 \leq K_{IVFS}(A, B) \leq 1$.

Property 3. Let D be an IVFS which is linearly related with two IVFSs A and B, i.e., $D=k_1A+k_2B$ where, $0 \leq k_1,k_2 \leq 1$, such that $M_D(x)= k_1M_A(x)+k_2M_B(x)$ and $M_D{:}X \rightarrow D[0,1]$.

Then (i) $C_{IVFS}(A, D)=k_1C_{IVFS}(A, A)+k_2C_{IVFS}(A, B)$,

(ii) $C_{IVFS}(D, D)=k_1^2C_{IVFS}(A, A)+2k_1k_2C_{IVFS}(A, B)+k_2^2C_{IVFS}(B, B)$.

Property 4. Let A and B be two IVFSs and U and V be another two IVFSs such that $U=k_1A$ and $V=k_2B$, where $0<k_1,k_2 \leq 1$.

Then, $K_{IVFS}(U,V)=K_{IVFS}(A,B)$.

Property 5. For any three IVFSs A, B and C, $C_{IVFS}(A, B+C)=C_{IVFS}(A, B)+C_{IVFS}(A,C)-C_{IVFS}(A, B.C)$.

Property 6. For any two IVFSs A, B,

(i) $2C_{IVFS}(A, A@B)=C_{IVFS}(A, A)+C_{IVFS}(A, B)$,

(ii) $C_{IVFS}(A\$B, A\$B)=C_{IVFS}(A, B)$,

(iii) $4C_{IVFS}(A@B, A@B)=C_{IVFS}(A, A)+2C_{IVFS}(A, B)+C_{IVFS}(B, B)$.

5. Partial and Multiple Correlation Coefficient Between IVFSs

Suppose we have three IVFSs A, B and C drawn from the universal set X. Let $M_A(x)$, $M_B(x)$ and $M_C(x)$ be the degree of membership of A, B and C respectively.

Here we are interested in finding the relationship between two IVFSs A and B, where the IVFS C is also account for the observed relationship between A and B. This relationship is called partial correlation between IVFSs A and B with the influence of IVFS C partialed out.

The simple correlation coefficient is calculated easily between each pair of IVFSs A, B and C. Then we can calculate $K_{IVFS}(A, B)$, $K_{IVFS}(A, C)$ and $K_{IVFS}(B, C)$. The formula to compute partial correlation coefficient between IVFSs A and B holding IVFS C as constant is defined as $K_{IVFS}(A, B.C)$

$=\{K_{IVFS}(A, B)-K_{IVFS}(A, C).K_{IVFS}(B, C)\}/\{(1-K_{IVFS}^2(A, C))^{1/2}(1- K_{IVFS}^2(B,C))^{1/2}\}$.

The following results follows from the definition of partial correlation coefficient.

(i) For any three IVFSs A, B and C, $K_{IVFS}(A,B.C)=K_{IVFS}(B,A.C)$ as $K_{IVFS}(A,B) = K_{IVFS}(B,A)$,

(ii) If $A=kB$ then $K_{IVFS}(A,B.C)=1$ where $0<k \leq 1$.

Property 7. (i) For any three IVFSs A, B and C, $-1 \leq K_{IVFS}(A, B.C) \leq 1$.

(ii) If $A=k_1C$ or $B=k_2C$ then $K_{IVFS}(A, B.C)$ does not exist.

(iii) Let A, B and C be three IVFSs and U, V and W be another three IVFSs such that $U=k_1A$, $V=k_2B$ and $W=k_3C$, where $0 < k_1,k_2,k_3 \leq 1$.

Then $K_{IVFS}(A, B.C)=K_{IVFS}(U, V.W)$.

Here we are interested in finding the correlation between A and the joint influence of B and C. This relation is known as multiple correlation and the coefficient is defined as $K_{IVFS}(A. B,C)=[K_{IVFS}^2(A, B)+K_{IVFS}^2(A, C)-K_{IVFS}(A,B).K_{IVFS}(A,C).K_{IVFS}(B,C)\}/\{1-K_{IVFS}^2(B,C)\}]^{1/2}$.

Property 8. (i) Since $K_{IVFS}(A.,B)= K_{IVFS}(B,A)$ so $K_{IVFS}(A.B,C)= K_{IVFS}(A.C,B)$.

(ii) If $B=kC$ then $K_{IVFS}(B,C)=1$ and $K_{IVFS}(A.B,C)$ does not exist.

(iii) For any IVFSs A, B, C, $0 \leq K_{IVFS}(A.B,C) \leq 1$.

6. An Illustration

Let X be the set of all people of a particular locality. The people may have different fuzzy characteristic such as tall, intelligent, smart, un-smart, etc. In our example we consider three IVFSs A, B and C, where A stands for tall people, B for intelligent people and C for smart people. For simplicity, we consider fifteen people only. In general, most of the intelligent people are smart but they may be tall or short. Tall people are smart but they may or may not be intelligent. Here we want to measure the amount of correlation among the people that are tall and intelligent or smart. That is, we want to find the correlation between each pair of this characteristic of the people. Here we consider the membership values of tallness, intelligence and smartness of a people as an interval. Because the determination of membership values as a point is a difficult job for a decision maker, but determination as an interval is simple and most realistic.

We calculate the multiple correlation of a characteristic with the joint influence of other two characteristics, i.e., we calculate the correlation of any one set, joint influenced by other two sets.

Let $X=\{x_1, x_2, x_3, \ldots, x_{15}\}$ be the universal set. Let A, B, C be three IVFSs represent respectively, the tall people, intelligent people and smart people.

As an illustration we consider the following data.

	x_1	x_2	X_3	x_4	x_5	x_6	x_7	x_8
A	[0.7,0.9]	[0.5,0.6]	[0.3,0.5]	[0.4,0.6]	[0.8,0.9]	[0.7,0.8]	[0.4,0.5]	[0.6,0.8]
B	[0.3,0.5]	[0.4,0.6]	[0.5,0.8]	[0.5,0.7]	[0.3,0.6]	[0.6,0.8]	[0.8,0.9]	[0.4,0.5]
C	[0.3,0.4}	[0.6,0.8]	[0.6,0.7]	[0.4,0.5]	[0.6,0.8]	[0.5,0.6]	[0.6,0.9]	[0.5,0.7]

	x_9	x_{10}	x_{11}	x_{12}	x_{13}	x_{14}	x_{15}
A	[0.5,0.7]	[0.6,0.9]	[0.5,0.8]	[0.3,0.5]	[0.4,0.5]	[0.8,0.9]	[0.5,0.7]
B	[0.7,0.8]	[0.5,0.7]	[0.6,0.8]	[0.4,0.5]	[0.7,0.9]	[0.6,0.7]	[0.2,0.4]
C	[0.4,0.6]	[0.5,0.6]	[0.8,0.9]	[0.5,0.6]	[0.3,0.4]	[0.7,0.9]	[0.6,0.7]

We calculate the simple correlation of each pair of IVFSs A, B and C. The simple correlations are $K_{IVFS}(A, B)=0.896$, $K_{IVFS}(A, C)=0.949$ and $K_{IVFS}(B, C)=0.941$.

The first result shows that tall people and intelligent people are positively and highly correlated with the strength of 0.896. Second and third results shows that tall and intelligent people are almost smart. The strength of their correlation is respectively 0.949 and 0.941.

For the above sets A, B, C the multiple correlations are $K_{IVFS}(A.B,C)=0.950$, $K_{IVFS}(B.A,C)=0.941$, $K_{IVFS}(C.A,B)=0.964$. The partial correlation coefficients are $K_{IVFS}(A,B.C)=0.028$. $K_{IVFS}(A,C.B)=0.701$. $K_{IVFS}(B,C.A)=0.649$.

7. Information Energy

In general, the membership value of a set is determined by the users subjectively. That is, the value of the membership specified for the same concept by different persons may vary. For an IVFS the membership value is an interval, which is a sub-interval of [0,1] because the decision maker cannot predict the exact value of membership easily. Every membership value presents the fuzziness of the corresponding IVFSs. At present, there are several typical methods used to measure the fuzziness of a fuzzy set [3, 8]. Here, we utilize the concept of information energy to indicate the fuzzyness of an IVFS. In this section, we also study some properties of information energy.

Definition 4. Let A be an IVFS. Then A is called an IVFS with fuzzy degree is maximum if $M_A(x)$=[0.5, 0.5]. And A is called an IVFS with fuzzydegree is minimum if $M_A(x)$=[0,0]. In this cases, A is called a degenerate ordinary fuzzy set.

For any two numbers a and b, we define two operations $\wedge$ and $\vee$ as follows:

$a \wedge b = min\{a, b\}$ *and* $a \vee b = max\{a, b\}$.

Definition 5. Let A be an IVFS. Then $E(M_{AL}(x_i)) = (2M_{AL}(x_i)-1)^2$ and $E(M_{AU}(x_i)) = (2M_{AU}(x_i)-1)^2$. Therefore $E_{xi}(A) = [E(M_{AL}(x_i)) \wedge E(M_{AU}(x_i)), E(M_{AL}(x_i)) \vee E(M_{AU}(x_i))]$. We call $E_{xi}(A)$ the interval-valued information energy of A at a point $x_i \in X$.

Definition 6. Let A be an IVFS. We define $E_X(M_{AL}(x)) = 1/n\sum E(M_{AL}(x_i))$ and $E_X(M_{AU}(x)) = 1/n\sum E(M_{AU}(x_i))$. Then, $E_X(A) = [E_X(M_{AL}(x)) \wedge E_X(M_{AU}(x)), E_X(M_{AL}(x)) \vee E_X(M_{AU}(x))]$.

Property 9. (i) $[0,0] \leq E_{xi}(A) \leq [1,1]$ for all $x_i \in A$ where A is an IVFS,

(ii) If $A \subset X$ be any IVFS then $[0,0] \leq E_X(A) \leq [1,1]$.

Property 10. Let A be an IVFS. Then, $E_X(A)$ = [0,0] iff A is an IVFS with fuzzy degree is maximum.

Property 11. (i) $E_X(A)$ reaches its maximum if $M_A(x)$=[0,0] or $M_A(x)$=[1,1].

(ii) $E_X(A) = E_X(A^c)$ where A^c is the complement of IVFS A.

References

1. K.T. Atanassov & G. Gargov (1989), Interval valued intuitionistic fuzzy sets, *Fuzzy Sets and Systems*, 31: 343-349.
2. H. Bustince & P. Burillo (1995), Correlation of interval-valued intuitionistic fuzzy sets, *Fuzzy Sets and Systems*, 74: 237-244.
3. D.Dumitrescu (1993), Entropy of a fuzzy process, *Fuzzy Sets and Systems*, 55: 169-177.
4. T. Gerstenkorn & J. Manko (1991), Correlation of intuitionistic fuzzy sets, *Fuzzy Sets and Systems*, 44: 39-43.
5. B. Gorzalczany (1983), Approximate inference with interval-valued fuzzy sets-an outline, in: *Proc. Polish Symp. on Interval and Fuzzy Math.*, Poznan: 89-95.
6. D. H. Hong & S. Y. Hwang (1995), Correlation of intuitionistic fuzzy sets in probability space, *Fuzzy Sets and Systems*, 75: 77-81.
7. N. Jana & M. Pal (2002), Some operations on interval-valued intuitionistic fuzzy sets, in: *Proc. International Conference on Analysis and Discrete Structure*, Dec.22-24, Indian Institute of Technology, Kharagpur, India.
8. De Luca & S. Termini (1972), A definition of non probabilistic entropy in the setting of fuzzy theory, *Inform. and Control*, 20: 301-312.

Fuzzy Logic and its Application in Technology and Management
Editors: Debjani Chakraborty, S. Nanda and D. Dutta Majumder

Basic Properties of Rough sets and Applications to Distributed Knowledge Systems

B.K. Tripathy and M.R. Patra

Deptt. of Computer Science, Berhampur University, Berhampur-760007
E-mail: tripathybk@rediffmail.com, manasrpbu@yahoo.com

Abstract: Design of distributed knowledge base systems is a critical task where one must ensure that the integration of knowledge partitions must reproduce the original knowledge base. In perusing this goal we observe that the concept of *Rough sets* can greatly contribute to the solution of this design problem. In this article we establish some basic properties of rough sets and illustrate the application of these results to the knowledge partitioning problem in distributed knowledge systems.

Keywords: Rough sets, lower approximation, upper approximation, knowledge base and equivalence relation.

1. Introduction

The fundamental concepts of sets have been extended in many directions in recent past. The notion of fuzzy sets by L.A.Zadeh ([4]), its generalization and the notion of rough sets by Z.Pawlak ([2]) being not worthy among them. While fuzzy set deals with approximate membership of elements in a set rough set captures indiscernibility of elements in a set. These two concepts have been shown to be independent ([1]).

The idea of the rough set consists of the approximation of a set by a pair of sets, called the lower and upper approximation of this set.

In this section we define the concepts and introduce the notation to be used in the article.

Let U ($\neq\Phi$) be a finite set of objects, called the universe and R be an equivalence relation over U. Then by U/R we denote the family of equivalences classes of R (or classification of U) referred to as categories or concepts of R and $[x]_R$ denotes a category in R containing an element $x\in U$.

By a knowledge base, we understand a relational system K=(U, R), where U is as above and R is a family of equivalence relation over U. For any subset $P \subseteq R$ and $P\neq\Phi$, P is the intersection of all equivalence relation and is denoted by IND (P). IND (P) is called the *indiscernibility relation* over P. The equivalence classes of IND (P) are called P-basic knowledge about U in K. For any $Q\in R$, Q is called a Q elementary knowledge about U in K and equivalence classes of Q are called Q-*elementary concepts* of knowledge R.

The family of all P-basic categories for all $\Phi \neq P \subseteq R$ will be called the family of basic categories in knowledge base $K=(U,R)$.

By IND (K) we denote the family of all equivalence relations defined in K. Symbolically,

$$IND(K) = \{IND(P) : \Phi \neq P \subseteq R\}.$$

For any $X \subseteq U$ and an equivalence relation $R \in IND(K)$, we associate two subsets

$$\underline{R}X = \cup \{Y \in U/R: Y \subseteq X\} \text{ and } \overline{R}X = \cup \{Y \in U/R: Y \cap X = \Phi\}$$

called the R-*Lower* and R-*Upper approximation* of X respectively.

The R-*boundary* of X is denoted by $BN_R(X)$ and is given by $BN_R(X) = \overline{R}X - \underline{R}X$. The elements of $\underline{R}X$ are those elements of U which can be classified as elements of X certainly with the knowledge of R and $\overline{R}X$ is the set of elements of X which can be classified as elements of X employing knowledge of R. The borderline region is the undecidable area of the universe.

We say X is *rough* with respect to R if and only if $\underline{R}X \neq \overline{R}X$, equivalently $BN_R \neq \Phi$. X is said to be R-*definable* if and only if $\underline{R}X = \overline{R}X$ or $BN_R(X) = \Phi$. So, a set is rough with respect to R if and only if it not R-definable.

2. Properties

In this section we state some properties of rough sets and prove them. It is well known that ([3]) for any two sets $X, Y \subseteq U$,

$$\underline{R}(X \cup Y) \supseteq \underline{R}(X) \cup \underline{R}(Y) \text{ and} \tag{2.1}$$

$$\overline{R}(X \cap Y) \subseteq \overline{R}(X) \cap \overline{R}(Y) \tag{2.2}$$

The above two inclusions are proper ([3]) and can be extended to a finite number of sets. As mentioned by Pawlak, these results confirm to the obvious observation that in general the knowledge included in a distributed knowledge base is less than the integrated one. That is, in general dividing the knowledge base into smaller fragments causes loss of information. This leads to the interesting problem of determining the exact circumstances under which there will be no loss of information even if one distributes the knowledge base or in other words under what circumstances there will definitely be loss of information. This leads to determining the necessary and sufficient conditions for the equalities to hold in (2.1) and (2.2). We provide answer to this in the following results:

Therom 2.1. Let $\{E_1, E_2, ----\}$ be a partition of the Universe U. Then for any two subsets X_1 and X_2, $\underline{R}(X_1) \cup \underline{R}(X_2) \subset \underline{R}(X_1 \cup X_2)$, if and only if there exists at least one F_i such that

$$X_1 \cap E_i \subset E_i,\ X_2 \cap E_i \subset E_i \text{ but } X_1 \cup X_2 \supset E_i \tag{2.3}$$

Proof: The sufficiency follows from the fact that $E_i \not\subset \underline{R}(X_1)$ and $E_i \not\subset \underline{R}(X_2)$ but $E_i \subseteq \underline{R}(X_1 \cup X_2)$.

Conversely, suppose $\underline{R}(X_1) \cup \underline{R}(X_2) \subset \underline{R}(X_1 \cup X_2)$. As $\underline{R}(X)$ for any X contains only E_j s , there is at least one E_i such that $E_i \subseteq \underline{R}(X_1 \cup X_2)$ but $E_i \not\subset \underline{R}(X_1) \cup \underline{R}(X_2)$. So, $E_i \subseteq X_1 \cup X_2$ and $E_i \not\subset \underline{R}(X_1)$ and $E_i \not\subset \underline{R}(X_2)$.

Hence $E_i \subseteq X_1 \cup X_2$, $E_i \not\subset X_1$ and $E_i \not\subset X_2$. So, $X_1 \cap E_i \subset E_i$, $X_2 \cap E_i \subset E_i$ and $E_i \subseteq X_1 \cup X_2$.

This completes the proof.

Note 2.1. This result can be extended to union of a finite number of sets $X_1, X_2, \ldots, X_n$ in a trivial way as "A necessary and sufficient condition for

$\cup^n_{i=1} \underline{R}(X_i) \not\subset \underline{R}(\cup^n_{i=1} X_i)$,

is that there exists at least one E_j such that

$X_i \cap E_j \subset E_j$, for j=1,2,…n and

$\cup^n_{i=1} X_i \supset E_j$.

Corollary 2.1. A necessary and sufficient condition for

$\cup^n_{i=1} \underline{R}(X_i) = \underline{R}(\cup^n_{i=1} X_i)$

is that there exists no E_j such that $X_i \cap E_j \subset E_j$ for all i=1,2,..n and $\cup^n_{i=1} X_i \supset E_j$.

Theorem 2.2. A necessary and sufficient condition for

$\overline{R}(\cap^n_{i=1} X_i) \subset \cap^n_{i=1} \overline{R}(X_i)$

is that there exists at least one E_j such that

$X_i \cap E_j$ for i=1,2,..n and

$\cap^n_{i=1} X_i \cap E_j = \phi$.

Proof: The sufficiency is clear as

$E_j \not\subset \overline{R}(\cap^n_{i=1} X_i)$ where as $\overline{R}(X_i) \supseteq E_j$ for all i=1,2,…n.

Conversely, suppose the conclusion is true. Then for some E_j, $E_j \subseteq \cap^n_{i=1} \overline{R}(X_i)$ but

$E_j \not\subset \overline{R}(\cap^n_{i=1} X_i)$.

So, $E_j \subseteq \overline{R}(X_i)$ for i=1,2,…n and $E_j \cap (\cap^n_{i=1} X_i) = \phi$

That is, $E_j \cap X_i \neq \phi$ for i=1,2,…n and $E_j \cap (\cap^n_{i=1} X_i) = \phi$.

This completes the proof.

Corollary 2.2. $\overline{R}(\cap^n_{i=1} X_i) = \cap^n_{i=1} \overline{R}(X_i)$ if and only if there is no E_j such that

$X_i \cap E_j \neq \phi$ for i=1,2,…n and

$\cap^n_{i=1} X_i \cap E_j = \phi$.

Note 2.2. It is known that $X \subseteq Y$ implies $\overline{R}(X) \subseteq \overline{R}(Y)$ and $\overline{R}(X) \subseteq \overline{R}(Y)$. The question arises whether $X \subset Y$ implies $\underline{R}(X) = \underline{R}(Y)$ and $R(X) = R(Y)$?

The answer is 'Yes'. We provide an example.

Let K=(U,R), where $U = \{x_1, x_2, \ldots x_8\}\}$ and the partition $\{E_1, E_2, E_3, E_4\}$ is such that $E_1 = \{x_1, x_4, x_8\}, E_2 = \{x_2, x_5, x_7\}, E_3 = \{x_3\}$ and $E_4 = \{x_6\}$.

Let $X = \{x_3, x_5\}$ and $Y = \{x_3, x_5, x_7\}$. Then $\underline{R}(X) = E_3$, $\underline{R}(Y) = E_3$. So, $X \subset Y$. But $\underline{R}(X) = \underline{R}(Y)$.

Again, $\overline{R}(X) = E_2 \cup E_3 = \overline{R}(Y)$.

The rough sets can be characterized through the notion of the lower and the upper approximations. There are four different kinds of rough sets defined as follows:

1. If $\underline{R}X \neq \Phi$ and , $\overline{R}X \neq U$, then we say that X is *roughly* R-*definable.*
2. If $\underline{R}X = \Phi$ and , $\overline{R}X \neq U$, then we say that X is *internally* R-*undefinable.*
3. If $\underline{R}X \neq \Phi$ and , $\overline{R}X = U$, then we say that X is *externally* R-*undefinable.*
4. If $\underline{R}X = \Phi$ and , $\overline{R}X = U$, then we say that X is *totally* R-*undefinable.*

We have the following results on the union and intersection of two elements of any one class of the above type.

Theorem 2.3.

I. If X and Y are internally R-definable then $X \cap Y$ is internally R-definable.

II. If X and Y are internally R-undefinable then $X \cup Y$ is externally R-undefinable.

III. If X and Y are roughly R-definable then $X \cap Y$ may be roughly R-definable or internally R-undefinable..

IV. If X and Y are roughly R-definable then $X \cup Y$ may be roughly R-definable or externally R-undefinable.

V. If X and Y are internally R-definable then $X \cup Y$ can be in any one of the four classes.

VI. If X and Y are externally R-undefinable then $X \cap Y$ can be in any one of the four classes.

VII. If X and Y are totally R-undefinable then $X \cap Y$ can be internally R-undefinable or totally R-undefinable.

VIII. If X and Y are totally R-undefinable then $X \cup Y$ can be externally R-undefinable or totally R-undefinable.

Proof: All the above results depend upon the following four properties of lower and upper approximation of a set.

$$\underline{R}(X \cap Y) = \underline{R}X \cap \underline{R}Y$$
$$R(X \cap Y) \subseteq \bar{R}X \cap \bar{R}Y$$
$$\underline{R}(X \cup Y) \supseteq \underline{R}X \cup \underline{R}Y \text{ and}$$
$$R(X \cup Y) = \bar{R}X \cup \bar{R}Y$$

As in case of (v) and (vi) there are four possibilities and in case of cases (iii), (iv), (vii) and (viii) there are two possibilities each. But using our Theorems 2.1 and 2.2 above and their corollaries we can reduce the number of cases to two in (v) and (vi) and unique solution in cases (vii) and (viii), under suitable conditions, as follows from the corresponding hypothesis. We omit the statements under space constraints.

3. Applications

The results of this paper are quite relevant to handle design issues in distributed knowledge base systems. One of the key objectives in the design of such systems is to partition knowledge into logical components based on certain predefined criteria and allocate them to suitable physical sites to support distributed applications. But many-a-times the distributed components undergo changes for performance reasons; such as to reduce communication overheads, increase knowledge availability at a site etc. In such occasions there is every possibility of loss of information if the criteria for the new partitions are different from the original criteria. The above Theorems (2.1 and 2.2) and their corollaries provide conditions that must be satisfied in order to avoid possible loss of information during the design process. The conditions and their bearing on the knowledge partitioning process are outlined below.

Rough sets can be classified into four types by employing the notion of lower and upper approximations, as mentioned in the previous section. These types have intuitive meaning attached with them.

(1) A set X is roughly R-definable means that we are able to decide for some elements of U whether they belong to X or –X.

(2) A set X is internally R-undefinable means that we are able to decide whether some elements of U belong to –x, but we are unable to indicate one element of X.

(3) A set X is externally R-undefinable means that we are able to decide some elements of U whether they belong to X but we are unable to indicate one element of –x.

(4) A set X is totally R-undefinable means that we are unable to decide for any element of U whether it belongs to X or –x.

A natural question is that what happens if the union/intersection of two or more number of any one of the above classes are formed. This has been settled to a certain extent in Theorem 2.3 by using our results of Theorem 2.1 and Theorem 2.2.

Reference

1. D. Dubois & H. Prade (1990), Rough fuzzy sets and Fuzzy rough sets, *International journal general systems* 17: 191-208.
2. Z. Pawlak (1982), Rough sets, *International journal Computer and Information science*, 11: 341-356.
3. Z. Pawlak (1991), *Rough sets – Theoretical aspects of reasoning about data*, Kluwer academic publishers.
4. L.A. Zadeh (1965), Fuzzy sets, *Information and Control*, 8: 338-353.

Fuzzy Logic and its Application in Technology and Management
Editors: Debjani Chakraborty, S. Nanda and D. Dutta Majumder

On Soft Relation and Fuzzy Soft Relation

Tanmoy Som

Department of Mathematics, Assam University, Silchar - 788 011
E-mail: som_tanmoy@yahoo.co.in

Abstract: In this physical world, we come across many complex problems pertaining to the areas of Engineering, Medical Science, Environmental Science, Economics, Social Science etc, which involve data that are not always crisp and precise. These problems have various types of uncertainties, some of which can be dealt with using the existing theories, viz., theory of probability, theory of fuzzy sets, theory of rough sets, theory of vague sets, or the theory of approximate reasoning etc. However, all these techniques lack in parameterization of the tools, due to which these could not be applied successfully in tackling such problems. Taking into account of this fact, Molodtsov (1999) introduced the concept of soft set, which is free from above difficulty, and has a rich potential for application in many directions, some of which are shown by him in his pioneer work and later by Maji et. al. (2001, 2003). With the motivation of this new concept, in this paper we define soft relation and fuzzy soft relation, which are certain extension of crisp and fuzzy relations respectively. Further we apply these concepts in solving decision-making problems.

1. Soft-relation (A classical approach)

Definition 1.1. A soft relation may be defined as a soft set over the power set of the cartesian product of two crisp sets. In other words if X and Y are two non-empty crisp sets (or crisp subsets of some universal set) and E is a set of parameters, then a soft relation, denoted as (R, E) is defined as a mapping from E to $P(X \otimes Y)$.

Example: Let U = Set of people in a social gathering

X = Set of male members in $U = \{m_1, m_2 ... m_7\}$ (say)

Y = Set of female members in $U = \{w_1, w_2 ... w_6\}$ (say)

Let E_1 and E_2 be two sets of parameters given by

E_1 = { is father of, is husband of, is son of, is nephew of}

E_2 = { is mother of, is wife of, is daughter of, is niece of)

Then a soft relation R over $P(X \otimes Y)$ corresponding to E_1 may be given as

$(R, E_1) = \{R$ (is father of) $= \{(m_1, w_1), (m_2, w_3), (m_4, w_5), (m_7, w_6)\}$,

R (is husband of) $= \{(m_1, w_2), (m_3, w_5), (m_6, w_6)\}$,

R (is son of)$= \{(m_2, w_1), (m_5, w_1), (m_3, w_4)\}$,

R (is nephew of) = $\{(m_1, w_3), (m_5, w_4)\}\}$

Another soft relation R over $P(X \otimes Y)$ corresponding to E_2 may be given by

$(R, E_2) = \{R$ (is mother of) = $\{(w_1, m_2), (w_1, m_5), (w_4, m_3)\}$,

R (is wife of) = $\{(w2, m1), (w5, m3), (w6, m6)\}$,

R (is daughter of)=$\{(w1, m1), (w3, m2), (w5, m4), (w6, m7)\}$,

R (is niece of) = $\{(w_2, m_4), (w_5, m_7)\}\}$

Note: Clearly R (is father of) and R (is daughter of) are derivable from each other.

Similarly $\{R$ (is husband of), R (is wife of)$\}$ and $\{$(is son of), (is mother of)$\}$ are other derivable pairs. There may be many other soft relations over $P(X \otimes Y)$.

Each approximate value set in the above soft relation (R, E_1) or (R, E_2) can be expressed in the parameterised matrix form as

	m_1	m_2	m_3	m_4	m_5	m_6	m_7
w_1	1	0	0	0	0	0	0
w_2	0	0	0	0	0	0	0
w_3	0	1	0	0	0	0	0
w_4	0	0	0	0	0	0	0
w_5	0	0	0	1	0	0	0
w_6	0	0	0	0	0	0	1
etc.							

R (is father of) etc.

1.1 Fuzzy soft relation

A fuzzy soft relation may be defined as soft set over the fuzzy power set of the cartesian product of two crisp sets. In other words if $P(X \otimes Y)$ is the fuzzy powerset $X \otimes Y$, X, Y are two crisp sets and E is the set of parameters, then a function $R: E \rightarrow P(X \otimes Y)$ is called a fuzzy soft relation. For each $\varepsilon \in E$ each ordered pair in $R(\varepsilon)$ has a degree of membership in the fuzzy soft relation R, indicating the strength of ε-parametric relationship present between the elements of the ordered pairs in R.

Example 1.1. Let, X = {Delhi, Kolkata, Mumbai} and Y = {Chennai, Kanpur, Bangalore} be two sets of cities, E = {far, very far, near, very near, crowded, well planned}. Let R be the fuzzy soft relation over the sets X and Y given by (R, E) = {R(far)={(Delhi, Chennai)/.6, (Delhi, Kanpur)/0.15, (Delhi, Bangalore)/.5, (Kolkata, Chennai)/.55, (Kolkata, Kanpur)/.2, (Kolkata, Bangalore)/.65, (Mumbai, Chennai)/.45, (Mumbai, Kanpur)/.5, (Mumbai, Bangalore)/.3} etc.

This can be conveniently represented in form of three-dimensional membership array matrix as given below:

	Chennai	Kanpur	Bangalore
Delhi	.6	.15	.5
Kolkata	.55	.2	.65
Mumbai	.45	.5	.3
		far	etc.

Note : It is evident from the above that a fuzzy soft relation may be considered as a parameterised fuzzy relation.

1.2. Relation on two fuzzy soft sets

Let (F, A) and (G, B) be two fuzzy soft sets over a common universal set U. Then a relation R of (F, A) on (G, B) may be defined as a mapping $R: A \times B \to P(U^2)$ such that for each $e_i \in A$, $e_j \in B$ and for all $u_l \in F(e_i)$, $u_k \in G(e_j)$, the relation R be characterised by the membership function given by

$$\mu_R(u_l, u_k) = \mu_{F(ei)}(u_l) \times \mu_{G(ej)}(u_k), \quad \text{where } u_l \in F(e_i), u_k \in G(e_j) \qquad (*)$$

Clearly the above mapping is well defined. Higher the value of the membership grade in the relation R for a pair, stronger is the parametric character present between the pair.

Example 1.2. Let $U = \{h_1, h_2, h_3, h_4, h_5\}$ be a set of houses, A = {costly, cheap}, B = {beautiful, in green surrounding}. Let (F, A) and (G, B) be two soft sets given by

F (costly) = $\{h_1/1, h_2/.7, h_3/.8, h_4/.5, h_5/.9\}$

F (cheap) = $\{h_1/.1, h_2/.25, h_3/.2, h_4/.6, h_5/.15\}$

G (beautiful) = $\{h_1/.6, h_2/1, h_3/.8, h_4/.5, h_5/.8\}$

G (in green surrounding) = $\{h_1/.6, h_2/.7, h_3/.8, h_4/.5, h_5/.4\}$

Then the relation $R: A \times B \to P(U^2)$ is given by the following membership matries :

	h_1	h_2	h_3	h_4	h_5
h_1	.6	1	.8	.5	.8
h_2	.42	.7	.56	.35	.56
h_3	.48	.8	.64	.40	.64
h_4	.30	.5	.40	.25	.40
h_5	.54	.9	.72	.45	.72

R (costly, beautiful)

	h_1	h_2	h_3	h_4	h_5
h_1	.06	.1	.08	.05	.08
h_2	.15	.25	.20	.125	.20
h_3	.12	.2	.16	.10	.16
h_4	.36	.6	.48	.30	.48
h_5	.09	.15	.12	.075	.12

R (cheap, beautiful) etc.

Note : The membership function (*) of R may be defined in other suitable ways also.

2. Another approach to Soft relation and Fuzzy Soft Relation

Definition 2.1. If (F, A) and (G, B) are two soft sets over a common universe U, then a soft subset (R, C) of $(F, A) \times (G, B)$ is called a soft relation of (F, A) and (G, B), where $C \subset A \times B$ and for every $(x, y) \in C$, $R(x, y)$ and $O(x, y)$ are identical approximations where $O(x, y) = F(x) \cap G(y)$.

Then clearly (R, C) is also a soft set and therefore the basic concepts such as union, intersection, complement etc. can be applied without any modification to soft relation.

Example 2.1. Let $U = \{c_1, c_2, c_3, c_4, c_5\}$ be the set of cars, E = {cheap, costly, fuel efficient, produced by firm B_1, produced by firm B_2, produced by firm B_3} be a set of parameters. Let (F, A) = {cheap cars = $\{c_1, c_2, c_3\}$, costly cars = $\{c_4, c_5\}$, fuel efficient cars = $\{c_1, c_3, c_5)$ and (G, B) = {cars produced by firm $B_1 = \{c_1, c_3\}$, cars produced by firm $B_2 = \{c_2, c_3, c_4\}$,cars produced by firm $B_3 = \{c_2, c_5\}$ be two soft sets over U. Then a soft relation (R, C) of all cheap and fuel efficient cars produced by firms B_1 and B_3 is given by (R,C) = {R(cheap, produced by B_1) = $\{c_1, c_3\}$, R(fuel efficient, produced by B_1) = $\{c_1,c_3\}$, R(cheap, produced by B_3) = $\{c_2\}$, R(fuel efficient, produced by B_3) = $\{c_5\}$}. A Generalization Here first we have to give generalization of AND and OR operators as given below:

2.2. AND Operator: The AND operator can be generalized for a family of n fuzzy soft sets $\{(F_i, A_i)/i \in N\}$ and may be denoted either by $(F_1, A_1) \wedge (F_2, A_2) \wedge \ldots \wedge (F_n, A_n)$ or by $\wedge_{i \in N}(F_i, A_i)$ and given by $\wedge (F_i, A_i) = (H, \times A_i) = (H, A_1 \times A_2 \times \ldots \times A_n)$ where $H (x_1, x_2, \ldots\ldots x_n) = F_1(x_1) \cap F_2(x_2) \cap \ldots \cap F_n(x_n)\ \forall (x_1, x_2 \ldots x_n) \in A_1 \times A_2 \times \ldots \times A_n$.

2.3. OR Operator: The OR Operator can also be generalized for a family of n fuzzy soft sets $\{(F_i, A_i)/i \in N\}$, denoted either by $(F_1, A_1) \vee (F_2, A_2) \vee \ldots \vee (F_n, A_n)$ or by $\vee (F_i, A_i)$ and given by $\vee (F_i, A_i) = (O, \times A_i)$, where $O(x_1, x_2 \ldots x_n) = F_1(x_1) \cup F_2(x_2) \cup \ldots \cup F_n(x_n) \forall (x_1, x_2 \ldots x_n) \in A_1 \times A_2 \times \ldots \times A_n$.

Definition 2.4.: If (F_1, A_1), (F_2, A_2) be n soft sets then the soft subset (R,C) of $(F_1,A_1) \times (F_2,A_2) \times \ldots \times (F_n,A_n)$ is called an n-ary soft relation. Here $C \subset A_1 \times A_2 \times \ldots \times A_n$ and $\forall (x_1, x_2 \ldots x_n) \in A_1 \times A_2 \times \ldots \times A_n$, $R(x_1, x_2 \ldots x_n)$ and $O(x_1, x_2 \ldots x_n)$ are identical approximations where $O(x_1, x_2 \ldots x_n) = F_1(x_1) \cap F_2(x_2) \cap \ldots \cap F_n(x_n)$.

Definition 2.5.: If (F, A) and (G, B) are two fuzzy soft sets then a fuzzy soft subset (R, C) of $(F, A) \times (G, B)$ is called a fuzzy soft relation. Here $C \subset A \times B$ and $\forall (x,y) \in A \times B$, $R(x, y)$ is a fuzzy subset of $O(x, y)$ where $O(x, y) = F(x) \cap G(y)$.

Example 2.2. Let $U = \{c_1, c_2, c_3, c_4, c_5\}$, be a set of cars, (F,A) be a fuzzy soft set which describes "the cost of the cars" and (G, B) be a fuzzy soft set which describes "the attractiveness of the cars" where A= {costly, moderate, cheap} and B= {fuel efficient, beautiful, having metallic colour }.

Let F(costly) $= \{c_1/.5, c_2/.8, c_3/0, c_4/.1, c_5/1\}$, F(moderate) $= \{c_1/.2, c_2/.4, c_3/.5, c_4/.6, c_5/.5\}$, F(cheap) $= \{c_1/.5, c_2/.1, c_3/1, c_4/.9, c_5/0\}$,$G$ (fuel efficient) $= \{c_1/.4, c_2/.6, c_3/.8, c_4/1, c_5/.2\}$, G(having metallic color) $= \{c_1/1, c_2/0, c_3/0, c_4/1, c_5/0\}$ and G(beautiful) $= \{c_1/.8, c_2/0, c_3/.5, c_4/.7, c_5/.9\}$.

Then a fuzzy soft relation R of all cheap, fuel efficient and beautiful cars is given by (R, C) ={R (cheap, fuel efficient)=$\{c_1/.4, c_2/.1, c_3/.8, c_4/.9, c_5/0\}$, R (cheap, beautiful) = $\{c_1/.5, c_2/0, c_3/.5, c_4/.7, c_5/0\}$}.

A Generalization: The fuzzy soft relation can be generalized to n fuzzy soft sets $\{(F_i, A_i)/i \quad N\}$ in the following manner.

Definition 2.6. The fuzzy soft set (R, C) of $\times (F_i, A_i)$ is called and n-ary fuzzy soft relation, here $C \subset A_1 \times A_2 \times ... \times A_n$ and $\forall (x_1, x_2 ... x_n) \in A_1 \times A_2 \times ... \times A_n$, $R(x_1, x_2 ... x_n) \subset O$ where $O(x_1, x_2 ... x_n) = F_1(x_1) \cap F_2(x_2) \cap ... \cap F_n(x_n)$.

A relation between two fuzzy soft sets is called a binary relation. Relation defined on *n*-soft sets is called an *n*-ary or *n*-dimensional relation.

3. An Application of Fuzzy Soft Relation in Decision Making Problems

The concept of fuzzy soft relation and its generalization can be used effectively for solving decision-making problems without much computational effort. This is illustrated by an example.

We first consider the problem solved by Maji et. al. [5]and solve it by using fuzzy soft relation. The problem considered is as follows:

Example 3.1. Let $U=\{h_1, h_2, h_3, h_4, h_5, h_6\}$ be a set of six houses and E={expensive, wooden, beautiful, cheap, in green surroundings, concrete, moderately beautiful, by the roadside} be a set of parameters. Let (F_1, A_1) be the fuzzy soft set which describes the cost fo the houses given by $(F_1, A_1) = \{F_1(\text{cheap}) = \{h_1/1, h_2/0, h_3/1, h_4/.2, h_5/1, h_6/.2\}, F_1(\text{expensive}) = \{h_1/0, h_2/1, h_3/.1, h_4/.9, h_5/.3, h_6/1\}\}$.

Let (F_2, A_2) be the fuzzy soft set which describes the attractiveness of the houses given by $(F_2, A_2) = \{F_2(\text{beautiful}) = \{h_1/1, h_2/.4, h_3/1, h_4/.4, h_5/.6, h_6/.8\}, F_2(\text{moderatly beautiful}) = \{h_1/.3, h_2/.7, h_3/.5, h_4/.6, h_5/.2, h_6/.3\}\}$.

Let (F_3, A_3) be the fuzzy soft set which describes the physical trait of the houses given by

$(F_3, A_3) = \{F_3(\text{wooden}) = \{h_1/.2, h_2/.3, h_3/1, h_4/1, h_5/1, h_6/0\}, F_3(\text{concrete}) = \{h_1/.7, h_2/.9, h_3/0, h_4/.1, h_5/.3, h_6/.8\}\}$.

Let (F_4, A_4) be the fuzzy soft set which describes the characteristics of the place where the houses are located given by $(F_4, A_4) = \{F_4(\text{in green surroundings}) = \{h_1/1, h_2/.1, h_3/.5, h_4/.3, h_5/.2, h_6/.3\}$ F_4 (near the road side) $=\{h_1/.2, h_2/.7, h_3/.8, h_4/1, h_5/.5, h_6/.9\}\}$.

Suppose that Mr. X is interested in buying a house on the basis of his choice of parameters beautiful, wooden, cheap, in green surroundings. This means that out of the houses available in U, he is to select the house that qualifies with all the parameters of his choice.

The above problem of Mr. X can be solved by using definition 2.6, a fuzzy soft relation (R, C) among the fuzzy soft sets (F_1, A_1), (F_2, A_2), (F_3, A_3) and (F_4, A_4) of the houses of U which are cheap, wooden beautiful and in green surroundings'. Clearly by the definition the fuzzy soft relation, (R, C) is given by

$(R,C) = \{R(\text{cheap, beautiful, wooden, in green surroundings}) = \{h_1/.2, h_2/0, h_3/.5, h_4/.2, h_5/.2, h_6/0\}$.

Evidently the house, which best satisfies the characteristics of Mr. X's choice is the house, which has the largest membership value in the relation. Here, h_3 has the largest membership value of 0.5, so Mr. X will buy the house h_3.

Remark 3.1. The solution of the above problem obtained by Maji et. al. [5,6] requires computing of row-sum, column-sum and score for each house. This requires more computational time compared to the solution obtained by using fuzzy soft relation. So the method used above is more efficient and economic.

Example 3.2. We now consider another decision-making problem of assigning a particular job to a suitable candidate who best fulfils the requirements of the job.

Let $U = \{p_1, p_2, p_3, p_4, p_5\}$ be the crisp set of available candidates for the job. Let E ={enterprising, average, confident, confused, willing to travel, un-willing to travel} be a set of parameters.

Let $(F_1, A_1) = \{F_1(\text{enterprising}) = \{p_1/.5, p_2/.7, p_3/.3, p_4/.1, p_5/.8\}, F_1(\text{average}) = \{p_1/.3, p_2/.1, p_3/.5, p_4/.8, p_5/.05\}\}$ be the soft set describing the enterprising qualities of the candidates.

$(F_2, A_2) = \{F_2(\text{confident}) = \{p_1/.6, p_2/.8, p_3/.5, p_4/.2, p_5/.9\}, F(\text{confused}) = \{p_1/.3, p_2/.1, p_3/.7, p_4/.9, p_5/.5\}\}$ $(F_3, A_3) = \{F_3(\text{willing to travel extensively}) = \{p_1/.7, p_2/.8, p_3/.5, p_4/2, p_5/.6\}, F_3(\text{unwilling to travel extensively}) = \{p_1/.3, p_2/.07, p_3/.65, p_4/.95, p_5/.1\}\}$ be the fuzzy soft set describing the willingness level to travel extensively.

Let us assume that the particular job requires an enterprising confident person who is willing to travel extensively. Our problem is to find the candidate who best suits the requirements of the job. To solve this we use definition 2.6 a fuzzy soft relation (R, C) of the fuzzy soft sets (F_1, A_1), (F_2, A_2), (F_3, A_3) of all candidates who are enterprising, confident, willing to travel extensively. By definition (R, C) is given by

$$(R, C) = \{p_1/.4, p_2/.5, p_3/.3, p_4/.1, p_5/.6\}.$$

From the above relation it is evident that the most suitable candidate for the job is p_5 who possesses the greatest membership value in the relation (R, C).

References

1. K. Atanassov (1986), Intuitionistic fuzzy sets, *Fuzzy Sets and Systems*, 20: 87-96.
2. K. Atanassov (1994), Operators over interval valued intuitionistic fuzzy sets, *Fuzzy sets and Systems*, 64: 159-174.
3. W.L. Gau & D.J. Buchrer (1993), Vague sets, *IEEE Tran. System Man Cybernet*, 23: 610-614.
4. M.B. Gorzalzany (1987), A method of inference in approximate reasoning based on interval valued fuzzy sets, *Fuzzy Sets and Systems*, 21: 1-17.
5. P.K. Maji, R. Biswas and A.R. Roy (2003), On Soft Set Theory, *Computers and Math with Appl*, 41:1-8.
6. P.K. Maji, R. Biswas and A.R. Roy (2001), Fuzzy Soft Sets, *The Journal of Fuzzy Mathematics*, 9(3): 589-602.
7. D. Molodostov (1999), Soft set theory-first results, *Computers Math. with Appl.* 37: 19-31.
8. Z. Pawlak (1999), Rough sets, *Int. J. Infor. & Comp. Sc.* 11: 341-356.

9. Z. Pawlak (1994), Hard sets and Soft sets, *ICS Research Report, Inst. of Comp. Sc.*, Poland.
10. H. Thielle (1999), On The concepts of qualitative fuzzy sets, *IEEE Inter. Sym. On multivalued logic*, Japan, 1999.
11. Y.Y. Yao (1998), Relational interpretations of neighbourhood operators and rough set approximation operators, *Infor. Sc.* 111: 239-259.
12. L.A. Zadeh (1965), Fuzzy sets, *Infor and Control* 8: 338-353.

Fuzzy Logic and its Application in Technology and Management
Editors: Debjani Chakraborty, S. Nanda and D. Dutta Majumder

Fuzzy Sets and Writer Shibram Chakraborty

Mihir Kumar Chakraborty and Sobhakar Ganguli

Department of Pure Mathematics, University of Calcutta, Kolkata
Email: mihirc99@cal3.vsnl.net.in

Abstract: In this article it has been focused that the basic philosophy of Fuzzy Set Theory was the theme of one of the short stories of eminent Bengali writer Shibram Chakraborty (1903 – 1980). The story was written sometime in the mid-thirties of the past century.

Keywords: Fuzzy set, Vagueness

The objective of this short article is to show that the core philosophical issue underlying Fuzzy set theory, a much publicized area of mathematics of the present day having extensive applications was addressed in one of his short stories by noted Bengali writer, and novelist Shibram Chakraborty, widely acclaimed for unparallel punning of words and humor. Looking for a similarity between a writing of Shibram and fuzzy set theory might seem to be far fetched. But it will be observed that not only the philosophical essence of the theory did he touched upon, the 'solution' to the problem towards which he pointed at in a casual story-telling manner converges principally with the current prescription suggested by the fuzzy set theorists to address vagueness. The story Shibram wrote in the 30's of the past century, exact date could not be ascertained. And we all know, Professor Lotfi Zadeh of Berkeley University published his first article on the theory in 1965. I firmly believe that Shibram deserves a proper place in the history of fuzzy set theory.

The title of his story is '*When mathematics is joined with literature*'. Main characters: Rajib, a decent boy and the author himself. The way Rajib or his father or elder brother answers Shibram's questions on daily affairs constitutes the background of the story. For example, Shibram asks, '*Hay, how are you?*' Rajib's brother replies, '*Carrying on, somehow.*' Or, when he asks '*Hallo, how are you Mr. Mukherjee?*' Rajib's father answers '*As the almighty wishes*'. Shibram then asked the maid servant accompanying Mr. Mukherjee to the market, '*How are you, dear?*' She melted, '*As you have blessed me with*'. Next Mahendra, the old friend, '*Oh Mahendra, we meet after such a long gap! How do you do?*' He replied, '*So, so*'. Utterly confused with all these fuzzy answers Shibram on his way back home met Rajib. '*How is your time passing Mr. Rajib?*' The answer was, '*First class*'. These are all usual ways of conversing in daily life. Could we not express ourselves in more concrete terms? Extremely bitter, angry and unhappy Shibram narrates –

'*I was flabbergasted with my experience – is everybody under the sun eccentric – or is there some problem in our language itself? Only poets can be happy with such mysticism in answers – my scientific mind was rather irritated and concerned and I decided to give a thought on a correct form of our usual style of expression*'.

These irreplaceable vague words and phrases of natural language are not by and large accepted in the discourses of science. But we may recall here Zadeh's insightful dictum that as the complexity of a system increases our ability to have a precise and yet significant description of it decreases until a threshold is reached beyond which precision and significance become almost mutually exclusive.

The solution to the problem that Shibram came up with is the following –

'*How is it if our language be tied up with some rules of mathematics especially the* ***adjectives and verbs****? There is nothing to err in the dictates of mathematics. 'Figures do not lie.'– thus goes a saying...... If anything is associated with numbers we get relieved and ambiguity of the language is removed.*' (bold face, ours : MKC , SG).

To represent vague terms Zadeh took recourse to numbers, quite understandably, since he is a computer scientist. But, surprisingly, Shibram also took steps to the same path. We may look into what he had suggested. '*Let the number 100 represent the whole or totality of everything relating to our body or mind — our beauty, intelligence, qualities, erudition and so on; we have to express everything proportionality to the highest number 100, that will clarify the difference of one state form another*'.

Fuzzy set theory takes 1 for the whole (totality) instead of 100.

Equipped with the theory Shibram decided to meet Rajib's father immediately to have a final settlement about his line of thought and to have his end gained. Let us follow his own narration again —

'*I ran with 82 - speed; as I was very close to Rajib's home, I found him flying a kite with 98 - attention*'.

At one point in the subsequent story-line Rajib with his eyes '*100-fixed*' at the kite stepped backward abruptly with '*92-speed*' and dashed with Shibram. Now, follow the description of how he takes Rajib to task —

'*I had an idea that you are 42 - intelligent but now do I find that you are not so, you are only 13 but 3 X 13 =39 – no, 49-precocious.*

Rajib was '*72- amazed*'.

'*How nonsense are you talking?*' said he.

'*Now I should revise myself to believe that you have attained 89- precociousness*' said Shibram.

Then there was a crowed gathering and Rajib complained to the people about the peculiar, mathematical, abusive language that Shibram was using against him. What then follows is a mass hysteria. The hilarious narration cannot be summarized one has to go through it.

Another sample from Shibram's text during the formative stage of his theory, before proceeding towards Rajib's home –

'*If I go and ask Rajib's father – how are you, O.K.? With 100 to express the highest state of good health, he shall have to think a while and then answer: okay, now I am 52 well, the day before yesterday, I had stomach problems and I was then 13- well, yesterday I was 7- well because of dental pain, when the dentist pulled them off, it reached the lowest level and I was only 2 - well — at best 4. I was almost gone. Since then I am feeling very weak but now 52......If I would express compassion for him he might say — I was 52-well before I met you; but your sympathy, has helped me...I am 53-well now.'*

It should be noted that the core philosophy of fuzzy set theory is basically the same. Application of vague words does not always permit us to give results '*yes*' (1) or '*no*' (0). The applicability degree in such cases is determined by some fraction between 0 and 1. Modern version of Shibram's example would be

'I am now .53-well'

or 'The degree of well-ness of mine is .53'.

Another point to be noted is that it does not matter if there is a little discrepancy in determination of the applicability-degree of vague terms. The theory allows space for such tolerance. This was present in Shibram's ideas too. Examine the above quotation. Sibram started with 52-well and ended up with 53-well applied to Rajib's father, the same person at the same time more or less. In his own words '*Oi janha bahanno taha tippanno*' (it does not matter much whether it is 52 or 53).

Yet another point. Recall the animated exchange between Rajib and Shibram after the dash and Shibram calling him '*89-precocious*'. At this, Rajib retorted, '*And you are 500-idiot'* — in fact, without any understanding of the mathematical scheme that Shibram had in his mind and that he had been applying all through. So we may notice how as a 'fuzzy set theorist' did Shibram react to the mistake that Rajib committed. In his own language,

'*I became 97- angry, took hold of his 45-ear with my fingers (I could not get hold of the ear entirely) – how do you say so and call me 500-idiot? It is difficult to understand my mathematics in which there is no number beyond 100!*'

The story ends with the narration of plight that Shibram had to undergo in the hands of the mob who took him to be hysteric and insane since he was scolding Rajib by using mathematical numbers. But Shibram, inspite of the physical abuse meted out to him by the crowd, was more concerned with the mistake people had been committing – '*their 72-idiocy made me 92- dumbfounded*'.

Concluding Remarks

Like most of his other writings Shibram probably created the present short story – an unparallel one indeed – out of his extraordinary appreciation – capability, sense of humor and passion for the magic of words. But while doing so, he has pointed his finger at one important aspect of the philosophy of language which is a very deep issue discussed now a-days by researchers in an area of knowledge that overlaps mathematics, logic, computer science and philosophy. And that he did much earlier, mostly as a comic and keeping himself outside the centre of mainstream knowledge-domain.

References

1. M.K. Chakraborty (2006), *Ganiter Dharapat Ó Galposalpo* (a collection of essays on the philosophy of mathematics), 3rd edition, Nandimukh Samsad.
2. Shibram Chakraborty, Anko Sahityer Jogfal (When Mathematics is joined with Literature), in: *Shibram Collected Works*, Annapurna Prakashani, Kolkata, 1985, 32-37.
3. L.A. Zadeh (1965), Fuzzy Sets, *Information and Control* 8: 338-353.

Part 2

Fuzzy Data Mining and Information Processing

Fuzzy Logic and its Application in Technology and Management
Editors: Debjani Chakraborty, S. Nanda and D. Dutta Majumder

Datamining & Knowledge Discovery Using A Fuzzy Mathematical Approach, in the Context of Indian Agricultural System Management

Dwijesh Dutta Majumder[1] *and Debasis Chanda*[2]

[1]Professor Emeritus, ISI & Secretary & Director, Institute of Cybernetics Systems and Information Technology, Kolkata
Email: ddmdr@hotmail.com
[2]Business Development Manager (Systems Integration), CMC Ltd.
Email: cdebasis03@yahoo.co.in

Abstract: Biologically evolved associative rules indicate that knowledge; data and storage are associatively related. Datamining deals with the discovery of *hidden knowledge*, *unexpected patterns* and *new association rules* from large databases (agricultural databases, in the context of this paper). This paper attempts to explore a Fuzzy Mathematical Model of datamining, which is *fast*, *scalable* and extracts *realistic* hidden patterns.

1. Introductions and Perspective

Indian *agriculture* is synonymous with the small farmer, who cultivates on an average less than 1.5 hectares. It is an accepted prospect that the *knowledge* residing in the Indian agricultural research system, if effectively transferred to the 156 million hectares of farmland in India, spread across rich and diverse agro climatic regions, has the potential to not only meet the growing needs of the domestic market, but also be a competitive source of many attractive global markets of the WTO era.

It is now universally acknowledged that no long term economic growth agenda for India can be feasible without including in its fold the agricultural sector, which is home to 72% of the population and 60% of the nation's workforce. Indeed, while targeting a real growth of over 8% in India's GDP, the Planning Commission observed that" agricultural development must be viewed as a core element of the Plan, since growth in this sector is likely to lead to the widest spread of benefits, especially to the rural poor including agricultural labour".

The objective would, therefore, be to maximize production of agricultural goods and services at least cost. This would call for optimal utilization of resources. That is, agricultural social profits in the form of value added are maximized only by getting the best from investments and by effective control of operational costs.

The *datamining* model, using a fuzzy mathematical approach, being proposed in this paper aims to discover *knowledge* in *agricultural databases*. This would on the one hand, facilitate delivery of customized knowledge to farmers, and on the other hand, enable managers / decision makers to take right decisions and manage effectively.

Datamining deals with *Knowledge Discovery in Databases (KDD)*. The six stages of the knowledge discovery process are

1. Data selection (where data relevant to the analysis task are retrieved from the database)
2. Data cleaning (to remove noise and inconsistent data)
3. Data transformation (where data is transformed or consolidated into forms appropriate for mining)
4. *Datamining* (an essential process where intelligent methods are applied in order to extract data patterns)
5. Pattern evaluation (to identify the truly interesting patterns representing knowledge based on some interestingness measures)
6. Knowledge presentation (where visualization and knowledge representation techniques are used to present the mined knowledge to the user)

Our paper will focus on the process of *Datamining*.

2. The Mathematical Approach

1. A crisp relation $R_c(X,X)$ is reflexive iff $\langle x,x\rangle \in R_c$ for each $x \in X$, that is, if every element of X is related to itself. A fuzzy relation $R_f(X, X)$ is reflexive iff $R_f(x, x) = 1$.
2. A crisp relation $R_c(X, X)$ is symmetric iff $\langle x,y\rangle \in R_c$, it is also the case that $\langle y,x\rangle \in R_c$, where x, y $\in X$. Thus, whenever an element x is related to an element y through a symmetric relation, y is also related to x. A fuzzy relation is symmetric iff $R_f(x, y) = R_f(y, x)$
3. A crisp relation $R_c(X, X)$ is transitive iff $\langle x,z\rangle \in R_c$, whenever both $\langle x,y\rangle \in R_c$, $\langle y,z\rangle \in R_c$, for at least one $y \in X$. In other words, the relation of x to y and of y to z implies the relation of x to z in a transitive relation. A fuzzy relation $R_f(X,X)$ is transitive if $R_f(x, z) \geq \max_{y \in Y} \min [R_f(x,y), R_f(y,z)]$ is satisfied for each pair $\langle x,z\rangle \in X^2$.
4. The transitive closure of a crisp relation $R_c(X, X)$ is defined as the relation that is transitive, contains $R_c(X, X)$, and has the fewest possible members. For fuzzy relations, this last requirement is generalized such that the elements of the transitive closure have the smallest possible membership grades that still allow the first two requirements to be met.

Given a relation $R(X, X)$, its transitive closure $R_T(X, X)$ can be determined by a simple algorithms that consists of the following three steps:

1. $R' = R \cup (R \circ R)$
2. If $R' \neq R$, make $R = R'$ and go to step 1.
3. Stop: $R' = R_T$

5. A fuzzy binary relation that is reflexive, symmetric, and transitive is known as a fuzzy equivalence relation or similarity relation.
6. A fuzzy binary relation that is reflexive and symmetric is called a fuzzy compatibility relation.

7. α -cut of a fuzzy set A is the crisp set α_A that contains all the elements of the Universal set X whose membership grades in A are greater than or equal to the specified value of α .

$$\alpha_A = \{x \mid A(x) \geq \alpha \}.$$

3. The Proposed Mathematical Model

The observations

i. The agricultural year comprises of four seasons based on the harvesting of crops, i.e. Kharif (July to October), Winter(October to January), Rabi (January to April) and Summer (April to June). It may immediately be observed that the concept of season itself is a fuzzy concept.

ii. The agricultural output is measured in terms of productivity (yield per hectare). The agricultural productivity is dependant on many factors (attributes) which are environmental in nature e.g. temperature, humidity, pressure, rainfall. Natural soil conditions also play important roles in agricultural productivity. The key attributes of soil content are Nitrogen, Potassium and Phosphorus. Each of the attributes referred to in the foregoing may be represented by fuzzy linguistic variables, viz. very low, low, medium, high and very high. The agricultural productivity is also dependent on a number of inputs like fertilizers, irrigation, etc. Fertilizer consumption, irrigated area may also be represented by the above indicated fuzzy linguistic variables.

iii. There are various classification of crops e.g. cereals, pulses, oilseeds, fibre as indicated below

Cereals	rice, wheat, barley, bajra
Pulses	gram, arhar, maskalai, masur, matar, khasuri
Oilseeds	til, rapeseed, mustard, linseed, groundnut, castor, sunflower
Fibre	jute, mesta, cotton, sunhemp

Thus crops under each class are similar to each other, and may be represented by fuzzy compatibility relationships.

3.1. Cluster Analysis

Clustering is the process of grouping data into classes or clusters, so that objects within a cluster have high similarity in comparison to one another, but very dissimilar to objects in other clusters. Dissimilarities are assessed based on the attribute values describing the objects.

Cluster Analysis focuses mainly on distance based cluster analysis, objective being to reduce the intra-cluster distance and increase the inter-cluster distance.

3.1.1.Cluster Analysis in the context of Indian Agriculture

The linguistic variables {very low, low, medium, high and very high} may be denoted by a set of integers such as $\{1,2,3,4,5\}$. Each of the records may, therefore, be represented as points in a multi-dimensional space, where each dimension is a member of the integer set $\{1,2,3,4,5\}$.

It may also be observed that dimensions, which are not represented by the linguistic variables {very low, low, medium, high, very high}, such as season_id and crop_id, may also be denoted by the integer sets:

The set representing season_id i.e. {kharif, winter, rabi, summer} may be denoted by (1,2,3,4}.

In the case of crop_id, the same may be represented by the integer set {0,1,2,3,4,5) by the following sequential process:

i. Representing each crop_id by an unique integer in ascending sequence starting from 1, i.e. 1,2,3, ...
ii. Dividing each integer (allocated to each crop), by the highest integer allocated, thus yielding a real number>0, in each case.
iii. Multiplying each such real number obtained by 5, and rounding of the same to the nearest integer.

A similar process may be followed for denoting district_ids by the integer set {1,2,3,4,5}.

Thus, a point in a multi-dimensional space may represent each record. The groups or clusters, thus formed, reflect in a natural way the structure of given data.

Every fuzzy equivalence relation (a relation that is reflexive, symmetric and max-min transitive) induces a crisp partition in each of its α-cuts. The fuzzy clustering problem can thus be viewed as the problem of identifying an appropriate fuzzy equivalence relation on given data. Although this cannot be done directly, we can readily determine a fuzzy compatibility relation (reflexive and symmetric) in terms of an appropriate distance function applied to given data.

Let us define a fuzzy compatibility relation *R*, on our data set (n *p*-tuples of $\mathbb{R}^p$), in terms of appropriate distance function of the Minkowski class by the formula given below, for all pairs $\langle x_i, x_k \rangle \in X$ where $q \in \mathbb{R}^+$ (the set of all non-negative real

$$R(x_i, x_k) = 1 - \delta(\sum_{j=1}^{p} |x_{ij} - x_{kj}|^q)^{1/q}$$

numbers) and δ is a constant that ensures that $R(x_i, x_k) \in [0,1]$. Clearly δ is the inverse value of the largest distance in X.

To illustrate the clustering method based on fuzzy equivalence relations, let us use a small data set consisting of the following five points in $\mathbb{R}^2$.

K	1	2	3	4	5
X_{k1}	1	2	3	4	5
X_{k2}	1	2	4	2	1

Where say X_{k1} represents crop_id, and X_{k2} represents productivity.

As the first step, we perform the analysis for $q = 2$, which corresponds to the Euclidean distance. First, we need to determine the value of δ in the above equation. Since the largest Euclidean distance between any pair of given data points is 4 (between x_1 and x_5), we have $\delta = 1/4 = 0.25$. Now, we can calculate memberships grade of *R* by the above equation. For example,

$$R(x_1, x_3) = 1 - 0.25\ \{(3-1)^2 + (4-1)^2\}^{0.5} = 0.1$$

When determined relation *R* may conveniently be represented by the matrix

$$R = \begin{bmatrix} 1 & .65 & .1 & .21 & 0 \\ .65 & 1 & .44 & .5 & .21 \\ .1 & .44 & 1 & .44 & .1 \\ .21 & .5 & .44 & 1 & .65 \\ 0 & .21 & .1 & .65 & 1 \end{bmatrix}$$

The relation is not max-min transitive; its transitive closure is

$$R_T = \begin{bmatrix} 1 & .65 & .44 & .5 & .5 \\ .65 & 1 & .44 & .5 & .5 \\ .44 & .44 & 1 & .44 & .44 \\ .5 & .5 & .44 & 1 & .65 \\ .5 & .5 & .44 & .65 & 1 \end{bmatrix}$$

This relation induces four distinct partitions of its α -cuts:

$\alpha \in [0, .44] : \{ \{x_1, x_2, x_3, x_4, x_5\}\}$

$\alpha \in (.44,.5] : \{ \{x_1, x_2, x_4, x_5\}, \{x_3\} \}$

$\alpha \in (.5,.65] : \{ \{x_1, x_2\},\{x_3\} , \{x_4, x_5\} \}$

$\alpha \in (.65, 1] : \{ \{x_1\},\{x_2\},\{x_3\}, \{x_4\}, \{x_5\} \}$

The clustering tree for Euclidean distance is shown below

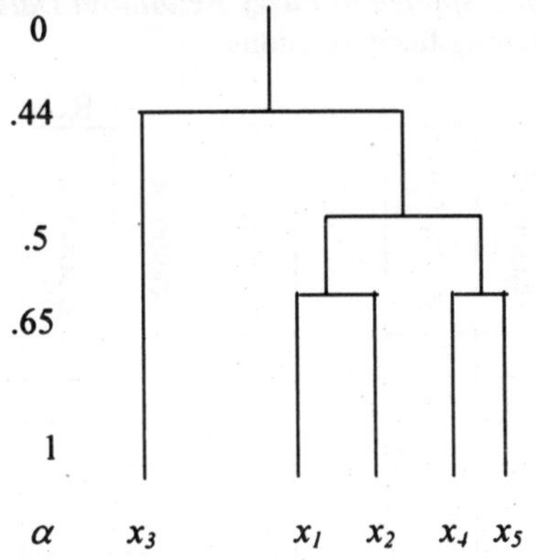

Thus we have *multidimensional classification* based on cluster analysis. Multidimensional classification implies similarity of concerned records in a cluster, or in other words, *close association.* The strength of association may be defined by means of a threshold value. In a multidimensional space, the threshold may be defined by the means of the radius of a sphere, having its centre as the centre of gravity of the cluster. Thus, for example, we may have strong *association rules* that suggest

District_id 4, season_id4, humidity high, fertilizer_consumption very high $\Rightarrow$ crop_id1.

And we may have a weak association rule that suggests that district_id4, season_id4, humidity high, fertilizer_consumption very high $\Rightarrow$ crop_id2.

3.1.2. The benefits of the proposed method

1. *Very fast response time*: In the proposed method, we are dealing only with exactly five integers during the datamining process, i.e. 1,2,3,4 and 5. Thus the fuzzy Euclidean distances being calculated are based on the above 5 integers only, thus obviating the need for calculations with extremely large numbers. Agricultural databases are huge, and in absence of the proposed method, would lead to calculations with extremely large numbers, thus leading to increased response times.
2. *Scalability*: The proposed method is extremely scalable. It is obvious from the proposed method that, no matter what the number of records and no matter what the extent of record addition, the representation of dimensions of such records never exceeds the integer set {0,1,2,3,4,5}.
3. *Extraction of realistic hidden patterns*: The proposed method is closer to real world expressions as compared to existing classical methods, e.g. it would be more realistic to convey that there is strong association between (high productivity, high rainfall and low temperature) instead of say (106.2 tonnes/hectare of productivity, 112 mm of rainfall and 5°C).
4. *Outlier analysis*: There may exist data objects, which do not comply with the general behaviour or model of the data. Such data objects, which are grossly different from or inconsistent with the remaining set of data, are called outliers (or exceptions), and may be of particular interest.

3.2. Datamining Methods Applied to Fuzzy Relational Databases

Let us consider the following fuzzy relations

Relation: *productivity*

Crop_id	district_id	year_id	season_id	productivity
1	1	1	1	5
2	1	1	2	4
2	1	1	3	4
4	2	1	4	2
5	4.	1	1	3

Relation: *season*

year_id	season_id	district_id	temp	rainfall	Humidity
1	1	1	5	4	5
1	2	1	4	3	4
1	3	1	4	5	2
1	4	1	2	1	3
1	1	2	3	4	4

Relation: *soil*

District_id	N content	K content	P content
1	5	5	2
2	4	3	5
3	2	3	3
4	3	4	2
5	2	5	5

Relation: *statistics*

District_id	Year_id	Irrigated area	Fertilizer consumption
1	1	2	2
2	1	5	5
3	1	3	3
4	1	2	2
5	1	5	5

Relation: *utility*

crop_id	utility1	utility2	Utility3
1	5	1	4
2	2	3	2
3	1	5	2
4	4	2	1
5	3	3	4

It may be noted that the relation utility has been defined keeping the food utility value of the concerned crop in mind, which will vary from region to region. Thus region-wise utility values have been shown as utility1, utility 2, utility 3, etc. From this relation we are, therefore, gives us in insight into the demographic character of the country based on food habits.

We may thus have a fuzzy compatibility relation defined for either of the domains utility1, utility2, utility 3, etc. on the domain universal set {1,2,3,4,5}, e.g. for the domain utiliy 1, we may have the following fuzzy compatibility relation.

	5	4	3	2	1
	1	.8	.6	.4	.2
4	.8	1	.8	.6	.4
3	.6	.8	1	.8	.6
2	.4	.6	.8	1	.8
1	.2	.4	.6	.8	1

Similarly, it would also be possible to define fuzzy compatibility relations on the domains productivity, district, temperature, humidity, rainfall, irrigated area, fertilizer consumption, Nitrogen content, Potassium content, Phosphorus content. For example, in case of districts the fuzzy compatibility relation may be defined on the nearness of districts to the base district.

Thus, it would be possible to query the above database with the following question: "Which similar crops have similar productivity in district 1?"
The answer is obtained by accomplishing a relational join on the relations "productivity " and "utility" on the attribute crop_id, selecting the tuples where distric_id=1 and then projecting the crop_ids and productivity as sets, where threshold productivity ≥ .6 and threshold utility 1 ≥ .8.

On performing the relational join, we have the following relation (relation 1)

crop_id	productivity	utility1
1	5	5
2	4	2
2	4	2

Now, we observe that productivities for crops 1 and 2 are similar in the sense that both are > the threshold value of 0.6. However, only crop 1 > the threshold utility value of .8. Thus, for the criteria defined by us, only crop 1 qualifies.
The result is shown below

Crop_id	Productivity
1	5

It may, therefore, be immediately observed that fuzzy relational databases provide us with insights (*knowledge*), which would otherwise not be possible with crisp relational databases, where we would not be able to define *similar* crops, *similar* productivity or *threshold* values. Also, search time is faster, as we are dealing with a few integers only when considering the attributes (domains) e.g., productivity, soil content, rainfall, humidity, etc. Moreover, the above methodology also provides us with mechanisms of arriving at strong or weak *association rules*, based on the threshold values.

As observed, the above methodology leads to *knowledge discovery*, and is, therefore, an effective *datamining* process that may be applied on relational databases. Moreover, multi-dimensional datamining on fuzzy relational databases, is also possible by sequential relational joins with relations containing the relevant dimensions (attributes). For example, we may accomplish a relational join between the relation 1 above and the relation season, and may obtain further insights (knowledge), into the database under consideration.

4. Conclusion

The mathematical model discussed in this paper leads to *Management Intelligence*. Discovery of *hidden knowledge*, *patterns* and *new association rules* enable the Manager to get the best out of investments, implement effective control of operational costs and manage agricultural resources optimally. Relevant intelligence and customized knowledge may then be passed onto the farmer for implementation, thereby leading to maximization of production of agricultural goods and services at least cost.

Investment decisions taken as a result of such Management Intelligence may be measured in terms of *Return on Investment* (*RoI*). In the context of Agricultural System Management, increase in total production in a particular agricultural year may be one of the yardsticks by which the RoI may be measured, vis-à-vis the total investment made in that particular year.

References

1. D Dutta Majumder and Debasis Chanda (2004), Study on a Framework for Agricultural Forecasting Systems An application of Information Technology & Datamining Techniques in the Indian Scenario, International Conference on *Recent Trends & New Directions of Research in Cybernetics & Systems Theory,* IASST, Guwahati in January 2004
2. Speech by Mr. Y C Deveshwar, Chairman ITC Ltd., at the 92nd AGM held on 25. 07.2003
3. S.K. Pal & D Dutta Majumder (1986), *Fuzzy Mathematical Approach to Pattern Recognition*, Wiley, New York.
4. George J. Klir & Bo Yuan (1995), *Fuzzy Sets and Fuzzy Logic Theory and Applications*, Prentice-Hall, Englewood Cliffs, NJ.
5. Pieter Adriaans & Dolf Zantinge (1997), *Datamining,* Addison-Wesley, Boston, MA, USA.
6. Jiawei Han & Micheline Kamber (2000), *Datamining Concepts and Techniques,* Morgan Kaufmann, April 2000.
7. Abraham Silberschatz, Henry F. Korth & S. Sudarshan (2001), *Database Systems Concepts*. McGraw-Hill Higher Education, 4th edition.

Fuzzy Logic and its Application in Technology and Management
Editors: Debjani Chakraborty, S. Nanda and D. Dutta Majumder

Compositional Rule of Inference: A Fuzzy Linear Regression Approach

Chandan Chakraborty and Debjani Chakraborty

Department of Mathematics, Indian Institute of Technology Kharagpur
Email: {chandan, debjani}@maths.iitkgp.ernet.in

Abstract: The paper aims to compress the linguistic information provided by a block of fuzzy rules. Rule represents the real world scenario within a logical framework. In view of this, a new composition rule of inference is being introduced using fuzzy linear regression in this paper. Also an objective function is defined here based on fuzzy distance measure. The proposed technique is applied to a given fuzzy rule base of a knowledge base evaluator for building a fuzzy model.

Keywords: Fuzzy Linear Regression, Fuzzy Distance Measure, Least Square Method, Knowledge base.

1. Introduction

The object-oriented modeling concept is developed to represent any real world situation more precisely. Some of the features of an entity need to be described fuzzily to make the system more flexible and informative along with the quantitative features. In this regard, fuzzy 'information binding or encapsulation' is very important to increase the processing efficiency of the system. The knowledge about the entity could be represented truly in terms of fuzzy rule-base (FRB). In order to build a compositional rule of inferencing model the inherent mathematical/logical framework of FRB is needed to identify on the basis of which a proper decision can be achieved. To encapsulate the whole set of fuzzy information here '*IF-THEN*' type of fuzzy rule for a real world entity is being considered with k attributes/features. In a FRB, a fuzzy rule is described as

$$R_i : \text{IF } X_1 \text{ is } \tilde{x}_{i1}, X_2 \text{ is } \tilde{x}_{i2},\ldots, X_k \text{ is } \tilde{x}_{ik} \text{ THEN } Y \text{ is } \tilde{y}_i \ \forall i=1,..,n \qquad (1)$$

The objective of this work is to develop a methodology for knowledge extraction by which the inference about the consequent can be drawn for a given set of antecedent.

$$\text{IF } \ (\tilde{x}^*_{i1}, \tilde{x}^*_{i2}, \ldots, \tilde{x}^*_{ik}) \text{ THEN } \ \tilde{Y}^* = ? \qquad (2)$$

For any FRB system we expect that there may exist a functional bond between the antecedent and consequent. Here the fuzzy regression technique is adopted to investigate the interrelationship that would be applicable to infer consequent of any fuzzy query rule. The statistical regression analysis deals with the precise data where the fuzzy linear regression analysis [1, 2, 5] has been studied to capture a large spectrum of information in the linguistic domain. Fuzzy regression analysis is a fuzzy data oriented modeling technique, which enables us to set up a relationship between independent variables and dependent variable in fuzzy environment. In classical regression model the difference between the observed values and expected values are supposed to be due to random error, which can be minimized by using several statistical methods to estimate the model parameters. But sometimes it may happen that the difference is due to imprecise observations or incompleteness / infiniteness of the system. In such situation the uncertainty arises due to its vagueness, not randomness. To solve the above problem we use the novel concept of fuzzy linear regression analysis for fuzzy rule base. A new concept on fuzzy distance [6] has been proposed that is used to construct an objective function in this paper.

Section 2 describes the methodology of compositional rule for fuzzy ineferencing and a model is developed. Also the fuzzy distance measure is briefly described in Subsection 2.1. The methodology is applied to a given fuzzy rule base of a knowledge base evaluator [10, 11] in Section 4.

2. Composition Rule for Fuzzy Inferencing

Suppose an entity is described by its '*k*' qualitative attributes called as fuzzy variables on the basis of which the decision has to be made. The antecedent part of each fuzzy rule of a FRB consists of '*k*' fuzzy variables whereas action as consequent. In the KB, there may exist an interrelationship between antecedent and consequent of a rule. To retrieve the inherent interrelationship existing in the FRB it is needed to capture the whole FRB within the following mathematical structure where the antecedent and consequent are connected as follows:

$$\tilde{y}_j = a_0 + a_1 \tilde{x}_{1j} + \ldots\ldots\ldots \;\; \ldots + a_k \tilde{x}_{kj} \quad \forall j. \tag{3}$$

The crisp parameters a_i 's of eq. (2.1) are to be estimated on the basis of the given FRB. In doing so, let us assume all input-output parameters as trapezoidal fuzzy numbers as follows: $\tilde{y}_j = (m^{(1)}_{y_j}, m^{(2)}_{y_j}, \alpha_{y_j}, \beta_{y_j})_{LR}$ and $\tilde{x}_{ij} = (m^{(1)}_{x_{ij}}, m^{(2)}_{x_{ij}}, \alpha_{x_{ij}}, \beta_{x_{ij}})_{LR}$ respectively. A fuzzy number is said to be *LR*-type trapezoidal fuzzy number, say $\tilde{M} = (m^{(1)}, m^{(2)}, \alpha, \beta)_{LR}$ with the following membership function

$$\mu_{\tilde{M}}(x) = \begin{cases} L\left(\dfrac{m^{(1)} - x}{\alpha}\right) & \text{for } x \le m^{(1)} \\ 1 & \text{for } m^{(1)} \le x \le m^{(2)} \\ R\left(\dfrac{x - m^{(2)}}{\beta}\right) & \text{for } x \ge m^{(2)} \end{cases} \qquad (\alpha > 0, \beta > 0)$$

where $m^{(1)}$, $m^{(2)}$ ($> m^{(1)}$) α and β are two centers, left and right spreads of $\tilde{M}$ respectively and $L(x) = R(x) = max(0, |1 - x|)$. In the above model eq. (2.1) it can be noted that the fuzzy consequent is approximated by a linear combination of fuzzy antecedents. Then the model eq. (2.1) can be approximated as follows

$$\tilde{y}_i = a_0 + a_1 \tilde{x}_{i1} + \ldots\ldots\ldots \;\; \ldots + a_k \tilde{x}_{ik} \cong (m_i^{(1)}, m_i^{(2)}, \alpha_i, \beta_i)_{LR} \tag{4}$$

where,

$$m_i^{(1)} = a_0 + \sum_{j=1}^{k} a_j m_{x_{ij}}^{(1)}, \; m_i^{(2)} = a_0 + \sum_{j=1}^{k} a_j m_{x_{ij}}^{(2)}, \; \alpha_i = \sum_{j=1}^{k} a_j \alpha_{x_{ij}} \text{ and } \beta_i = \sum_{j=1}^{k} a_j \beta_{x_{ij}}.$$

2.1. Fuzzy Distance Measure

Many research works [2, 7-9] have been done on defining the distance measure for fuzzy numbers. P. Diamond [7] and Yang et al. [2] has been defined a measure in the Euclidean space. In [8], Tran et al. introduced the distance concept based on the interval. In [9] a fuzzy distance for Gaussian type fuzzy numbers is defined that includes a normalization factor. In the literatures, though the vagueness is inherent within the distance measures but it's not reflected as a fuzzy number. The fuzzy points may be two non-numeric / linguistic values expresses in natural language [3], blurred regions in satellite image, hazy spots in medical image and pattern recognition. Since a fuzzy number is a collection of different points and therefore the distance between two fuzzy numbers is nothing but the distances of pair wise elements of the respective fuzzy numbers. In our opinion, consideration of infimum or supremum or any value from the distance set is crude oversimplification. Rather it would be more reliable to consider the distance between two fuzzy / blur / hazy points as fuzzy number. In this paper, a fuzzy valued distance measure [6] is briefly described that satisfies the necessary metric properties softly. In regression analysis, the estimation of model parameters is the most important issue, which is done here by means of minimizing the fuzzy distance measure. Based on the fuzzy distance measure, the distance between $\left(\tilde{y}_i, \sum_{j=0}^{k} a_j \tilde{x}_{ij}\right)$ is written as

$$\tilde{d}_{LR}^{i}(\tilde{y}_i, \sum_{j=0}^{k} a_j \tilde{x}_{ij}) = (d_{m^{(1)}}^{i}, d_{m^{(2)}}^{i}, d_{\alpha}^{i}, d_{\beta}^{i})_{LR}$$

where
$$\begin{cases} d_{m^{(1)}}^{i} = (2\lambda - 1)(m_{y_i}^{(1)} - \sum_{j=0}^{k} a_j m_{x_{ij}}^{(1)}) \\ d_{m^{(2)}}^{i} = (2\lambda - 1)(m_{y_i}^{(2)} - \sum_{j=0}^{k} a_j m_{x_{ij}}^{(2)}) \\ d_{\alpha}^{i} = \frac{1}{2}\left\{\lambda\left(\alpha_{y_i} + \sum_{j=0}^{k} a_j \beta_{x_{ij}}\right) - (\lambda - 1)(\beta_{y_i} + \sum_{j=0}^{k} a_j \alpha_{x_{ij}})\right\} \\ d_{\beta}^{i} = \frac{1}{2}\left\{\lambda(\beta_{y_i} + \sum_{j=0}^{k} a_j \alpha_{x_{ij}}) - (\lambda - 1)\left(\alpha_{y_i} + \sum_{j=0}^{k} a_j \beta_{x_{ij}}\right)\right\} \end{cases}$$
and

$$\lambda = \begin{cases} 1 & if \quad m_{y_i} \ge m_i \\ 0 & if \quad m_{y_i} < m_i \end{cases}.$$

The error inherent in the model is explained by vagueness. Therefore the center of the fuzzy distance actually represents the error and spreads as vagueness. Hence the model parameters are estimated by minimizing the objective function based on least square approach as follows:

$$\text{Min}\left(\sum_{i=1}^{n} d_m^{i\;2}\right) = (2\lambda-1)^2 \,\text{Min}\sum_{i=1}^{n}\left[\left(m_{y_i}^{(1)} - \sum_{j=0}^{k} a_j m_{x_{ij}}^{(1)}\right)^2 + \left(m_{y_i}^{(2)} - \sum_{j=0}^{k} a_j m_{x_{ij}}^{(2)}\right)^2\right].$$

The system of simultaneous equations can be represented in the matrix form as

$$\begin{bmatrix} \sum_{i=1}^{n}\left(m_{y_i}^{(1)} + m_{y_i}^{(2)}\right) \\ \sum_{i=1}^{n}\left(m_{y_i}^{(1)} m_{x_{i1}}^{(1)} + m_{y_i}^{(2)} m_{x_{i1}}^{(2)}\right) \\ \vdots \\ \sum_{i=1}^{n}\left(m_{y_i}^{(1)} m_{x_{ik}}^{(1)} + m_{y_i}^{(2)} m_{x_{ik}}^{(2)}\right) \end{bmatrix}_{(k+1)\times 1} = \text{X} \begin{bmatrix} a_0 \\ a_1 \\ \vdots \\ a_k \end{bmatrix}_{(k+1)\times 1} \quad \text{where}$$

$$X = \begin{bmatrix} 2n & \sum_{i=1}^{n}\left(m_{x_{i1}}^{(1)} + m_{x_{i1}}^{(2)}\right) & \cdots & \sum_{i=1}^{n}\left(m_{x_{ik}}^{(1)} + m_{x_{ik}}^{(2)}\right) \\ \sum_{i=1}^{n}\left(m_{x_{i1}}^{(1)} + m_{x_{i1}}^{(2)}\right) & \sum_{i=1}^{n}\left(m_{x_{i1}}^{(1)2} + m_{x_{i1}}^{(2)2}\right) & \cdots & \sum_{i=1}^{n}\left(m_{x_{i1}}^{(1)} m_{x_{ik}}^{(1)} + m_{x_{i1}}^{(2)} m_{x_{ik}}^{(2)}\right) \\ \vdots & \vdots & \ddots & \vdots \\ \sum_{i=1}^{n}\left(m_{x_{ik}}^{(1)} + m_{x_{ik}}^{(2)}\right) & \sum_{i=1}^{n}\left(m_{x_{ik}}^{(1)} m_{x_{i1}}^{(1)} + m_{x_{ik}}^{(2)} m_{x_{i1}}^{(2)}\right) & \cdots & \sum_{i=1}^{n}\left(m_{x_{ik}}^{(1)2} + m_{x_{ik}}^{(2)2}\right) \end{bmatrix}_{(k+1)\times(k+1)}$$

And λ values are verified with $\hat{m}_i^{(1)}$ and $\hat{m}_i^{(2)}$ and $m_{y_i}^{(1)}$ and $m_{y_i}^{(2)}$ for all i. Therefore the fitted fuzzy regression model is as

$$\hat{\tilde{y}} = \hat{a}_0 + \hat{a}_1 \tilde{x}_1 + \ldots\ldots\ldots\ \ldots. + \hat{a}_k \tilde{x}_k .$$

Hence the predicted consequent of the fuzzy query rule is given by

$$\hat{\tilde{y}}^{*}_{Expected} = \hat{a}_0 + \hat{a}_1 \tilde{x}_1^{*} + \ldots\ldots\ldots\ \ldots. + \hat{a}_k \tilde{x}_k^{*} .$$

3. Knowledge Base Evaluator

A set of fuzzy rules are extracted from the Knowledge Base Evaluator (KBE) [10, 11], which is an expert system that evaluates whether or not the expert system is suitable for a domain. In the FRB, there are five fuzzy variables: EMPLOYEE_ACCEPTANCE, SOLUTION_ AVAILABLE, EASIER_SOLUTION, TEACHABILITY, RISK. The decision is made about the SUITABILITY of the system. The antecedents and consequent assign perception-based values to capture a large spectrum of information in decision-making.

Table 3.1 Original fuzzy rules of KBE

Rule Antecedents					***Consequent***
Employee_ Acceptance	*Solution_ Available*	*Easier Solution*	*Teachability*	*Risk*	*Output (Suitability)*
Neutral	Adequate	Complete	Difficult	High	Poor
Negative	None	Partial	Frequent	Low	Poor
Negative	Partial	None	Difficult	Moderate	Fair
Positive	Partial	Complete	Frequent	High	Poor
Positive	Partial	Partial	Possible	Low	Good
Positive	Adequate	None	Frequent	Low	Fair
Negative	Partial	None	Frequent	High	Fair
Positive	None	Complete	Difficult	Moderate	Poor
Neutral	Adequate	Complete	Frequent	Low	Fair
Neutral	None	None	Difficult	Low	Fair
Positive	Adequate	None	Difficult	High	Poor
Negative	Adequate	Partial	Frequent	High	Poor
Negative	Partial	Complete	Possible	Low	Fair
Negative	None	Partial	Difficult	High	Fair
Neutral	Adequate	Partial	Difficult	Low	Poor
Neutral	None	None	Difficult	Low	Fair

In this example the decision maker fuzzifies the different linguistic terms for various fuzzy antecedents in the interval [0, 1] as follows:

$$\begin{cases} Term\_1 = (0.00, 0.20, 0.0, 0.10)_{LR} \\ Term\_2 = (0.35, 0.65, 0.1, 0.05)_{LR} \quad \text{where} \\ Term\_3 = (0.80, 1.0, 0.10, 0.0)_{LR} \end{cases}$$

Table 3.2 Term sets for antecedents and consequent defined on [0, 1]

Terms	*Employee Acceptance*	*Solution_ Available*	*Easier_ Solution*	*Teachability*	*Risk*	*Suitability*
Term_1	Negative	Adequate	None	Difficult	Low	Poor
Term_2	Neutral	Partial	Partial	Possible	Moderate	Fair
Term_3	Positive	None	Complete	Frequent	High	Good

A PC-aided intelligent system could be designed in such a way that the given large set of fuzzy rules is compressed by means fuzzy model building. A user-interface for fuzzy inference can be framed as follows for the web users.

IF	*Employee_Acceptance*	is	________	&
	Solution_Available	is	________	&
	Easier_Solution	is	________	&
	Teachabilty	is	________	&
	Risk	is	________	
THEN				
	Suitabilty	is	________	

Based on the KB in table 3.1 and 3.2 the parameters are estimated based on the simultaneous equations (see Appendix) and hence the fitted fuzzy regression model is written as follows:

$$\textit{Suitability} = 0.182 + 0.12 \times \textit{Employee\_Acceptance} + 0.189 \times \textit{Solution\_Available} - 0.134 \times \textit{Easier\_Solution} + 0.247 \times \textit{Teachability} - 0.085 \times \textit{Risk}.$$

The prediction of suitability for a given fuzzy query rule can be inferred using the above fitted model. And the predicted value needs to be mapped into the linguistic domain defined by decision maker for approximate reasoning about consequent.

4. Conclusion

Every system needs information compression tool as the data becomes large due to the complexity of situation and incorporation of more decision makers in practice. In this paper basically a fuzzy regression model is built as a fuzzy information processing method, which is also able to infer the consequent of a fuzzy query rule on the basis of synthesized prior fuzzy rule base. The model would give better result for consistent fuzzy rule base that must avoid redundancy, subsumed and conflicting rules [10]. Two rules are considered to be redundant if both of them have same antecedents and consequents. One rule is subsumed by other one when they have same consequents; but one rules' antecedents include the other's. Two rules are considered to be conflict if their antecedents are the same, but their consequents are different.

Acknowledgement

Authors acknowledge the support given by Council of Scientific & Industrial Research, New Delhi, India (Project No. 25(0119)/01/EMR-II).

References

1. M. S. Yang & T. S. Lin (2002), Fuzzy least-squares linear regression analysis for fuzzy input-output data, *Fuzzy Sets and Systems*, 126: 389-399.
2. M. S. Yang & C. H. Ko (1997), On Cluster-Wise Fuzzy Regression Analysis, *IEEE Transactions on Systems, Man and Cybernetics – Part B: Cybernetics*, 27 (1): 1-13
3. Debjani Chakraborty (2001), Structural quantization of vagueness in linguistic expert opinions in an evaluation programme, *Fuzzy Sets and Systems*, 119: 171-186
4. L.A. Zadeh (1965), Fuzzy Sets, *Information & Control*, 8: 338-353.
5. Chiang Kao & Chun-Lu Chyu (2002), A fuzzy linear regression model with better explanatory power, *Fuzzy Sets and Systems*, 126: 401-409.
6. Chandan Chakraborty & Debjani Chakraborty (2004), Aggregation of experts' opinions through clustering in group decision making environment, *Proc. of an Int. Conf. on Recent Trends & New Directions of Research in Cybernetics & Systems Theory*, IASST, Guwahati, India
7. P. Diamond (1988), Fuzzy least squares, *Information Science*, 46: 41-157.
8. L. Tran & L. Duckstein (2002), Comparison of fuzzy numbers using a fuzzy distance measure, *Fuzzy Sets and Systems*, 130 (3): 331-341
9. R. Xu & C. Li (2001), Multidimensional least-squares fitting with a fuzzy model, *Fuzzy Sets and Systems*, 119: 215-223.

10. H.M. Lee, J.M. Chen, C.L. Liu (2002), Inconsistency Resolution and Rule Insertion for Fuzzy rule-Based Systems, *Journal of Information Science and Engineering*, 18: 187-210.
11. R. Keller (1987), *Expert System Technology Development and Application,* Jordon Press, Englewood Cliffs, NJ.

Appendix

The system of simultaneous equations is as follows:

$$
\begin{aligned}
32\,a_0 + 15.2\,a_1 + 15.2\,a_2 + 15.2\,a_3 + 16\,a_4 + 14.4\,a_5 &= 11.2 \\
15.2\,a_0 + 11.165\,a_1 + 6.98\,a_2 + 7.945\,a_3 + 7.67\,a_4 + 6.87\,a_5 &= 5.665 \\
15.2\,a_0 + 6.98\,a_1 + 11.165\,a_2 + 7.305\,a_3 + 8.635\,a_4 + 6.555\,a_5 &= 6.305 \\
15.2\,a_0 + 7.945\,a_1 + 7.305\,a_2 + 11.165\,a_3 + 8.32\,a_4 + 7.51\,a_5 &= 5.015 \\
16\,a_0 + 7.67\,a_1 + 8.635\,a_2 + 8.32\,a_3 + 12.85\,a_4 + 6.28\,a_5 &= 6.985 \\
14.4\,a_0 + 6.87\,a_1 + 6.55\,a_2 + 7.51\,a_3 + 6.28\,a_4 + 11.25\,a_5 &= 4.265
\end{aligned}
$$

This system has a unique solution:

a_0 = 0.1817742723284198, a_1 = 0.1198380690661583, a_2 = 0.1888305673942709, a_3 = -0.1340613210910273, a_4 = 0.24727621749686, a_5 = -0.08522412417181284.

Fuzzy Logic and its Application in Technology and Management
Editors: Debjani Chakraborty, S. Nanda and D. Dutta Majumder

Fast and Robust Learning Through Fuzzy Linear Proximal Support Vector Machines for Multicategory Data Classification

Reshma Khemchandani

Department of Mathematics
Indian Institute of Technology,Hauz Khas, New Delhi.
Email: mar030007@ccsun50.iitd.ac.in

Abstract: Given a dataset, each point of which labeled by one of k labels, we propose fuzzy extension of multicategory proximal support vector machines, where a fuzzy membership is assigned to each data points. Patterns are classified by assigning them to nearest of k parallel planes, which are proximal to one of the k categories and separates one category from the rest. The algorithm is simple and fast as no quadratic or linear programming problem is solved, effectively, only solution of system of equation is needed. This work is extension of author's previous work on fuzzy proximal support vector machines for binary data classification.

Keywords: Learning, Support Vector Machines, Fuzzy Methods.

1. Introduction

In the field of pattern classification, Support Vector Machines (SVMs) have played a vital role [2,3], its motivation lies in maximizing the ability to generalize well from a small number of training samples. SVMs assign data points to one of two half-spaces, either in the pattern space or in a transformed higher-dimensional space. The most popular SVM is the "maximum margin classifier", which aims at minimizing an upper bound on the generalization error through maximizing the margin between two disjoint half spaces [2,3]. The maximal margin classifier represents the classification problem as a convex optimization problem: minimizing a quadratic function under linear inequality constraints.

Fung and Mangasarian recently proposed Proximal SVMs (PSVMs) [5], where each class of points are assigned to the closest of two parallel planes, that are pushed apart as far as possible. PSVMs require the solution of a single set of linear equations, and can thus be considerably faster than traditional SVMs.

In general, real life data sets are usually corrupted with noise and are associated with fuzziness in the membership of data points to a class. A classifier, which is able to utilize information regarding this fuzziness, can improve its performance and lessen the effect of outliers. Traditional SVM based classifiers lack such a mechanism.

Recently, fuzzy proximal support vector machines (FPSVMs) [7], were implemented, wherein each data points has been associated with fuzzy membership value to a class. This formulation, which can also be interpreted as regularized least squares, leads to extremely fast and simple algorithm for generating linear classifier which is obtained by solving a single system of linear equations, which requires the inversion of a matrix in a dimension of input space, which leads to the computational advantage. It seems logical, to apply this simple 2-class FPSVM classifier to k-category classification by using a one from rest (OFR) separation for each class. In comparison, SVMs k class classifier, which require the solution of either a large single, or k smaller quadratic or linear programs we here require solution of linear equations. In this paper, we propose a FPSVMs for multicategory data classification, by incorporating a membership matrix S that indicates the fuzzy membership of data points to their respective classes. We derive results to show that the solution of Fuzzy PSVMs (FPSVMs) for multicategory data classification also requires the solution of a reduced system of linear equations.

The paper is organized as follows: Section 2 briefly dwells on SVMs, PSVMs. In section 3 we briefly review the 2-category FPSVMs and then introduce our Multicategory FPSVMs (MFPSVMs). Section 4 contains experimental results and concluding remarks.

2.a. Support Vector Machines

Let the patterns to be classified be denoted by a set of m row vectors A_i, $(i = 1,2 \ldots m)$ in the n-dimensional real space $\Re_n$, where $A_i = (A_{i1}, A_{i2} \ldots A_{in})$. Also, let $f_i \in \{1, -1\}$ denote the class to which the i-th pattern belongs. We first consider the case when the patterns belonging to the two classes are strictly linearly separable. Then, we require to determine $w \in \Re^n$ and $\gamma \in \Re$ such that

$$\begin{aligned} A_i w &\geq \gamma + 1 \quad \text{for } f_i = 1 \\ A_i w &\leq \gamma - 1 \quad \text{for } f_i = -1. \end{aligned} \tag{1}$$

The plane described by

$$w^T x = \gamma \tag{2}$$

lies midway between the bounding planes given by

$$x^T w = \gamma + 1 \text{ and } x^T w = \gamma - 1 \tag{3}$$

and separates the two classes from each other with a margin of $1/\|w\|_2$ on each side. In other words, the margin of separation between the two classes is given by $2/\|w\|_2$.

Data samples that lie on the planes given by (3) are termed as support vectors. The maximum margin classifier, which is the standard SVM, is obtained by maximizing this margin, which is equivalent to the following problem

$$\text{Minimize} \frac{1}{2} w^T w \tag{4}$$

subject to the constraints (1).

When the two classes are not strictly linearly separable, there will be an error in satisfying the inequalities (1) for some patterns, and we can modify (1) to

$$A_i w + y_i \geq \gamma + 1 \quad \text{for } f_i = 1,$$
$$A_i w - y_i \leq \gamma - 1 \quad \text{for } f_i = -1, \tag{5}$$
$$y_i \geq 0, \; i = 1, 2, \ldots m,$$

where y_i denotes an error variable for the ith data sample. In this case the classifier is termed as a "soft margin" one, and it approximately classifies points into two classes with some error. The classification of a given test sample x is obtained by determining the sign of $w^T x$ - γ. The soft margin depends on the value of the non-negative error variables y_i . In this case, one needs to choose a trade-off between the margin and the error, and the standard SVM formulation for classification of the data points with a linear kernel is given by

$$\text{Minimize } ce^T y + \frac{1}{2} w^T w \tag{6}$$

subject to the constraints (5). Here, $y = (y_1, y_2 \ldots y_m)^T$, and c denotes a scalar whose value determines the trade-off; a larger value of c emphasizes the classification error, while a smaller one places a larger significance on the classification margin. The standard formulation of the SVM "soft-margin" classifier is thus (6), subject to constraints (5), which can be summarized in the following form

$$D(Aw - e\gamma) + y \geq e, \quad y \geq 0 \tag{7}$$

where D is a diagonal matrix whose entries are given by $D_{ii} = f_i$, i.e.

$$D = diag(f_1, f_2, \ldots f_m) \tag{8}$$

and e is a vector of ones, i.e. $e = (1, 1 \ldots 1)^T$.

2.b. Proximal Support Vector Machines

Mangasarian and Musicant first suggested [9] that the norm of the error variables in (6) be changed from the L_1 norm to the L_2 norm. This leads to the following optimization problem for determining the classifier

$$\text{Minimize } c \cdot \frac{1}{2} y^T y + \frac{1}{2}\left(w^T w + \gamma^2\right) \tag{9}$$

subject to the constraints

$$D(Aw - e\gamma) + y \geq e. \tag{10}$$

Observe that the non-negativity constraint on the error variables is no longer required, because if any y_i is negative, then (9) can be reduced by letting y_i to be equal to zero, while still satisfying constraints (10). Fung and Mangasarian [5] claim that the formulation in (9)-(10) is not only as good as the classical SVM formulation, but in fact has added advantages such as strong convexity of the objective function.

The idea of Proximal SVMs (PSVMs) evolves from (9)-(10) with only a simple change in the constraints; viz. (10) is replaced by

$$D(Aw - e\gamma) + y = e. \tag{11}$$

Surprisingly, this small variation in the formulation changes the nature of the optimization problem in a significant manner. The planes $x^T w = \gamma + 1$ and $x^T w = \gamma - 1$ now no longer represent bounding planes, but corresponds to planes around which points of the corresponding class are clustered. The objective function

(9) read along with constraints (11) can also be interpreted [5] as a regularized least-squares solution of the system

$$D(Aw - e\gamma) = e, \tag{12}$$

and is not based on maximizing the margin, which is a key feature of traditional SVMs [3, 9,10].

3. Fuzzy Proximal SVMs

A key difficulty with real data is that noise tends to corrupt the samples, and a classifier which tries to learn from noisy data needs to ignore outliers, and to assign reduced importance to samples with which one has less confidence. Often, the samples may not be representative of a class, in which case one can treat them as if there is an uncertainty with regard to the class to which they belong. For example, a person showing some symptoms of a disease may have characteristics that overlap with both perfectly healthy patients as well as unhealthy ones.

Therefore, the membership of the class to which a sample belongs tends to be fuzzy, with a fuzzy membership given by (18)-(19), indicating the extent to which the sample may be said to belong to one class or the other. Samples with a higher membership value can be thought of as more representative of that class, while those with a lower membership value should be given less importance when building a classifier. Before introducing the idea of fuzzy proximal support vector machines for multicategory data classification, we begin with a brief description of the two-category linear FPSVM which is a fuzzy extension of proximal support vector machine [5]. We consider the problem of classifying m points in the n-dimensional real space $\Re_n$, represented by the $m \times n$ matrix A, according to the membership of each data point A_i $(i=1,2...m)$ in the class $A(+)$ or $A(-)$ as specified by a given $m \times m$ diagonal matrix D with plus ones or minus ones along the diagonal. For the problem, the FPSVM with linear kernel [7] is given by following optimization problem with parameter $c > 0$ and linear equality constraints.

$$\text{Minimize} \frac{c}{2}\|Y\|^2 + \frac{1}{2}\left(w^T w + \gamma^2\right) \tag{15}$$

subject to the constraints

$$D(Aw - e\gamma) + y = e, \tag{16}$$

where $Y=Sy$, and S denotes a diagonal matrix $S=\text{diag}(s_1, s_2, .. s_m)$, whose diagonal entries correspond to the membership values of the data samples. In practice, the membership values are associated with a lower bound, i.e.

$$\sigma_i \le s_i \le 1,\ i = 1, 2, \ldots m \tag{17}$$

where σ_i denotes the lower bound for the ith sample.

The fuzzy memberships of data sample for the two classes can be computed in a following manner: Given a x_i belonging to class 1, Chen and Chen suggest using the approach of Keller and Hunt [8] to compute fuzzy memberships of x_i for both classes1 and 0. These are given by

$$s_{class1} = 0.5 + \frac{[\exp(f(d_0(x_i) - d_1(x_i))/d] - \exp(-f)}{2(\exp(f) - \exp(-f))} \tag{18}$$

$$s_{class0} = 1 - s_{class1}$$

Given a x_i belonging to class 0, the fuzzy membership values of x_i for both classes1 and 0 are given by

$$s_{class0} = 0.5 + \frac{[\exp(f(d_1(x_i) - d_0(x_i))/d] - \exp(-f)}{2(\exp(f) - \exp(-f))} \tag{19}$$

$$s_{class1} = 1 - s_{class0}$$

Here, f is a constant that determines the rate at which memberships decrease towards 0.5; $d_1(x_i)$ and $d_0(x_i)$ are the distances between x_i and the means of classes 1 and 0, respectively, and d is the distance between the centers of gravity of the two classes. Solution of the problem FPSVM is briefly explained in [7]. Instead of solving the quadratic problem, we solve the system of linear equations which is equivalent to the inversion of matrix of dimension $(n+1)\times(n+1)$, as against one of dimension $m\times m$ required in the case of SVMs.

To extend this formulation to k classes, all we need to redefine the following matrix A:

$$A = \left[A^1 \cdots A^k\right]^T$$

$$A+ = A^r,$$

$$A- = \left[A^1 \cdots A^{r-1} \; A^{r+1} \cdots A^k\right]^T$$

$$r \in \{1,2,\ldots k\}$$

where $A^l \in \Re^{m l \times n}$ represents m^l points in class l. In short, we assign class 1 to $A(+)$ and rest to $A(-)$. We then define for $m=m^1+m^2+\ldots+m^k$ the $m\times m$ diagonal matrix D of $\pm$ 1 as follows:

$$D_{ii} = 1 \quad for \; A_i \in A^r,$$

$$D_{ii} = -1 \; for \; A_i \notin A^r,$$

$$r \in \{1,2\cdots,k\} \; .$$

Once the k minimization problems of the type (15)-(16) are solved (with A and D defined as above) by solving the linear system of equations, k unique separating planes are generated:

$$w^T x = \gamma \tag{20}$$

A new given point $x \in \Re_n$ is assigned to class r depending on which of the k half spaces generated by the k planes (20) it lies close to, that is,

$$x^T w^r - \gamma^r = \max_{i=1,2..k} x^T w^i - \gamma^i \tag{21}$$

Note that instead of solving k quadratic program, the classifier is obtained by solving k linear system of equations, which require inverting matrix in the dimension of input space.

4. Experimental Results and Concluding Remarks

The FPSVM, PSVM for multicategory data classification methods are implemented in MATLAB [11] running on a PC with an Intel P4 processor (1.6 MHz), 256 MB RAM, with a Linux version 8 operating system. In order to illustrate the importance of the membership function values, we took a simple case of 1: k-1 classification, in which data samples belonging to class $A(+)$ are assigned a membership value v_1, while samples belonging to class $A(-)$ are assigned a membership value v_0. This is the representative of a situation in which data belonging to one class is associated with a higher degree of fuzziness. The FPSVM can clearly be applied in a situation where each sample is assigned a membership value, as in a typical fuzzy learning scenario. The MFPSVM network was trained on the iris data set, which is a part of the UCI Machine Learning Repository [1]. The implementation of above algorithm is still in progress.

Table 1 compares the performance of the Fuzzy Proximal SVM with that of a PSVM for multicategory data classification, on some benchmark data sets from the UCI Machine Learning Repository [1]. The data was divided into 5 parts, 4 of which comprised the training data, and the 5th part was used for testing the generalization ability of the classifier. This process was repeated 5 times, using a different part for testing on each occasion. The classification accuracy was computed by averaging across all the 5 trials.

Table1: Classification accuracy in (%) of MPSVM and MFPSVM

	Iris data		***Glass data***	
	Training	*Testing*	*Training*	*Testing*
MFPSVM	86.96	86.67	83.18	81.35
MPSVM	84.0	83.33	82.48	80.42

The experimental results demonstrate that the FPSVM for multicategory data classification provides improved generalization ability in comparison with Proximal SVMs for the same. For example, in the case of screening patients who are clearly healthy from those who require further examination, it is desirable to err on the side of caution.

In this paper, we extend a Fuzzy Proximal Support Vector Machine for binary case to multicategory data classification, which uses knowledge of the uncertainty associated with the membership of a data sample to a given class, to improve generalization. The solution to a k- FPSVM requires the inversion of a matrix whose size is the pattern dimension, as against the number of patterns. This is responsible for a significant speedup in comparison with traditional SVM-type classifiers.

References

1. C.L Blake & C.J. Merz, *UCI Repository for Machine Learning databases*, Irvine, CA: University of California, Department of Information and Computer Sciences. On-line at *http://www.ics.uci.edu/~mlearn/MLRepository.html.*
2. C. Burges (1998), A Tutorial on Support Vector Machines for Pattern Recognition, *Data Mining and Knowledge Discovery*, 2, No. 2: 121-167.

3. P.S. Bradley & O.L. Mangasarian (2000), Massive Data Discrimination via Linear Support Vector Machines, *Optimization Methods and Software*, 13: 1-10.
4. J-H. Chen and C-S. Chen (2002), Fuzzy Kernel Pecreptron, *IEEE Trans. Neural Networks*, 13(6): 1364-1373.
5. G. Fung & O.L. Mangasarian (2001), *Proximal Support Vector Machine Classifiers*, in D. Lee et. al (Eds.), Proc. KDD-2001: *Knowledge Discovery and Data Mining*, San Francisco, CA, Association for Computing Machinery, New York, pp. 77-86.
6. T. Inoue & S. Abe (2001), *Fuzzy Support Vector Machines for Pattern Classification*, Proc. International Joint Conference on Neural Networks, pp. 1449-1454.
7. Jayadeva, R. Khemchandani & S.Chandra Fast and Robust learning through linear fuzzy proximal support vector machines, Article in Press.
8. J.M. Keller & D.J. Hunt (1985), Incorporating Fuzzy Membership Functions into the Perceptron. Algorithm, *IEEE Trans. Pattern Analysis and Machine Intelligence*, 7: 693-699.
9. O.L. Mangasarian & D.R. Musicant (2000), *Lagrangian Support Vector Machines*, Technical Report 00-06, Data Mining Institute, Computer Sciences Department, University of Wisconsin, Madison, Wisconsin, June 2000.
10. O.L. Mangasarian (2000), *Generalized Support Vector Machines*, in: *Advances in Large Margin Classifiers*, A. Smola et. al (Eds.), MIT Press: Cambridge, MA, pp. 135-146.
11. http://www.mathworks.com

Fuzzy Logic and its Application in Technology and Management
Editors: Debjani Chakraborty, S. Nanda and D. Dutta Majumder

Lossless Join Property in Fuzzy Relational Data Model Using Fuzzy Function

R.N. Dutta[1], B.K. Tyagi[2] and A. Sharfuddin[3]

[1]*Central Road Research Institute, Mathura Road, New Delhi – 110 020*
[2]*B. K. Tyagi, A. R. S. D College, University of Delhi, New Delhi – 110 021*
[3]*Department of Mathematics, Jamia Millia Islamia, New Delhi – 110 025*

Abstract: It is often necessary to decompose a given relation into component of relations for design and query purposes in the theory of relational database system. The same classical relational approach has been extended for fuzzy relations using Zadeh's approach of imprecise data values. For maintaining database integrity it is mandatory that the join operation of the decomposed relations yield the original relation and such decomposition process is called lossless join decomposition. But the given join may not always recover the original relation. In this paper, we study the decomposition problem (necessary and sufficient condition) of fuzzy relations using the concept of well defined theory of equality of two fuzzy values in a fuzzy set using fuzzy functional dependencies [15] and restricted fuzzy functional dependencies.

Keywords: Fuzzy equality, Partial fuzzy function, Fuzzy functional dependency, Fuzzy relations & relational data model, Lossless join, Projection, Cylindrical extension, Formal system.

1. Introduction

Codd's [7] concept of relational data model was studied and extended further by many researchers [1, 2, 8, 16] and has been found a very useful tool for making a complete relational database system. The model has its own limitations in terms of data values it stores as atomic and sets of atomic values only. Therefore, inferences from these data as question answering are limited as it can answer for crisp values only. The information is dynamically added, modified, deleted, etc. from the database often include ambiguous, imprecise and non-deterministic objects. In order to deal with imprecisely known real world objects [17, 18], several extensions [4, 5, 6, 11, 13,] of the classical relational model have been proposed.

In particular, we shall consider Buckles & Petry's model [5] followed by Raju & Mazumdar's [11] and Bhuniya & Niyogi's [3] in our study for lossless join decomposition in fuzzy relational data model. We shall use the theory of "equality" of two fuzzy values in a fuzzy set and define "fuzzy equality" over fuzzy sets, which correspond to a partial fuzzy function. When a classical relation is generalized to fuzzy relation, the extended notion of fuzzy functional dependency (*ffd*) to fuzzy relation corresponds to a partial fuzzy function. In fact, all the inference rules for the classical functional dependencies hold with this new definition of fuzzy functional dependencies (*ffds*) and these rules form a complete formal system. For details of our work on "fuzzy functional dependency using fuzzy function", the reader is referred to [15]. Now, the in presence of integrity constraints, i.e. with this new definition of *ffds*, we shall study the lossless join property of fuzzy relational data model.

2. The classical relational data model

We are adopting all the standard notations and terms as used in [16] and need not again explicitly define these in this paper, due to space constraints.

3. Fuzzy relations

Let U be a classical set of objects, called universe of discourse. An element of U is denoted by u.

Definition 3.1: A fuzzy set F in a universe of discourse U is characterized by a membership function $\mu_F : U \to [0, 1]$ (1)

Where, $\mu_F(U)$ denotes the grade of membership of u in the fuzzy set F for each $u \in U$. Thus, a fuzzy set F is $F = \{ \mu_F(u_1)/ u_1,\ldots, \mu_F(u_n)/u_n\}$. (2)

Definition 3.2: A fuzzy relation r is a fuzzy subset of the Cartesian product of some universes. Thus an n-ary fuzzy relation is a fuzzy subset of $U_1 \times \ldots \times U_n$ which is characterized by membership function $\mu_r : U_1 \times \ldots \times U_n \to [0, 1]$. (3)

R is expressed as $r = \sum \mu_R(u_1,\ldots, u_n)/ (u_1,\ldots, u_n)$. (4)

Definition 3.3: A fuzzy relation r on a relation scheme $R = \{A_1,\ldots,A_n\}$ is a fuzzy subset of dom $A_1 \times \ldots \times$dom A_n, where dom A_i is called the domain of attribute name A_i, $i = 1,\ldots,n$. dom A_i may be a fuzzy subset or a set of fuzzy subsets and is associated with an universe of discourse U_i.

According to Raju & Mazumdar's definition [11], r is called a type-1 fuzzy relation if dom A_i is a fuzzy subset, and if dom A_i is a set of fuzzy subsets then r is called a type-2 fuzzy relation.

In classical relational scheme, the set of attribute names for a relation will be called the relation scheme. Let $A_1,\ldots,A_n$ be attributes of a relation scheme, then a fuzzy relation r is an instance of the relation scheme R such that r is a fuzzy subset of $U_1 \times \ldots \times U_n$, where $U_i =$ dom A_i. Thus, r may be expressed as

$$R = \sum \mu_r(a_1,\ldots,a_n)/ (a_1,\ldots, a_n), \quad (5)$$

where, $\mu_r(a_1, \ldots, a_n) \le \min(\mu_{A_1}(a_1),\ldots,\mu_{A_n}(a_n))$, a_i, $i = 1,\ldots,n$, are discrete numbers for type-1 fuzzy relation and for type-2 fuzzy relation, a_i, may be a set of scalars or discrete numbers.

3.1. Fuzzy Projections and natural join: Let an instance r be a fuzzy relation of the relation scheme $R = \{A_1,\ldots, A_n\}$. Let $X = \{A_{i_1},\ldots, A_{i_k}\}$ be a subset of the relation scheme R_i such that $\{i_1,\ldots,i_k\}$ is a subsequence of the sequence $(1,\ldots, n)$. Then, the projection of r on $R_i = P_{R_i}(r)$ is a k-ary fuzzy relation in dom X = dom $A_{i_1} \times \ldots \times$ dom A_{i_k}. Then the membership function of r_i is defined as

$$\mu_{r_i}(a_{i_1},\ldots,a_{i_k}) = \max_{\substack{\forall a_j \in u_j \\ j \notin (i_1,\ldots,i_k)}} [\mu_r(a_1,\ldots,a_n)] \qquad (6)$$

Let $r_1,\ldots, r_n$ be a fuzzy relation schemes obtained from the relation schemes $R_i = \{ A_{i_1}, \ldots, A_{i_k} \}$, where $i = 1, \ldots, n$ having membership function μ_{r_i}.

Let $R = \{A_1,\ldots, A_n\} = \cup_{i=1}^{n} R_i$. The fuzzy join of fuzzy relations $r_1,\ldots,r_n$ denoted by $\bar{r} = \rhd\lhd_{i=1}^{n} r_i$ is a fuzzy relation over $U_1 \times \ldots \times U_n$. The membership function of $\bar{r}$ is defined as $\mu_{\bar{r}}(a_1, \ldots, a_n) = \min_i [\mu_{\bar{r}_i}(a_1, \ldots, a_n)]$, (7)

where $\mu_{\bar{r}_i}(a_1,\ldots,a_n)$ is the cylindrical extension of $\mu_{r_i}(a_{i_1},\ldots,a_{i_k})$. The fuzzy relation $\hat{r}$ obtained by joining the projections of the original relation may not be identical with the original relation.

4. Fuzzy Functional Dependency

We now extend the concept of functional dependencies of classical data model [1, 16] to fuzzy relational database model as fuzzy functional dependencies. To obtain necessary and sufficient condition the concept of *restricted* fuzzy functional dependencies (*rffds)* are being used for lossless join decomposition. In order to capture integrity arising out of approximately equality of domain values, the type-1 and type-2 fuzzy relations, the 'fuzzy equality' as 'partial fuzzy function' is defined and used. The degree of equality between the two fuzzy values act as a fuzzy relation on fuzzy sets and is called a partial fuzzy function. We now introduce the notion of fuzzy functional dependencies (*ffds)* based on our earlier work [15] using fuzzy function.

Let $R = \{A_1,\ldots,A_n\}$ be a relation scheme. A fuzzy relation r over R is a fuzzy subset of dom $A_1 \times \ldots \times$dom A_n. The membership function of r is denoted by μ_r. A tuple $t(a_1,\ldots,a_n)$ in dom $A_1 \times \ldots \times$dom A_n is thus associated with grade of membership function $\mu_r(t)$ in r, i.e.,

$$r = \{ \mu_r(t)/t: t \in \text{dom } A_1 \times \ldots \times \text{dom } A_n\}. \qquad (8)$$

This fuzzy relation is called type-1 fuzzy relation as defined in Raju& Mazumdar's model [11]. Let $(i_1,\ldots, i_k)$ and $(j_1,\ldots, j_m)$ be subsequences of $(1, \ldots, n)$. Let $X = \{ A_{i_1}, \ldots, A_{i_k} \}$, $Y = \{ A_{j_1}, \ldots, A_{j_m} \}$, dom X = dom $A_{i_1} \times \ldots \times$ dom A_{i_k} and dom Y = $dom\ A_{j_1} \times \ldots \times dom\ A_{j_m}$. Then the fuzzy relation r induces a binary fuzzy relation over XY, denoted by $r[X, Y]$, which is a fuzzy subset of $dom\ X \times dom\ Y$, and is given by

$$r[X,Y] = \{\mu_r(t, t')/(t, t') : (t, t') \in dom\ X \times dom\ Y,$$
$$\text{and } \mu_{r[X,Y]}(t, t') = \sup_{t''} \{\mu_r(t'') : t''[X] = t \text{ and } t''[Y] = t'\}\}. \qquad (9)$$

In the second level generalization of fuzzy relation of type-2 the data values are allowed to be sets of fuzzy set. Let R={$A_1, \ldots, A_n$} be a relation scheme. Let $I^{\text{dom } A_i}$ be the set of all fuzzy subsets of dom A_i, i = 1,…,n. Then a fuzzy relation r over R is a fuzzy subset of $I^{\text{dom } A_1} \times \ldots \times I^{\text{dom } A_n}$. The definition of fuzzy relation r over R of type-1 can be considered as a special case of this type-2 definition, if we consider a $\in$ dom A_i to be a fuzzy set associated with a membership function μ_a defined as

$$\mu_a(x) = \begin{cases} 1, & \text{if } x = a, \\ 0, & \text{otherwise} \end{cases}.$$

Therefore, in what follows, we shall not distinguish between the two definitions of a fuzzy relation. Thus,

$$r = \{\mu_r(t)/t : t \in I^{\text{dom } A_1} \times \ldots \times I^{\text{dom } A_n}\}. \qquad (10)$$

A definition of type-2 also induces a fuzzy binary relation over XY, denoted by r[X, Y], which is a fuzzy subset of $I^X \times I^Y$, where $I^X = I^{\text{dom } A_{i1}} \times \ldots \times I^{\text{dom } A_{ik}}$ and $I^Y = I^{\text{dom } A_{j1}} \times \ldots \times I^{\text{dom } A_{jm}}$, and is given by

$$r[X, Y] = \{\mu_r(t, t')/(t, t') : (t, t') \in I^X \times I^Y, \text{ and},$$
$$\mu_{r[X,Y]}(t, t') = \sup_{t''}\{\mu_r(t'') : t''[X] = t \text{ and } t''[Y] = t'\}\}. \qquad (11)$$

To simplify the presentation, the fuzzy equality on X, I^X and Y, I^Y will be denoted by E_X and E_Y respectively.

Definition 4.1: A fuzzy relation r over a relation scheme R = {A_1,..., A_n} satisfies a *ffd* $X \rightsquigarrow Y$, where X, Y ⊆ R, if r[X, Y] is a partial fuzzy function, that is, if for all $t_1, t_2 \in$ dom X and for all $t_1', t_2' \in$ dom Y,

$$\mu_{r[X,Y]}(t_1, t_1') \wedge \mu_{r[X,Y]}(t_2, t_2') \wedge E_X(t_1, t_2) \le E_Y(t_1', t_2'). \qquad (12)$$

The inference rules of classical relations, namely Armstrong's axioms holds for integrity constraints induced by fuzzy functional dependencies in case of this definition also and form a sound and complete formal system. For details, reader may see [15]. The Definition 4.1 holds for both type-1 and type-2 fuzzy relations as defined in [11] and is nearest to the definition of Raju & Mazumdar's model [11].

In Raju & Mazumdar's model [11], the membership degrees of a tuple is restricted by the data values (which are fuzzy sets). These membership values once defined no longer play any part in their definition of *ffd*. The definition of *ffd* is then purely a matter of resemblance of data values in two tuples. Therefore, their approach does not capture the fuzzy association among attribute values reflected by the membership degrees of tuples. The following example shows that the lower membership degrees are responsible for a *ffd* to hold in our sense whereas the same *ffd* in the sense of [11] does not hold. We keep the example theoretical to simplify the calculations.

Example 4.1: Let R ={A, B} be a relation scheme. Let r be a fuzzy relation over R containing two tuples t_1 and t_2 defined as follows:

r :	A	B	$\mu_r(t)$
t_1 :	a	b_1	0.5
t_2 :	a	b_2	0.6

where $a \in I^{\text{dom } A}$ and $b_1, b_2 \in I^{\text{dom } B}$ are fuzzy sets. The membership function of fuzzy set a is

$$\mu_a(x) = \begin{cases} 1, & \text{if } x = u, u \in \text{dom } A, \\ 0, & \text{if } x \ne u. \end{cases}$$

Let $v_1, v_2 \in$ dom B. Let the membership functions of b_1 and b_2 be

$\mu_{b_1}(v_1) = 1$, $\mu_{b_1}(v_2) = 0.8$, $\mu_{b_1}(x) = 0$ if $x \ne v_1$ and $x \ne v_2$, and

$\mu_{b_2}(v_1) = 0.7$, $\mu_{b_2}(v_2) = 1$, $\mu_{b_2}(x) = 0$ if $x \neq v_1$ and $x \neq v_2$.

Then, $I(b_1, b_2) = \{x \in \text{dom } B : \mu_{b_1}(x) \leq \mu_{b_2}(x)\} = \text{dom } B - \{v_1\}$.

Similarly, $I(b_2, b_1) = \text{dom } B - \{v_2\}$.

Therefore, $[\tilde{\subseteq}(b_1, b_2)] = \bigwedge_{x \in \text{dom } B} (b_1^c \vee b_2)(x) = \bigwedge_{x \in \text{dom } B} (\mu_{b_1^c}(x) \vee \mu_{b_2}(x))$

$$= \bigwedge_{x \in \text{dom } B} ((1 - \mu_{b_1}(x)) \vee \mu_{b_2}(x)) = \mu_{b_2}(v_1) = 0.7.$$

Thus, $[\tilde{\subseteq}_*(b_1, b_2)] = [\tilde{\subseteq}(b_1, b_2)] \vee (\bigwedge_{x \in \text{dom } B} \psi_{I(b_1, b_2)}(x)) = 0.7 \vee 0 = 0.7$.

Similarly, $[\tilde{\subseteq}_*(b_2, b_1)] = [\tilde{\subseteq}(b_2, b_1)] \vee (\bigwedge_{x \in \text{dom } B} \psi_{I(b_2, b_1)}(x)) = \mu_{b_1}(v_2) = 0.8$.

Therefore, $E_B(b_1, b_2) = 0.7 \wedge 0.8 = 0.7$.

Now, it follows that

$$\mu_{r[A,B]}(t_1[A], t_1[B]) \wedge \mu_{r[A,B]}(t_2[A], t_2[B]) \wedge E_A(t_1[A], t_2[A])$$
$$= \mu_r(t_1) \wedge \mu_r(t_2) \wedge E_A(a, a)$$
$$= 0.5 \wedge 0.6 \wedge 1 = 0.5 < 0.7 = E_B(b_1, b_2) = E_B(t_1[B], t_2[B]),$$

so that *ffd* $A \rightsquigarrow B$ holds in r. It is easily seen that $A \rightsquigarrow B$ does not hold in r in the sense of [11]. But, if we consider Bhuniya & Niyogi's model [3] which is a modification of [11], then the *ffd* $A \rightsquigarrow B$ holds for $\mu_{EQX} - \mu_{EQY} = 1 - 0.7 = 0.3$ and $1 - \beta = 1 - 0.7$. Consequently, for the parameter value $\beta = 0.7$, the condition $\alpha < \beta < 1$ is satisfied.

Now, in view of [11], if we consider $\mu_{r[X,Y]}(t_1, t_1') > 0$ and $\mu_{r[X,Y]}(t_2, t_2') > 0$ to be arbitrary values, so that these membership values no longer plays any part in their definition of *ffds* (Definition 4.1) we restrict $E_X(t_1[X], t_2[X]) < 1$, if $t_1[X] \neq t_2[X]$ for $X \subseteq R = \{A_1, \ldots, A_n\}$ and $t_1, t_2 \in \text{dom } A_1 \times \ldots \times \text{dom } A_n$. Now, we define restricted fuzzy functional dependency (*rffd*).

Definition 4.1(a): A fuzzy relation r over a relation scheme $R = \{A_1, \ldots, A_n\}$ satisfies a *rffd* $X \rightsquigarrow Y$, where $X, Y \subseteq R$, if $r[X, Y]$ satisfies the following conditions, that is, if for all $t_1, t_2 \in \text{dom } X$ and for all $t_1', t_2' \in \text{dom } Y$,

$$E_X(t_1, t_2) \leq E_Y(t_1', t_2') \text{ and } E_Y(t_1', t_2') < 1. \quad (12a)$$

Definition 4.2: Let R be a relation scheme and F be a set of *ffds* over R, that is, F is a set of *ffds* of the form $X \rightsquigarrow Y$, where $X, Y \subseteq R$. Then F^+ is defined to be the set of all *ffds* implied by F, that is, $X \rightsquigarrow Y \in F^+$ if $X \rightsquigarrow Y$ holds in every fuzzy relation in which each of the *ffds* in F holds.

Similarly, we can define Closure over set of a set of attributes.

Definition 4.3: Let F be a set of *ffds* defined over a relation scheme R. The closure of a set of attributes $X \subseteq R$ with respect to F, written as X_F^+, is defined as

$$X_F^+ = \{A : X \rightsquigarrow A \in F^+\}.$$

5. Lossless join of fuzzy relations

In order to maintain the integrity of a database and the recovery of the original relation, it is necessary for a decomposition to have the lossless join property.

Let U be a given finite set of attributes and let A be set of fuzzy relational schemes $\{R_1,\ldots,R_k\}$, where $R=\cup_{i=1}^{k} R_i$. A fuzzy relational database is a set of fuzzy relations $r_1, \ldots, r_n$, where r is an instance of R and r_j is its projection on the attributes of the scheme R_i. The natural join $\bar{r} = \rhd\lhd_{i=1}^{k}\ r_i$ is performed on these relations according to (7).

Let $R_i = \{A_{i_1}, \ldots, A_{i_{k_i}}\}$ where $A_{i_j} \in \{A_1, \ldots, A_n\}$. Then the membership function μ_{r_i} of r_i as given in (6) can be rewritten as

$$\mu_{r_i}(a_{i_1}, \ldots, a_{i_{k_i}}) = \max\{\mu_r(a_1, \ldots, a_n), \mu_r(a_{1_1}, \ldots, a_{n_1}), \ldots, \mu_r(a_{1_{k_1}}, \ldots, a_{n_{k_n}})\}$$

Definition 5.1: (Lossless join). Let R be a fuzzy relation scheme and $\{R_1, \ldots, R_n\}$ be a decomposition of R such that $R = \cup_{i=1}^{n} R_i$. Then this decomposition is a lossless join decomposition with respect to a set of *ffds* F if for every fuzzy relation r of R that satisfies F, the following condition holds: $r = \rhd\lhd_{i=1}^{n} P_{R_i}(r)$.

6. Lossless join decomposition

In this section we study the properties of lossless join decomposition of a fuzzy relation with respect to a set of *rffds*.

Theorem 6.1: Let $R = \{A_1, \ldots, A_n\}$ be a relation scheme with a *rffd*: $X \rightsquigarrow Y$, where $X, Y \subseteq R$. Then the relation scheme R has a lossless join decomposition into two components $R_1(X, Y)$ and $R_2(X, Z)$, where $Z = R - XY$.

Proof: Let r be an instance of R and r_1 and r_2 be the projections of R_1 and R_2 over r respectively. Let $\hat{r}_i$ (i = 1, 2) be the cylindrical extensions of r_i and $\bar{r}$ be the natural join of r_1 and r_2.

Let t be a tuple in $D = \text{dom } A_1 \times \ldots \times \text{dom } A_n$. Then the given decomposition has lossless join if there exists at least one $i \in \{1, 2\}$ such that $\mu_{\hat{r}}(t) = \mu_r(t)$. For a tuple t in r with $\mu_r(t) > 0$, we have, $\mu_{\hat{r}_i}(t) \geq \mu_r(t)$ for i = 1, 2. Consider the case $\mu_{\hat{r}_2}(t) \geq \mu_r(t)$, then r must have at least another tuple t_1 with $\mu_{\hat{r}_2}(t_1) \geq \mu_r(t)$ such that

$t_1[XZ] = t[XZ]$ and $t_1[Y] \neq t[Y]$. (14)

So, $E_X(t_1[X], t[X]) = 1$, $E_Z(t_1[Z], t[Z]) = 1$ and $E_Y(t_1[Y], t[Y]) < 1$ (15)

Again, r satisfy the rffd: $X \rightsquigarrow Y$.

From Definition (4.1a), $E_X(t_1[X], t[X]) \leq E_Y(t_1[Y], t[Y])$ and $t_1[Y] \neq t[Y]$, and by using (14) and (15), we get $E_X(t_1[X], t[X]) \nleq E_Y(t_1[Y], t[Y])$. Hence the Definition (4.1a) for *rffds*: $X \rightsquigarrow Y$ is violated. Thus for a tuple t of r, we have $\mu_{\hat{r}_2}(t_1) = \mu_r(t)$, i.e., $\mu_{\hat{r}}(t) = \mu_r(t)$.

Now we show that for a tuple $t \in D$ which is definitely not in r, i.e., $\mu_r(t) = 0$, either $\mu_{r_1}(t[XY]) = 0$, or, $\mu_{r_2}(t[XZ]) = 0$, i.e., $\mu_{\bar{r}}(t) = 0$.

We will prove this result by contradiction. Suppose for a tuple $t \in D$, $\mu_r(t) = 0$, but both $\mu_{r_1}(t[XY]) \neq 0$ and $\mu_{r_1}(t[XZ]) \neq 0$. For $\mu_{r_1}(t[XY]) \neq 0$, it is required that there must exist a tuple t_1 in r with $t_1[XY] = t[XY]$ and $t_1[Z] \neq t[Z]$. Similarly, for

μ_{r_2} (t[XZ]) $\neq 0$, we have a tuple t_2 in r, such that $t_2[XZ] = t[XZ]$ and $t_2[Y] \neq t[Y]$. In other words the relation r must contain at least two tuples t_1 and t_2 such that $t_1[X] = t_2[X]$ and $t_1[Y] \neq t_2[Y]$. For t_1 and t_2, $E_X(t_1[X], t_2[X]) = 1$, $E_Y(t_1[Y], t_2[Y]) < 1$, by Definition (4.1a), this would violate the *rffd*: X⇝Y. We have, therefore, for any tuple $t \in D$, $\mu_{\bar{r}}(t) = \mu_r(t)$. Hence, R has lossless join decomposition.

Lemma 6.1: If $\rho = (R_1, R_2)$ is a decomposition of R and F is a set of *rffds*, then ρ has a lossless join with respect to F if and only if

$$(R_1 \cap R_2) \rightsquigarrow (R_1 - R_2) \text{ or } (R_1 \cap R_2) \rightsquigarrow (R_2 - R_1).$$

Proof: Let R = {R_1, ..., R_k} ba a relation scheme with two components R_1 = {X, Y}, R_2 = {X, Z} such that X, Y, Z ⊆ R. Let r be an instance of R, containing only tuples t and t_1 given by $t = (a_1, a_2, \ldots, a_n)$ and t_1 is defined by

$$t_1[A_i] = \begin{cases} a_i & \text{if } A_i \in X^+ \\ b_i & \text{otherwise,} \end{cases}$$

such that $E_{\text{dom } A_i}(a_i, b_i) < 1$, for $a_i, b_i \in \text{dom } A_i$, i = 1, ...,n. It can be shown that r satisfies F.

Then the join $\bar{r} = P_{R_1}(r) \bowtie P_{R_2}(r)$ contains two additional tuples t_2, t_3 such that

$$t_2[A_i] = \begin{cases} a_i & \text{if } A_i \in X^+ \cup Z \\ b_i & \text{for } b_i \in Y - X^+ \end{cases}, \qquad t_3[A_i] = \begin{cases} a_i & \text{if } A_i \in X^+ \cup Y \\ b_i & \text{for } b_i \in Z - X^+ \end{cases}.$$

Now, t_2 belongs to R if $Z \subseteq X^+$, i.e., X⇝Z is in F^+, where F^+ is a set of *rffds* that can be derived from F using inference rules. Similarly, t_3 belongs to R if $Y \subseteq X^+$, i.e., X⇝Y is in F^+. So, R satisfies the ffds X⇝Y, X⇝Z. In other words, $(R_1 \cap R_2) \rightsquigarrow (R_1 - R_2)$ or $(R_1 \cap R_2) \rightsquigarrow (R_2 - R_1)$.

It is given that, $(R_1 \cap R_2) \rightsquigarrow (R_1 - R_2)$ or $(R_1 \cap R_2) \rightsquigarrow (R_2 - R_1)$. This implies X ⇝ Y or X ⇝ Z. We have to show that $\rho = (R_1, R_2)$ is a lossless join decomposition. This can be proved following the steps in the proof of Theorem 6.1.

We next examine the problem of lossless join of fuzzy relations with respect to a set of fuzzy functional dependencies. Let r be an instance of a fuzzy relation scheme R, and U be a finite set of attributes present in R. Let ρ = {R_1, ..., R_k} denote a decomposition of the relation scheme R, i.e., each $R_i \subseteq R$ and $R = \cup_{i=1}^{n} R_i$. Further, let $Closure_F(R_i)$ denote the closure of the attributes of R_i with respect to F, i.e., the set of attributes derived from R_i using set of ffds F and the inference rules.

The algorithm for a dependency preserving decomposition for the lossless join property of fuzzy relation schemes in $O(n^2)$ time, which has been extended straight way from [8] and holds for fuzzy data model as well without any modification. Loizou et al [8] has proved the following algorithm gives the same result as of the tableau method of Aho[2]. Other properties of dependency preserving decomposition for losslessness were proved in [8] in the Theorem 1 and Theorem 2.

7. Conclusion

We have studied a necessary and sufficient condition for lossless join decomposition of a fuzzy relational database scheme. The algorithm to test lossless join decomposition developed by Loizou et al. [8] also holds true in our definition of *restricted* fuzzy functional dependency [15] applied in the fuzzy domain.

References

1. S. Abiteboul, R. Hull & V. Vianu (1995), *Foundation of Databases*, Addition-Wesley Publication Company Inc. USA.
2. A.V. Aho, C. Berri & J.D. Ullman (1979), The theory of join in relational databases, *ACM Transaction of Database Systems*, 4(3): 297-314.
3. B. Bhuniya & P. Niyogi (1993), Lossless join property in fuzzy relational databases, *Data and Knowledge Engineering* 11(2): 109 – 124.
4. P. Bosc, D. Dubois &H. Prade(1998), Fuzzy functional dependencies and redundancy elimination, *J. of the American Soc. for Information Sc.*, 49(3): 217 - 235.
5. B.P. Buckles & F.E. Petry (1982), A fuzzy representation of data for relational databases, *Fuzzy Sets and Systems*, 7: 213-226.
6. G. Chen, J.Vandenbulcke & E.E. Kerre(1991), A step towards the theory of fuzzy relational database design, *in Proc. of IFSA'91 World Congress*, Brussels.
7. F.E. Codd (1970), A relational model of data for large shared data banks, *Communications of the ACM* 13(6): 377-387.
8. G. Loizou & P. Thanisch (1983), Testing dependency preserving decomposition for losslessness, *Information Systems* 8(1): 25-27.
9. M. Demirci (1999), Fuzzy functions and their fundamental properties, *Fuzzy Sets and Systems* 105: 239-246.
10. M. Demirci (2000), Fuzzy functions and their applications, *Journal of Mathematical Analysis and Applications* 252: 495-517.
11. K.V.S.V.N Raju & A.K. Mazumdar (1988), Fuzzy Dependencies and Lossless Join Decomposition of Fuzzy Relational Database Systems, *ACM Transaction of Database Systems* 13 (2): 129-166.
12. M. Sasaki (1993), Fuzzy functions, *Fuzzy Sets and Systems*, 55:295-301.
13. P.C. Saxena & B.K. Tyagi (1995), Fuzzy functional dependencies and independencies in extended fuzzy relational database models, *Fuzzy Sets & Sys.* 69: 65-89.
14. A. Sostak (1985), On a fuzzy topological structure, *Supp. Rend. Circ. Mat. Palermo* 11: 89-103.
15. B.K. Tyagi, A. Sharfuddin, R.N. Dutta & D.K. Tayal (2005), A Complete Axiomatization of Fuzzy Functional Dependencies using Fuzzy Function, *Fuzzy Sets and Systems*, 151(2): 363 – 379.
16. J.D. Ullman (1980), *Principals of Database Systems*, Computer Sc. Press, USA.
17. L.A. Zadeh (1965), Fuzzy Sets, *Information and Control* 69: 338 - 353.
18. L.A. Zadeh (1981), PRUF - a meaning representation language for natural language, in: *Fuzzy Reasoning and its Applications*, E.H. Mamdani & B.R. Gaines, Eds., Academic Press, New York.

Part 3

Fuzzy Modeling and Decision Making

Fuzzy Logic and its Application in Technology and Management
Editors: Debjani Chakraborty, S. Nanda and D. Dutta Majumder

Fuzzy Preference based TOPSIS for Interval Multi-criteria Decision Making

Atanu Sengupta and Tapan Kumar Pal

Department of Applied Mathematics with Oceanology and Computer Programming
Vidyasagar University, Medinipur - 721 102, WB
Email: atanu_sgupta@yahoo.co.in

Abstract: In the decision-making literature, the approach based on interval analysis allows handling imprecise input data, which assume that the data of a decision-making problem are not well defined but may vary in given intervals. Giove (2002)'s I-TOPSIS (interval version of the traditional TOPSIS method) [9] used the *acceptability index* earlier suggested by us [18] for comparing crisp interval numbers. This paper brings an extension of Giove's work in the light of an improved work (further suggested by us) *Fuzzy Preference in interval order.* As DM's rationale and attitude is reflected in the later index, the present work logically yields an improved solution.

Keywords: MCDM problem, Interval analysis, I-TOPSIS, Acceptability index, Fuzzy Preference in interval order.

1. Introduction

In almost all real-world decision making problem multiplicity of criteria for judging the alternatives is pervasive and for many of such problems, the task of decision-making is merely to solve a Multiple Criteria Decision-Making (MCDM) problem, which can concisely be expressed in a matrix format as

$$
\begin{array}{c c} & \begin{array}{cccccc} C_1 & C_2 & \cdots & \cdots & C_n \end{array} \\
D = \begin{array}{c} A_1 \\ A_2 \\ \vdots \\ \vdots \\ A_m \end{array} & \begin{bmatrix} x_{11} & x_{12} & \cdots & \cdots & x_{1n} \\ x_{21} & x_{22} & \cdots & \cdots & x_{2n} \\ \vdots & \vdots & \cdots & \cdots & \vdots \\ \vdots & \vdots & \vdots & \vdots & \vdots \\ x_{m1} & x_{m2} & \cdots & \cdots & x_{mn} \end{bmatrix}
\end{array} \tag{1}
$$

$$W = [w_1 \;\; w_2 \;\; \cdots \;\; w_n]$$

where, $A_1, A_2 \ldots A_m$ are the possible alternatives among which a DM has to choose from, $C_1, C_2, \cdots C_n$ are the criteria over which alternative performances are measured, x_{ij} is the rating against alternative A_i with respect to the criterion C_j and w_i is the weight assigned to criterion C_j.

In classical MCDM methods, the problem data given in terms of ratings and weights are all known preciously [20, 8, 5]. The TOPSIS method (the Technique for Order Preference by Similarity to the Ideal Solution [13]) is one of the most acceptable approaches for solving classical MCDM problems. TOPSIS defines an index called *similarity* or *relative closeness to the ideal solution* by combining the proximity to the ideal solution and the remoteness from the negative-ideal solution. Hence, a decision making via TOPSIS approaches to select an alternative with maximum relative similarity to the hypothetical ideal one. However, there are many conditions, under which simple crisp data are not adequate to model real-life situations. Since human judgments and preferences are often vague and generally are not representable with exact numerical values, in the literature we have several MCDM approaches, which consider inexact values as the problem data [1, 3, 4, 6, 7, 10, 11, 12].

In the decision-making literature, the approach based on interval analysis allows handling imprecise input data. This approach consists in assuming that the data of a decision-making problem are not well defined but may vary in given intervals. Giove [9] proposed an interval version of the traditional TOPSIS method. This paper describes I-TOPSIS (Interval TOPSIS) approach using the *acceptability index* introduced in [18] for comparing crisp interval numbers. The main weakness of *acceptability index* is that, as it is only a value-based ranking index, it can be applicable only partially, for selection of the best alternative from a pessimistic (or a risk averse) DM's point of view. So, to take a broad view, enhancing the applicability of *acceptability index* as a pessimistic DM's preference index too and for attaining more flexibility and efficiency of DM's interval ranking strategy, *Fuzzy Preference* in interval ordering has been proposed as a generalized ranking index, which is given in [17].

The present work is an extension of Giove's work [9] in the light of *Fuzzy Preference in interval order*. As DM's rationale and attitude is reflected in the later index, the present work logically yields an improved solution.

2. Relevant preliminaries

***2.1. The original* TOPSIS:** Originally the TOPSIS is presented by Chen et al [5], with reference to Hwang & Yoon [13]. The basic principle is that the chosen alternative should have the shortest distance from the ideal solution and the farthest distance from the negative-ideal solution.

2.2. A note on Giove (2002)'s Bag based Interval-TOPSIS

Rebai [16] introduced the BB-TOPSIS method (Bag based TOPSIS) in the context of an MCDM problem in which both numerical and logical data appear in the criteria definitions. Here we recall the main points only keeping our notations as far as consistent to the original one. A fuzzy bag $B(X)$ is defined over a finite set X, $card(X) = n$ as the following function $B_X : X \to \Re^{2n}$:

$$B_X = \{C_1(X)/w_1, C_2(X)/w_2, \ldots\ldots, C_n(X)/w_n\} = \left\{\prod_{j=1}^{n} C_j(X)/w_j\right\}, \tag{2}$$

where X is an alternative described against the criterion $\{C_j(X)\}$, and w_j is the weight of the jth criterion, while the symbol $\prod$ denotes the Cartesian product. Rebai [16] defined two bags B_X and B_Y over X and Y, with $card(X) = card(Y)$ as follows

$$\left.\begin{array}{l} cardinality: |B_X| = \sum_{j=1}^{n} C_j(X) w_j \\ symmetric\ difference: B_X \Delta B_Y = \left\{ \prod_{j=1}^{n} |C_j(X) - C_j(Y)| w_j \right\} \\ distance: d(B_1, B_2) = |B_X \Delta B_Y| \end{array}\right\} \tag{3}$$

Among other distance functions the above one was chosen as a suitable one (refer [9] and [16] for detail). A bag can then be assigned to each alternative taking the values for each criterion and relative weights. Given an alternative, dominance value for a criterion defines a superiority (or inferiority or non-inferiority, depending the nature of the criterion) score over other alternative values under the same criterion. For instance, the superiority score for the alternative X under j^{th} criterion is given by

$$S_j(X) = card\{Y : C_j(Y) < C_j(X),\ Y \in A\}, \tag{4}$$

where A is the set of alternatives. Then following bag defines the Multi-attribute Canonical Fuzzy Bag with the superiority score as

$$MCFB^{>}(X) = \left\{ \prod_{j=1}^{n} S_j(X)/C_j \right\}. \tag{5}$$

The ideal and anti-ideal alternatives are defined as

$$MCFB^{+}_{>}(X) = \vee\, MCFB^{>}(X) = \left\{ \prod_{j=1}^{n} \vee\, S_j(X)/C_j \right\}, \tag{6}$$

$$MCFB^{-}_{>}(X) = \wedge\, MCFB^{>}(X) = \left\{ \prod_{j=1}^{n} \wedge\, S_j(X)/C_j \right\}, \tag{7}$$

where the symbols $\vee$ and $\wedge$ stand for the operators maximum and minimum respectively. Next for each of the given alternatives $X \in A$, the algorithm computes two sets of distances:

Set 1: distance between $MCFB^{>}(X)$ and the positive ideal bag $MCFB^{+}_{>}(X)$ and

Set 2: distance between $MCFB^{>}(X)$ and the anti-ideal bag $MCFB^{-}_{>}(X)$.

The two sets of distances are given by

$$d^{+}_{>} = d\left(MCFB^{>}(X),\ MCFB^{+}_{>}(X)\right) \tag{8}$$

$$d^{-}_{>} = d\left(MCFB^{>}(X),\ MCFB^{-}_{>}(X)\right) \tag{9}$$

and the separation index, which is similar to [5], is given as

$$C_>(X) = \frac{d_>^-}{d_>^- + d_>^+}. \tag{10}$$

Finally, the alternatives $X \in A$, are ranked according to the value of the index $C_>(X)$, as is done in the classical TOPSIS method.

Giove [9] extended Rebai [16]'s aforesaid BB-TOPSIS in the case of interval data. Interval analysis is a simple and intuitive way to treat data uncertainty for complex decision problems and is used for a lot of practical applications. Giove used $\mathscr{A}$-index for measuring 'how much an interval is greater than any other interval number' but in a slightly changed version as follows

$$\xi(X_1 > X_2) = \frac{m(X_1) - m(X_2)}{w(X_1) + w(X_2) + 1}, \tag{11}$$

where, X_1 and X_2 be any two interval numbers, and $m(X_1)$, $m(X_2)$, $w(X_1)$ and $w(X_2)$ be their means and widths respectively. Here the index value can be positive or negative depending on whether $m(X_1)$ is greater than $m(X_2)$ or not. There is another change in the denominator of the index; a value equal to (+ 1) is added here and in Giove's opinion [9] this has been done "to avoid loss of significance where both the spreads are null". However this has not been explained further whether it improves the comparability of the index as a whole. (Refer [17, 18] for different properties and results of $\mathscr{A}$-index.) In this regard we state our views in the next section of this paper.

According to Giove's proposal [9], for a set of interval data, the superiority score for the alternative $X \in \{X_i | i = 1, 2, ..., m\}$, where $X_i = X_{ij} = [(x_{ij})_L, (x_{ij})_R]$ under j^{th} criterion is given by

$$IS_j(X) = \sum_{\substack{i=1 \\ i \notin g(X)}}^{m} \xi_j(X > X_{ij}), \quad \text{for any } j = 1, 2, ..., n, \tag{12}$$

where $g(X)$ is the index function for the set of alternatives $\{X_1, X_2, ..., X_m\}$ such that $g(X_i) = i$ and $IS_j(X)$ is defined as a replacement for the interval extension of $S_j(X)$ given in (4).

In Giove's paper [9], the description of rest of the procedure is almost similar to that of Rebai's BB-TOPSIS [16]. Expressions (5) to (10) are rewritten as expressions (13) to (18) with changes in the variable names suitable for extension of BB-TOPSIS with interval data.

The interval-valued Multi-attribute Canonical Fuzzy Bag with the superiority score is defined as

$$IMCFB^>(X) = \left\{ \prod_{j=1}^{n} IS_j(X)/w_j \right\}. \tag{13}$$

Subsequently, the ideal and anti-ideal bags are defined as

$$IMCFB_{>}^{+}(X) = \vee IMCFB^{>}(X) = \left\{ \prod_{j=1}^{n} \vee IS_j(X)/w_j \right\}, \tag{14}$$

$$IMCFB_{>}^{-}(X) = \wedge IMCFB^{>}(X) = \left\{ \prod_{j=1}^{n} \wedge IS_j(X)/w_j \right\}, \tag{15}$$

The two sets of distances and the separation index are given by

$$id_{>}^{+} = d\left(IMCFB^{>}(X),\ IMCFB_{>}^{+}(X)\right) \tag{16}$$

$$id_{>}^{-} = d\left(IMCFB^{>}(X),\ IMCFB_{>}^{-}(X)\right) \tag{17}$$

$$IC_{>}(X) = \frac{id_{>}^{-}}{id_{>}^{-} + id_{>}^{+}}. \tag{18}$$

Finally, the alternatives $X \in A$, are ranked according to the value of the index $IC_{>}(X)$.

3. Fuzzy Preference Ordering of Intervals in I-TOPSIS

Giove's modification [9] in the basic structure of $\mathscr{A}$-index as given in (11) has the following two limitations:

(i) The main objective of $\mathscr{A}$-index is to rank any two interval-numbers with the help of a DM's objective preference. As a value judgment index it performs efficiently and reveals thorough consistency to our intuition. Findings of this index are also not contradictory to the existing main approaches in the literature (for example, Ishibuchi & Tanaka [14]'s strict preference relations and Kundu's fuzzy leftness relation [15] for interval ranking). Moreover it has shown some advantages over the other approaches.

Whereas Giove's modified $\mathscr{A}$-index [9] neither can explain nor can set apart the basic cases like definitely acceptable or possibly acceptable situation of 'one interval is superior (or inferior) than the other'. For example, let B_1, B_2 and B_3 be three intervals such that $B_1 = [1,2] = <1.5, .5>$, $B_2 = [2,3] = <2.5, .5>$ and $B_3 = [2,10] = <6,4>$, then using the modified index (11), we have,

$$\begin{aligned} \xi(B_2 > B_1) &= .5 \\ \xi(B_3 > B_1) &= .82 \end{aligned} \tag{19}$$

Here both of intervals B_2 and B_3 are weakly mutually exclusive to a same interval B_1. But Giove's modification does not show any resemblance between the comparing pairs. Whereas $\mathscr{A}$-index shows that

$$\begin{aligned} \mathscr{A}(B_1 \odot B_2) &= 1 \\ \mathscr{A}(B_1 \odot B_3) &= 1 \end{aligned} \tag{20}$$

$\mathscr{A}$-index interprets that both of intervals B_2 and B_3 are equally and definitely preferred to interval B_1 as the maximizing alternatives. Can the modified index be explained in a better way if and unless it is treated as some sort of 'difference between the intervals'?

(ii) If the modified index is treated as some sort of 'difference between the intervals', then the incompatibility of the following kind can arise.

If the intervals $B_1 = [1, 2] = <1.5, .5>$ and $B_2 = [2, 3] = <2.5, .5>$ are multiplied by scalars $k = 1, 5, 10$, then comparing the interval pairs (B_{1^k}, B_{2^k}), $\forall k$, such that $B_{t^k} = k.B_t$, $t = 1, 2$, (for example, $B_{2^{10}} = 10.B_2$ $= [20, 30] = <25, 5>$), we get,

$$B_{2^1} - B_{1^1} = [0, 2] = <1, 1>$$
$$B_{2^5} - B_{1^5} = [0, 10] = <5, 5>$$
$$B_{2^{10}} - B_{1^{10}} = [0, 20] = <10, 10>$$

$\xi(B_{2^1} > B_{1^1}) = .5$ $\qquad$ $\mathscr{A}(B_{1^1} \odot B_{2^1}) = 1$

$\xi(B_{2^5} > B_{1^5}) = .833$ $\qquad$ $\mathscr{A}(B_{1^5} \odot B_{2^5}) = 1$

$\xi(B_{2^{10}} > B_{1^{10}}) = .9$ $\qquad$ $\mathscr{A}(B_{1^{10}} \odot B_{2^{10}}) = 1$

This example clearly shows that $\mathscr{A}$-index works more consistently than the modified index in defining DM's objective acceptability of whether an interval is superior or inferior to the other on the basis of their mean values and the levels of uncertainty. Moreover neither of the indices is suitable for being used as the concept of *difference between the intervals* as we see in real analysis. In our previous works [17, 18] the definition of $\mathscr{A}$-index clearly states that it does not work for comparing two real numbers. We see similar sort of limitation for defining division between two intervals (refer Alefeld & Herzberger [2]). In another argument, interpretation of $\mathscr{A}$-index does not conform with the plausibility of defining *the grade of acceptability of superiority of a real number to another real number*. Here an important point to be noted that the concept of $\mathscr{A}$-index is in no way be analogous to the concept of *difference between the intervals* as we see in real analysis.

However, as a complete preference index, limitations of $\mathscr{A}$-index is discussed in [17] and we introduce an advanced scheme, *fuzzy preference ordering* for comparing any two interval numbers which also takes into account a pessimistic or optimistic DM's attitude. The procedure for ranking many intervals simultaneously is given and used elaborately in [19]. Similar sort of approach is applied here in I-TOPSIS to get better version of normalization in the processing of interval data considering (4) and (12) in the procedures stated above.

For a maximization problem, any pair of intervals A and B are ordered as (A, B) if $m(A) \leq m(B)$ and are classified into two sets, S_1 and S_2 as follows:

$$\begin{aligned} &\text{(i)} \quad (A, B) \in S_1, \text{ if } w(A) \geq w(B), \\ &\text{(ii)} \quad (A, B) \in S_2, \text{ if } w(A) < w(B). \end{aligned} \tag{21}$$

Then,

- For $(A, B) \in S_1$, unless A and B are identical, *B is always the best choice*.
- For $(A, B) \in S_2$, a membership function of *fuzzy preferences between A and B* can be constructed.

We define a fuzzy set B' as the *rejection of B* compared to A for any pair (A, B), where, $B' = \{(A, B) \in S_2 \mid m(A) \le m(B), w(A) < w(B)\}$, then

$$\pi_{B'}(A, B) = \max\left\{0, \frac{m(A) - (b_L + w(A))}{m(B) - (b_L + w(A))}\right\}^p \tag{22}$$

where $\pi_{B'}(A, B) \in [0, 1]$ and the parameter p is the (non-fuzzy) level of pessimism for any decision maker and $p \in [(1/_M), M]$, where M is a finite large number. When the value of $p = 1/_M$, the decision maker is said to be of an absolutely pessimistic person who practically does not take any risk. Similarly, for an absolutely optimistic DM, we use $p = M$. The level of pessimism, at $p = 1$, may be regarded as the reflection of a moderate DM's attitude against the aforementioned extreme odds. (Refer [17, 19] for detail.)

We define another fuzzy set, A', *rejection of A* for $(A, B) \in S_2$, as the complement of B' as

$$\pi_{A'}(A, B) = 1 - \pi_{B'}(A, B).$$

We can interpret different values of $\pi_{B'}$ as follows:

if $\pi_{B'} = 1$, then B is definitely rejected and A is definitely accepted;

if $\pi_{B'} = 0$, then B is definitely accepted and A is definitely rejected and

if $\pi_{B'}$ is in between 0 and 1; then there exists different degrees of *rejection of B*.

Hence we can frame a computational routine as

1. if $A \equiv B$, then put $\pi_{A'}(A, A) = .5$. (23)
2. if $(A, B) \in S_1$, then put $\pi_{B'}(A, B) = 1 - \pi_{A'}(A, B) = 0$. (24)
3. if $(A, B) \in S_2$, then (taking $p = 1$) compute

$$\pi_{B'}(A, B) = 1 - \pi_{A'}(A, B) = \max\left\{0, \frac{m(A) - (b_L + w(A))}{m(B) - (b_L + w(A))}\right\}. \tag{25}$$

Therefore, for a set of interval data from the MCDM problem (1), the *superiority score* for the alternative $X \in \{X_i = [(x_i)_L, (x_i)_R] / i = 1, 2, ..., m\}$ under the jth criterion is given by

$$IS_j(X_i) = \sum_{k=1}^{m} \pi_{i'}^{j}(X_i, X_k), \quad k \in \{i = 1, 2, ..., n\}; \quad \text{for any } j = 1, 2, ..., n, \tag{26}$$

where, $IS_j(X_i)$ in (26) is based on fuzzy preference ordering for interval numbers and it replaces Giove's $IS_j(X)$ [9] given in (12).

Subsequently, the Interval-valued Multi-attribute Canonical Fuzzy Bag with the superiority score given in (13) is redefined as

$$IMCFB^{>}(X) = \left\{ \prod_{j=1}^{n} IS_j(X_i)/w_j \right\}. \tag{27}$$

Following (14) and (15), the ideal and anti-ideal bags are then redefined as

$$\begin{aligned} IMCFB_{>}^{+}(X) &= \vee\, IMCFB^{>}(X) \\ &= \left\{ \left(\operatorname*{Max}_{i=1}^{m} IS_1(X_i) \big/ w_1 \right), \left(\operatorname*{Max}_{i=1}^{m} IS_2(X_i) \big/ w_2 \right) \cdots \left(\operatorname*{Max}_{i=1}^{m} IS_n(X_i) \big/ w_n \right) \right\} \\ &= \left\{ \prod_{j=1}^{n} \left(\operatorname*{Max}_{i=1}^{m} IS_j(X_i) \big/ w_j \right) \right\} \end{aligned} \tag{28}$$

$$\begin{aligned} IMCFB_{>}^{-}(X) &= \wedge\, IMCFB^{>}(X) \\ &= \left\{ \left(\operatorname*{Min}_{i=1}^{m} IS_1(X_i) \big/ w_1 \right), \left(\operatorname*{Min}_{i=1}^{m} IS_2(X_i) \big/ w_2 \right) \cdots \left(\operatorname*{Min}_{i=1}^{m} IS_n(X_i) \big/ w_n \right) \right\} \\ &= \left\{ \prod_{j=1}^{n} \left(\operatorname*{Min}_{i=1}^{m} IS_j(X_i) \big/ w_j \right) \right\} \end{aligned} \tag{29}$$

The two sets of distances as are given in (16) – (17) and the separation index are given by

$$\begin{aligned} id^{+}(X_i) &= d\left(IMCFB^{>}(X),\ IMCFB_{>}^{+}(X) \right) \\ &= \sum_{j=1}^{n} \left(\operatorname*{Max}_{i=1}^{m} IS_j(X_i) - IS_j(X_i) \right) w_j \end{aligned} \tag{30}$$

$$\begin{aligned} id^{-}(X_i) &= d\left(IMCFB^{>}(X),\ IMCFB_{>}^{-}(X) \right) \\ &= \sum_{j=1}^{n} \left| \left(\operatorname*{Min}_{i=1}^{m} IS_j(X_i) - IS_j(X_i) \right) \right| w_j \end{aligned} \tag{31}$$

$$\zeta_i = \frac{id^{-}(X_i)}{id^{-}(X_i) + id^{+}(X_i)}. \tag{32}$$

Finally, the alternatives $X_i \in A$ are ranked according to the value of the index ζ_i.

4. An Example of the Algorithm

The following interval multi-criteria decision problem was used earlier by Giove [9]. We have given three sets of interval data on three criteria C_1, C_2 and C_3 with given weights 0.3, 0.5 and 0.2 respectively. The problem is to rank the alternatives X_1, X_2 and X_3.

Alternatives	Maximizing Criteria with weights		
	$C_1/0.3$	$C_2/0.5$	$C_3/0.2$
X_1	[5,7]	[4,8]	[1,9]
X_2	[4,8]	[2,6]	[5,7]
X_3	[6,10]	[3,7]	[2,6]

Table 1: The Interval MCDM problem

Table 2 denotes pair-wise fuzzy rejection grades for each $C_j, \forall j = 1,2,3,$ where $\pi_{i'}^{j}\{X_i, X_k\}$ represents rejection of X_i over X_k for the jth criterion. The pairs of intervals, $\{X_i, X_k\}$ are unordered pairs according to (21) and therefore made ordered pairs before applying the rules (23) – (25) for getting the superiority scores.

Alternatives	C_1	C_2	C_3
	$\pi_{1'}^{1}\{X_1, X_1\} = .5$	$\pi_{1'}^{2}\{X_1, X_1\} = .5$	$\pi_{1'}^{3}\{X_1, X_1\} = .5$
X_1	$\pi_{1'}^{1}\{X_1, X_2\} = 0$	$\pi_{1'}^{2}\{X_1, X_2\} = 0$	$\pi_{1'}^{3}\{X_1, X_2\} = 1$
	$\pi_{1'}^{1}\{X_1, X_3\} = 1$	$\pi_{1'}^{2}\{X_1, X_3\} = 0$	$\pi_{1'}^{3}\{X_1, X_3\} = .5$
	$\pi_{2'}^{1}\{X_2, X_1\} = 1$	$\pi_{2'}^{2}\{X_2, X_1\} = 1$	$\pi_{2'}^{3}\{X_2, X_1\} = 0$
X_2	$\pi_{2'}^{1}\{X_2, X_2\} = .5$	$\pi_{2'}^{2}\{X_2, X_2\} = .5$	$\pi_{2'}^{3}\{X_2, X_2\} = .5$
	$\pi_{2'}^{1}\{X_2, X_3\} = 1$	$\pi_{2'}^{2}\{X_2, X_3\} = 1$	$\pi_{2'}^{3}\{X_2, X_3\} = 0$
	$\pi_{3'}^{1}\{X_3, X_1\} = 0$	$\pi_{3'}^{2}\{X_3, X_1\} = 1$	$\pi_{3'}^{3}\{X_3, X_1\} = .5$
X_3	$\pi_{3'}^{1}\{X_3, X_2\} = 0$	$\pi_{3'}^{2}\{X_3, X_2\} = 0$	$\pi_{3'}^{3}\{X_3, X_2\} = 1$
	$\pi_{3'}^{1}\{X_3, X_3\} = .5$	$\pi_{3'}^{2}\{X_3, X_3\} = .5$	$\pi_{3'}^{3}\{X_3, X_3\} = .5$

Table 2: The normalization operation: Computation of pair-wise superiority scores for each criterion.

Table 3 gives the superiority Score for alternative X_i for the jth criterion

$$IS_j(X_i) = \sum_{k=1}^{3} \pi_{i'}^{j}(X_i, X_k), \quad k \in \{i = 1,2,3\}; \quad \text{for any } j = 1,2,3,$$

For each criterion the column maximum, S_j^{max} and the column minimum, S_j^{min} are computed as

$$S_j^{\max} = \operatorname*{Max}_{i=1}^{3} IS_j(X_i) \qquad S_j^{\min} = \operatorname*{Min}_{i=1}^{3} IS_j(X_i)$$

Alternatives	$IS_1(X_i)$	$IS_2(X_i)$	$IS_3(X_i)$
X_1	1.5	.5 → Column Min	2 → Column Max
X_2	2.5 →Column Max	2.5 → Column Max	.5 → Column Min
X_3	.5 → Column Min	1.5	2

Table 3: The superiority Score for alternative X_i with reference to j^{th} criterion.

Next, according to (30) and (31), the TOPSIS distances i.e., the distances between the ideal solution set and the superiority score set for each X_i and the distances between the anti-ideal solution set and the superiority score set for each X_i, are computed as follows and given next in *Table 4*:

$$id^+(X_i) = \sum_{j=1}^{3} \left(S_j^{\max} - IS_j(X_i)\right).w_j$$

$$id^-(X_i) = \sum_{j=1}^{3} \left|S_j^{\min} - IS_j(X_i)\right|.w_j$$

Alternatives	$id^+(X_i)$	$id^-(X_i)$
X_1	1.3	.6
X_2	.3	1.6
X_3	1.1	.8

Table 4: The TOPSIS distances

Alternatives	ζ_i	Rank
X_1	.3158	III
X_2	.8421	I
X_3	.4210	II

Table 5: Separation index for ranking

Now, through the *separation index*,

$$\zeta_i = \frac{id^-(X_i)}{id^+(X_i) + id^-(X_i)},$$

we have the following rank of alternatives (*Table 5*), where rank I assigned to the alternative with highest preference, rank II gives the alternative with the next maximum preference, etc.

5. Conclusion

This paper presents a modified version of Giove's I-TOPSIS approach [9] for interval-valued MCDM problems. The modification is done using *Fuzzy Preference in interval order* instead of applying $\mathscr{A}$-index for comparing intervals. Obviously our result is different from Giove's result. A numerical comparison, however, is not practically possible. As DM's rationale and preference attitude is reflected in the fuzzy preference-ordering scheme, the present algorithm logically yields an improved solution. As a next step forward, we intend to extend this approach for large number of criteria having possible interactions among them.

References

1. M.A. Abo-Sinna & A.H. Amer (2005), Extensions of TOPSIS for multiobjective largescale nonlinear programming problems, *Applied Mathematics and Computation* 162 (1): 243–256.
2. G. Alefeld & J. Herzberger (1983), *Introduction to Interval Computations*, Academic Press, New York.
3. R.E. Bellman & L.A. Zadeh (1970), Decision-making in a fuzzy environment, *Management Science* 17 (4): 141 – 164.
4. C.T. Chen (2000), Extensions of the TOPSIS for group decision-making under fuzzy environment, *Fuzzy Sets and Systems* 114 (1): 1 – 9.
5. S.-J. Chen, C.-L. Hwang & F. P. Hwang (1992), *Fuzzy multiple attribute decision making*, Springer-Verlag, Berlin.
6. C.T. Chen, C.T. Lin & S.F. Huang (2006), A fuzzy approach for supplier evaluation and selection in supply chain management, *International Journal of Production Economics* 102 (2): 289-301.
7. M. Delgado, J.L. Verdegay & M.A. Vila (1992), Linguistic decision-making models, *International journal of Intelligent System* 7: 479 – 492.
8. J.S. Dyer, P.C. Fishburn, R.E. Steuer, J. Wallenius & S. Zionts (1992), Multiple criteria decision making and multiattribute utility theory: The next ten years, *Management sciences* 38 (5): 645-654.
9. Silvio Giove (2002), Interval TOPSIS for Multicriteria Decision Making, in: M. Marinaro & R. Tagliaferri (Eds.): WIRN VIETRI 2002, *Lecture Notes in Computer Science* 2486, Springer-Verlag Berlin Heidelberg: 56-63.
10. F. Herrera, E. Herrera-Viedma & J.L. Verdegay (1996), A model of consensus in group decision-making under linguistic assessments, *Fuzzy Sets and Systems* 78 (1): 73 – 87.
11. H.M. Hsu & C.T. Chen (1996), Aggregation of fuzzy opinions under group decision-making, *Fuzzy Sets and Systems* 79 (3): 279 – 285.
12. H.M. Hsu & C.T. Chen (1997), Fuzzy credibility relation method for multiple criteria decision-making problems, *Information Sciences* 96 (1): 79 – 91.
13. C.L. Hwang & K. Yoon (1981), Multiple Attribute Decision Making, *Lecture Notes in Economics and Mathematical Systems*, 186, Springer-Verlag, Berlin.

14. H. Ishibuchi & H. Tanaka (1990), Multiobjective programming in optimization of the interval objective function, *European Journal of Operational Research* 48 (2): 219 – 225.
15. S. Kundu (1997), Min-transitivity of fuzzy leftness relationship and its application to decision making, Fuzzy Sets and Systems 86 (3): 357–367.
16. A. Rebai (1993), BB-Topsis: A bag based technique for order preference by similarity to ideal solution, *Fuzzy Sets and Systems* 60: 143-162.
17. A. Sengupta & T.K. Pal (2000), On Comparing Interval Numbers, *European Journal of Operational Research* 127: 28–43.
18. A. Sengupta, T.K. Pal & D. Chakraborty (2001), Interpretation of Inequality Constraints Involving Interval Coefficients and a Solution to Interval Linear Programming, *Fuzzy Sets and Systems* 119, 129–138.
19. A. Sengupta & T.K. Pal (2006), Solving the Shortest Path Problem with Interval Arcs, *Fuzzy Optimization and Decision Making* 5: 71–89.
20. J. Teghem Jr., C. Delhaye & P.L. Kunsch (1989), An interactive decision support system for multicriteria decision-aid, *Mathematical and Computer Modelling* 12 (10/11): 1311 – 1320.

Fuzzy Logic and its Application in Technology and Management
Editors: Debjani Chakraborty, S. Nanda and D. Dutta Majumder

A Soft Computing Procedure for Multiobjective Chance-Constraint Programming with Continuous Random Variables

R.K. Jana[1] *and M.P. Biswal*

Department of Mathematics, Indian Institute of Technology, Kharagpur
Email: {rabin, mpbiswal}@maths.iitkgp.ernet.in
[1]Presently at: Systems Engineering and Operations Research Department,
George Mason University, Fairfax VA 22030, USA Email: rabin_iitkgp@yahoo.com

Abstract: A soft computing procedure for a multiobjective chance constraint programming problem is presented in this paper. The random variables appear in the chance constraints are assumed to follow any continuous distribution. The fuzzy goal programming formulation of the problem is developed using a stochastic simulation based genetic algorithm. In this process, the chance constraints are directly used and their feasibilities are guaranteed using stochastic simulation. Finally, the genetic algorithm is again used to achieve the highest membership value of each of the membership goals and thereby the most satisfactory decision is obtained. The proposed procedure is illustrated by a numerical example.

Keywords: Chance Constraint, Multiobjective Programming, Continuous Random Variable, Fuzzy Goal Programming, Stochastic Simulation, Genetic Algorithm.

1. Introduction

Chance Constraint Programming (CCP) is a major approach of stochastic programming. It can be used to solve problems involving constraints having finite probability of being violated. Charnes and Cooper [2] originally developed CCP. They formulated chance constraints individually for each stochastic constraint. Miller and Wagner [12] considered joint probabilistic constraints for independent random variables. The general case was studied by Pre'kopa [15]. Also Kataoka [11], Panne and Popp [14] and many others contributed in this field.

In CCP, the chance constraints are converted to their respective deterministic equivalents according to the specified confidence level. Then the deterministic model is solved by traditional nonlinear programming techniques. In case of normal distribution, deriving the deterministic equivalent is easier and well known [10]. Therefore, in most of the literature on CCP it is assumed that the random variables follow the normal distribution. Deterministic equivalents of the chance constraints with uniform random variables [5] and with exponential random variables [1, 5] are also obtained. But the deterministic equivalents of chance constraints in case of other random variables are yet to appear in the literature.

Due to the rapid development of computer technology and software in the last few years, genetic algorithm (GA), initiated by Holland [6], and GA based methods have been found to be suitable to deal problems involving complex mathematical approaches. Iwamura and Liu [8] used GA to solve CCP problems for the first time. Ruiqing et al. [17] have presented a GA based method to solve chance constraint integer programming and chance constraint goal programming problems. Very recently, Jana and Biswal [9] presented solution procedure of CCP problems with any continuous random variables. Their method is based on stochastic simulation and GA. Deriving the deterministic equivalent is not required in their method. Therefore, it is applicable to CCP problems with any continuous random variables. They checked the feasibility of the chance constraints by using the stochastic simulation technique [16].

A CCP problem involving more than one objective function is known as a multiobjective chance constraint programming (MOCCP) problem. Hulsurkar et al. [7] applied fuzzy programming technique to solve such problems in the case of normal random variables. But this technique is not applicable to MOCCP problems with any random variables. Therefore, it is better to use a GA based method that can be applicable for any continuous random variables.

Goal programming, introduced by Charnes and Cooper [3], can be used as a useful tool for solving problems involving multiple conflicting objectives. GA based approaches in the framework of conventional goal programming [18] and fuzzy goal programming (FGP) [4] have been investigated. But they are very few in numbers in the literature.

In this paper, a soft computing approach, based on GA within the framework of FGP, is proposed for solving MOCCP problems. In the model formulation of the problem, a GA is initially introduced to determine the fuzzy aspiration levels of objectives. Then, a FGP formulation of the problem for achievement of highest membership value, to the extent possible, of each of the membership goals defined for membership functions of the stated fuzzy goals of the problem is presented.

2. Multiobjective Chance Constraint Programming Problem

A multiobjective chance constraint programming problem can be stated as

Maximize: $$z^{(k)}(x) = \sum_{j=1}^{n} c_j^{(k)} x_j, k = 1,2,...,K \qquad (1)$$

subject to, $$P(\sum_{j=1}^{n} a_{ij} x_j \le b_i) \ge 1 - \alpha_i,\ \ i = 1,2,...,m \qquad (2)$$

$$0 \le x_j \le x_j^u, j = 1,2,...,n,\ \alpha_i \in (0,1), i = 1,2,...,m \qquad (3)$$

where a_{ij} and b_i are any continuous random variables, α_i are specified probabilities and x_j^u is the upper bound of the jth ($j = 1,2,...,n$) decision variable.

A multiobjective programming approach through the classical CCP technique is not applicable here. To overcome this situation, the proposed soft computing procedure is used. In this process, random number generation corresponding to different continuous probability distributions is an essential part. The details about algorithms for generating random numbers for continuous probability distributions can be found in [9,16]. Next we briefly discuss the technique of stochastic simulation for chance constraints.

2.1. Stochastic Simulation for Chance Constraints

Let us consider,

$$P[g_i(x,r) \le 0] \ge 1-\alpha_i, i=1,2,...,m \quad (4)$$

where $r=(R_1,R_2,...,R_t)$ is a t-dimensional continuous random vector and each R_t has a given continuous probability distribution. We generate N independent random vectors $r^s=(R_1^s,R_2^s,...,R_t^s), s=1,2,...,N$ from their probability distributions. Let N' be the number of occasions on which $g_i(x,r^s)\le 0, i=1,2,...,m$. Then by the definition of probability, equation (4) holds if $N'/N \ge (1-\alpha_i), i=1,2,...,m$.

2.2. Steps of Proposed GA

The GA used in the decision-making context is presented below:

Step 1. Representation and initialization: Let $V=\{x_1,x_2,...,x_n\}$ be the binary coded representation of chromosome in a population. The population size is defined by p_size and p_size chromosomes are initialized randomly in the search domain.

Step 2. Constraints check by stochastic simulation: Check the constraints of the problem employing stochastic simulation.

Step 3. Evaluation function: The evaluation function is defined as $eval(V_i)=Z^{(k)}$, $i=1,2,\dots,$ p_size, where $Z^{(k)}$ is given by (1). The best chromosome with largest fitness at each generation is obtained as V^* = argmax{ $eval(V_i) \mid i=1,2,\dots,$ p_size}.

Step 4. Selection: The roulette-wheel scheme [6] is used for selecting two parents for mating purposes.

Step 5. Crossover: Probability of crossover is defined as p_c and the arithmetic crossover operator is applied.

Step 6. Mutation: Probability of mutation is defined as p_m. A random number $r\in(0,1)$ is generated. A chromosome is selected for mutation if $r<p_m$. Then a bit is selected randomly and mutated.

Step 7. Termination: The execution of whole process terminates when the number of iterations is reached to the generation number specified in the genetic process.

2.3. Characterization of membership function

Let $Z_B^{(k)}$ and $Z_W^{(k)}$ be the best and worst values, respectively, of the kth ($k=1,2,...,K$) objective obtained by using the GA. Then the kth membership function can be constructed as -

$$\mu_{Z^{(k)}}(x)=\begin{cases} 1 , & Z^{(k)}(x) > Z_B^{(k)} \\ \dfrac{Z^{(k)}(x)-Z_W^{(k)}}{Z_B^{(k)}-Z_W^{(k)}}, & Z_W^{(k)} \le Z^{(k)}(x) \le Z_B^{(k)} \\ 0 , & Z^{(k)}(x) < Z_W^{(k)} \end{cases} \quad (5)$$

Now, the FGP formulation is presented in the following section.

3. FGP Formulation

In FGP, the membership functions are transformed into fuzzy goals. According to [4], the given MOCCP problem is equivalent to

$$\text{Maximize: } \sum_{i=1}^{K} w_k \lambda_k \tag{6}$$

$$\text{Subject to, } \lambda_k \le \mu_{Z^{(k)}}(x),\ 0 \le \lambda_k \le 1,\ k = 1,2,\ldots,K \tag{7}$$

along with the given set of constraints (2) and (3). In this formulation, w_k represent the relative importance of achieving the aspired levels of the respective fuzzy goals and are determined as [13]:

$$w_k = 1/(Z_B^{(k)} - Z_W^{(k)}),\ k = 1,2,\ldots,K \tag{8}$$

The above problem is a CCP problem with one objective function. Therefore, we again use the proposed stochastic simulation based GA to solve it.

4. Numerical Examples

To illustrate the potential use of the proposed procedure, we take the following CCP problem having two objective functions:

$$\text{Maximize } Z^{(1)} = 2x_1 + 3x_2 + 8x_3 \tag{9}$$

$$\text{Maximize } Z^{(2)} = 5x_1 + 6x_2 + 3x_3 \tag{10}$$

$$\text{subject to } \quad P(a_{11}x_1 + a_{12}x_2 + a_{13}x_3 \le b_1) \ge 0.95 \tag{11}$$

$$P(a_{21}x_1 + a_{22}x_2 + a_{23}x_3 \le b_2) \ge 0.80 \tag{12}$$

$$P(a_{31}x_1 + a_{32}x_2 + a_{33}x_3 \le b_3) \ge 0.90 \tag{13}$$

$$0 \le x_i \le 2, i = 1,2,3 \tag{14}$$

where (i) $a_{11}, a_{12}, a_{13}, b_1$ follow beta distribution with parameters 1and 50, 2 and 100, 4 and 90, 8 and 5, respectively, (ii) $a_{21}, a_{22}, a_{23}, b_2$ follow gamma distribution with parameters 2 and 0.01, 2 and 1, 3 and 0.1, 3 and 5, respectively, and (iii) $a_{31}, a_{32}, a_{33}, b_3$ follow Weibull distribution with parameters 2 and 3, 3 and 2, 3 and 1, 5 and 30, respectively.

To formulate the fuzzy goals of the objectives of the problem, the proposed GA is coded in C and the parameter values are set as: p_size = 100, $p_c = 0.8$, $p_m = 0.08$. At 25th generation, the best and worst objective values are obtained as follows:

Best value	(x_1, x_2, x_3)	Worst value	(x_1, x_2, x_3)
$Z_B^{(1)}$=25.9815	(1.9956,1.9986,1.9993)	$Z_W^{(1)}$=0.0056	(0.0022,0.0002,0.0001)
$Z_B^{(2)}$=27.9560	(1.9988,1.9940,1.9993)	$Z_W^{(2)}$=0.0268	(0.0030,0.0000,0.0039)

The membership functions for the goals are obtained as follows:

$$\mu_{Z^{(1)}}(x) = (2x_1 + 3x_2 + 8x_3 - 0.0056)/25.9759 \tag{15}$$

$$\mu_{Z^{(2)}}(x) = (5x_1 + 6x_2 + 3x_3 - 0.0268)/27.9292 \tag{16}$$

Then the FGP formulation of the above problem becomes

$$\text{Maximize: } \lambda_1 / 25.9759 + \lambda_2 / 27.9292 \quad (17)$$

$$\text{Subject to, } \lambda_1 \le (2x_1 + 3x_2 + 8x_3 - 0.0056) / 25.9759 \quad (18)$$

$$\lambda_2 \le (5x_1 + 6x_2 + 3x_3 - 0.0268) / 27.9292 \quad (19)$$

$$0 \le \lambda_1, \lambda_2 \le 1 \quad (20)$$

and constraints (11) – (14).

This problem is solved using the GA. Except the generation number, taken as 100 in this case, all other parameter values are kept same. The solution obtained is given below:

$$(x_1, x_2, x_3) = (1.9403, 1.8954, 1.9843)$$

with maximum value of $Z^{(1)}$ = 25.4412 and $Z^{(2)}$ = 27.0268. The corresponding membership values are, respectively, $\lambda_1 = 0.9752$ and $\lambda_2 = 0.9615$.

Note: The parameter values used in the proposed GA, such as p_size = 100, p_c = 0.8, p_m = 0.08, are obtained as the best suitable values after a rigorous computational experience.

5. Conclusions

In this paper, the FGP formulation of a MOCCP problem and the efficient use of GA to solve such a problem is presented. The main advantage of the proposed approach is that deriving the deterministic equivalent of chance constraints is not required here. The chance constraints are used directly. Therefore, the method is suitable for MOCCP problem with any continuous random variables. The result of the numerical problem shows that the proposed procedure is very efficient. The goal satisfaction levels of both the objectives are very close to unity. Hence the obtained solution is the most satisfactory solution in the decision-making context.

The proposed approach can be extended to different areas like multiobjective chance constraint bilevel programming, multiobjective chance constraint multilevel programming and multiobjective chance constraint fractional programming etc. with any continuous random variables in the chance constraints. The coefficients in the objective functions are not taken as random variables. Solving MOCCP problem with all the coefficients in the chance constraints as well as all the coefficients in the objective functions as any continuous random variable is a further topic of research. References

References

1. M.P. Biswal, N.P. Biswal & Li, Duan (1998), Probabilistic linear programming problems with exponential random variables, A technical note. *European Journal of Operational Research*, 111: 589-597.
2. A. Charnes & W.W. Cooper (1959), Chance constrained programming, *Management Science*, 6: 73-79.
3. A. Charnes & W.W. Cooper (1961), *Management models and industrial applications of linear programming*, Wiley, New York.
4. M.Gen, K. Ida, J. Lee & J. Kim. (1997), Fuzzy nonlinear goal programming using genetic algorithm, *Computers and Industrial Engineering*, 33 (1-2): 39-42.

5. Goicoechea, D.R. Hansen, & L. Duckstein, (1982), *Multi-objective decision analysis with engineering and business applications*, John Wiley and Sons, New York.
6. L. Holland (1975), *Adaptation in natural and artificial systems*. University of Michigan Press. Ann Arbor, MI.
7. Suwarna Hulsurksr, M.P. Biswal & S.B. Sinha (1997), Fuzzy programming approach to multi-objective stochastic linear programming problems, *Fuzzy Sets and Systems*, 88: 173-181.
8. K. Iwamura & B. Liu (1996), A genetic algorithm for chance constrained programming. *Journal of Information and Optimization Sciences*, 17(2): 409-422.
9. R.K. Jana & M.P. Biswal, (2004). Stochastic simulation-based genetic algorithm for chance constraint programming problems with continuous random variables. *International Journal of Computer Mathematics*, 81 (9): 1069-1076.
10. P. Kall & S.W. Wallace (1994), *Stochastic programming*. John Wiley & Sons.
11. Kataoka (1963), A stochastic programming model. *Econometrica*, 31: 181-196.
12. L.B. Miller & H. Wagner (1965), Chance-constrained programming with joint constraints. *Operations Research*, 13: 930-945.
13. B.B. Pal, B.N. Moitra & U. Maulik (2003), A goal programming procedure for fuzzy multiobjective linear fractional programming problem, *Fuzzy Sets and Systems*, 139: 395-405.
14. V.D. Panne & W. Popp (1963), Minimum cost cattle feed under probabilistic protein constraints, *Management Science*, 9: 405-430.
15. A. Pre'kopa (1973), Contribution to the theory of stochastic programming. *Mathematical Programming*, 4: 202-221.
16. R.Y. Rubinstein (1981), *Simulation and Monte Carlo method*, John Wiley & Sons.
17. Z. Ruiqing, K. Iwamura & B. Liu (1998), Chance constrained integer programming and stochastic simulation based genetic algorithm. *Journal of System Science and Systems Engineering*, 7(1): 96-102.
18. D.W. Zheng, M. Gen & K. Ida (1996), Evolution Program for nonlinear goal programming. *Computers and Industrial Engineering*, 31 (3/4): 907 – 911

Fuzzy Logic and its Application in Technology and Management
Editors: Debjani Chakraborty, S. Nanda and D. Dutta Majumder

Two-Storage Inventory Model with Variable Demand, Possibility Constraints, Fuzzy Unit Costs and Imprecise Resources

Manas Kumar Maiti[1] *and Manoranjan Maiti*[2]

[1]Department of Mathematics, Mahishadal Raj College, Mahishadal, Medinipur
[2]Department of Mathematics, Vidyasagar University, Medinipur, WB
[1]E-mail: manasmaiti@yahoo.co.in [2]E-mail: vidya295@sanchatnet.in

Abstract: An inventory model with two different storage facilities is developed for multi-item by considering inventory level dependent demand rate in fuzzy environment (purchase cost, investment amount and storehouse capacity are imprecise). The model is formulated with chance constraints and is solved using Fuzzy chance constraint programming (FCCPT) with (i) Generalized reduced gradient method (FCCPT- GRG) and (ii) Fuzzy simulation based genetic algorithm (FCCPT- FSGA).

Keywords: Chance constraint programming, Fuzzy Simulation, Genetic Algorithm.

1. Introduction

Now-a-days in the market places like super markets, municipality markets, etc., it is almost impossible to have a big showroom/shop due to the scarcity of space and very high rent. Normally, moderate and big business houses operates through two rented houses- one, smaller in size, is in the front of the market place and other with large capacities at the back of the place. During last two decades, these types of models has been developed and solved by many researchers [1-2].

Normally, in inventory control systems, resource constraints are assumed deterministic. In real life, it is not so. During controlling period of inventory, the resource constraints may be possibilistic in nature and it may happen that the constraints on resources will satisfy in almost all cases except possibly for a very few cases, where they may be allowed to violate. In stochastic environment for the solution of this type of problems Mohon [3] proposed a 'here and now' approach- i.e. the chance constraints programming approach in which a minimum probability level for satisfying each of the constraint is specified. Similarly possibilistic constraints may also be defined [4].

Here, an inventory model for multi-items is considered with displayed units dependent demand. Two storehouses are used for storage- one (say RW_1) at the front of market place and the other (say RW_2) at the back. At the beginning, RW_1 is filled up first and then items are stored at RW_2. The items are sold from RW_1 and are continuously released from RW_2 to RW_1 once a unit item is sold from RW_1.

Replacement of an item occurs when its inventory level reaches its reorder point. Here, the size of RW_1 is finite but that of RW_2 is imprecise. Similarly unit costs of the items and the capital for investment are also fuzzy in nature. Hence there are two constraints and these constraints will hold with at least some possibility α. With these assumptions, total profit for the inventory system is calculated and is maximized following (i) FCCPT-GRG and (ii) FCCPT-FSGA [5]. The model is illustrated with numerical data and the results from two methods are compared.

2. Possibility in Fuzzy Environment

Any fuzzy subset $\tilde{a}$ of $\Re$ (where $\Re$ represents set of real numbers) with membership function $\mu_{\tilde{a}} : \Re \to [0,1]$, is called a fuzzy number. Let $\tilde{a}$ and $\tilde{b}$ be two fuzzy quantities with membership functions $\mu_{\tilde{a}}$ and $\mu_{\tilde{b}}$ respectively. Then according to Liu and Iwamura [5],

$$Pos(\tilde{a} * \tilde{b}) = \sup\{\min(\mu_{\tilde{a}}(x), \mu_{\tilde{b}}(y)), \quad x, y \in \Re, \quad x * y\}$$

where the abbreviation Pos represent possibility, and * is any one of the relations >, <, =, $\leq, \geq$. Analogously if $\tilde{b}$ is a crisp number say b then

$$Pos(\tilde{a} * b) = \sup\{\mu_{\tilde{a}}(x), \quad x \in \Re, \ x * b\}$$

If $\tilde{a}, \tilde{b} \in \Re$ and $\tilde{c} = f(\tilde{a}, \tilde{b})$ where $f : \Re \times \Re \to \Re$ be a binary operation then membership function $\mu_{\tilde{c}}$ of $\tilde{c}$ is defined as

$$\mu_{\tilde{c}}(z) = \sup\{\min(\mu_{\tilde{a}}(x), \mu_{b}(y)), \ x, y \in \Re \ and \ z = f(x, y), \forall z \in \Re$$

Let $\tilde{a} = (a_1, a_2, a_3, a_4)$ be a trapezoidal fuzzy number with membership function

$$\mu_{\tilde{a}}(x) = \begin{cases} \dfrac{x - a_1}{a_2 - a_1} & \text{for } a_1 \leq x \leq a_2 \\ 1 & \text{for } a_2 \leq x \leq a_3 \\ \dfrac{a_4 - x}{a_4 - a_3} & \text{for } a_3 \leq x \leq a_4 \\ 0 & otherwise \end{cases}$$

$\tilde{c} = (c_1, c_2)$ is a fuzzy number with membership function $\mu_{\tilde{c}}$ given by $\tilde{c}$

$$\mu_{\tilde{c}}(x) = \begin{cases} 1 & \text{for } 0 \leq x \leq c_1 \\ \dfrac{c_2 - x}{c_2 - c_1} & \text{for } c_1 \leq x \leq c_2 \\ 0 & otherwise \end{cases}$$

and b be any crisp number then using above definitions following lemmas can easily be derived.

Lemma 1: $Pos(\tilde{a} \geq b) \geq \alpha \ \ if \ \dfrac{a_4 - b}{a_4 - a_3} \geq \alpha$

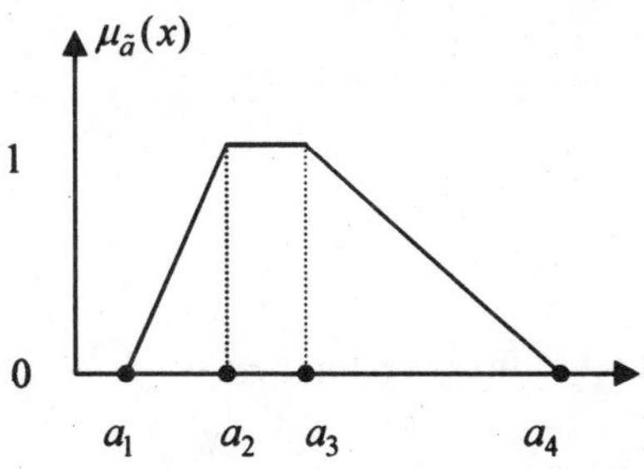

Fig 1: Trapezoidal fuzzy number

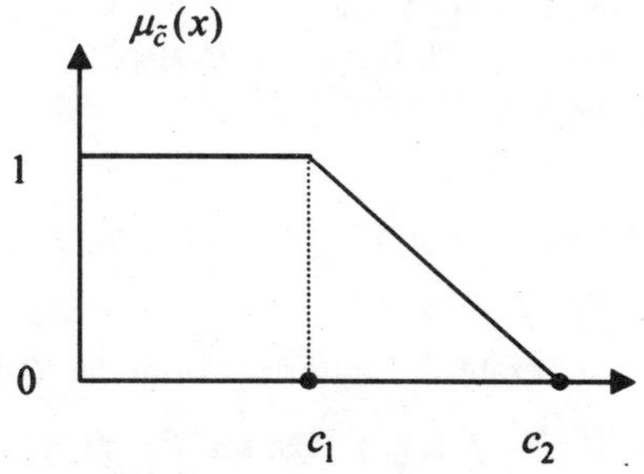

Fig 2: Pictorial representation of $\mu_{\tilde{c}}(x)$

Lemma 2: $Pos(\tilde{c} \geq b) \geq \alpha$ *if* $\dfrac{c_2 - b}{c_2 - c_1} \geq \alpha$

Lemma 3: $Pos(\tilde{c} \geq \tilde{a}) \geq \alpha$ *if* $\dfrac{c_2 - a_1}{a_2 - a_1 + c_2 - c_1} \geq \alpha$

3. Fuzzy Chance Constraint Programming

A chance constrained programming with fuzzy parameters may be written as [5]

$$\left.\begin{aligned} &Max\ \tilde{f} \\ &Subject\ to\ \ Pos\{\xi / f(x,\xi) \geq \tilde{f}\} \geq \beta \\ &\qquad\qquad Pos\{\xi / g_j(x,\xi) \leq 0\} \geq \alpha_j \quad j = 1,2,..n \end{aligned}\right\} \qquad (1)$$

where, x is a decision vector, ξ is a vector of fuzzy parameters, $f(x,\xi)$ is the objective function, $g_i(x,\xi)$ are constraint functions and β, α_j are predetermined confidence level for objective function and the constraints respectively.

***3.1. Fuzzy Simulation*:** The basic technique of chance constrained programming in a fuzzy environment is to convert the possibility constraints to their respective deterministic equivalents according to predetermined confidence level. However, the procedure is usually very hard and only successful for some special cases. Recently Liu & Iwamura [4] developed fuzzy simulation process to check feasibility of a solution x of the problem (1). Following Liu & Iwamura [5] the following two algorithms are developed to check feasibility of a solution x and to determine $\tilde{f}$.

***Algorithm 1*:** Algorithm to check $Pos\{\xi / g_j(x,\xi) \leq 0\} \geq \alpha_j, j = 1,2,....n$ for a particular value of decision vector x.

1. Set j=1.
2. Generate ξ_0 uniformly from the α_j cut set of fuzzy vector ξ.
3. If $g_j(x,\xi) \leq 0$ go to step 6.
4. Repeat steps 2 to 3 N times.

5. Return Infeasible.
6. Set $j = j+1$, if $j \leq n$ goto step 2.
7. Return feasible.
8. End algorithm.

***Algorithm 2*:** Algorithm to determine $\tilde{f}$

1. Set $\tilde{f} = -\infty$.
2. Generate ξ_0 uniformly from the β cut set of fuzzy vector ξ.
3. If $\tilde{f} < f(x, \xi_0)$ then set $\tilde{f} = f(x, \xi_0)$.
4. Repeat step 2 and 3, N times.
5. Return $\tilde{f}$.
6. End algorithm.

***3.2. Fuzzy Simulation based Genetic Algorithm*:** Using the above two algorithms problem (1) can be solved using following genetic algorithm.

1. Set probability of crossover P_c and probability of mutation P_m.
2. Set iteration counter $i = 1$.
3. Generate Initial population set of solutions $P(i)$ of size N, where feasibility of each solution is checked by algorithm1.
4. Determine objective function value due to every solution using algorithm2. These values are assumed as fitness of each solution.
5. Select solutions from $P(i)$ for evaluation. Let this set of solutions be $P'(i)$.
6. Select solutions from $P'(i)$ for crossover.
7. Made crossover on selected solutions to get population $P_1(i)$.
8. Select solutions from $P_1(i)$ for mutation.
9. Made mutation on selected solutions to get new population $P(i+1)$.
10. If average fitness of $P(i+1)$ > average fitness of $P(i)$ then set $i \leftarrow i+1$
11. If termination conditions does not hold go to step 4.

Take best solution from $P(i)$ as the solution of the problem.

4. Assumptions and Notations

Following assumptions and notations are used for the proposed model:

(i) $W_1, \tilde{W}_2$ are the inventory area of RW_1 and RW_2 respectively and $\tilde{R}$ is investment limit.

(ii) α_1, α_2, α_3 denote the confidence levels for profit function, storage capacity constraint and price constraint respectively.

(iii) f is the minimum profit with confidence level α_1.

(iv) The inventory system involves N items, and for i^{th} item following notations are used.

(v) R_i is reorder level, Q_i is order quantity, Q_{is} is inventory level at RW_1, where $Q_{is} = \lambda_i W_1 / ar_i$, $\sum_{i=1}^{N} \lambda_i = 1$ and ar_i is the area required to store one unit.

(vi) Here R_i, Q_i, λ_i are decision variables.

(vii) c_{0i} is the ordering cost, $\tilde{p}_i$ is per-unit purchase cost in \$, m is mark-up of selling price and h_1, h_2% of purchase cost is assumed as holding cost at RW_1 and RW_2 respectively and H_i is the holding cost per cycle.

(viii) T_i is the cycle time and $\tilde{Z}_i$ is the average profit, $q_i(t)$ inventory level at time t and demand D_i is of the form:

$$D_i = \begin{cases} A_i Q_{is}^{\gamma_i} & \text{if } q_i \geq Q_{is} \\ A_i q_i^{\gamma_i} & \text{if } Q_{i0} \leq q_i \leq Q_{is} \\ A_i Q_{i0}^{\gamma_i} & \text{if } q_i \leq Q_{i0} \end{cases}$$

Where A_i, γ_i are scale and shape Parameters of demand function

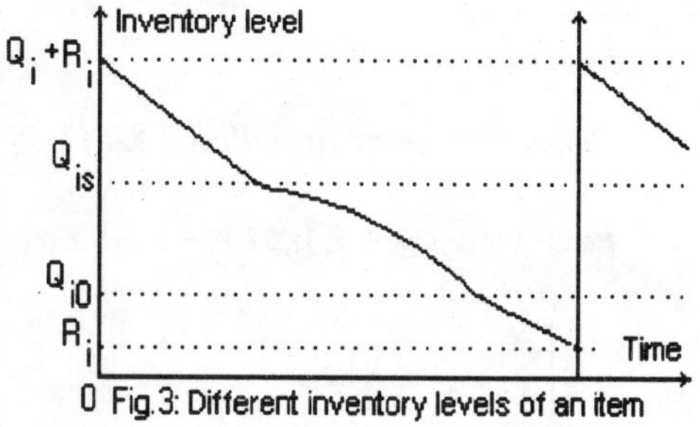

Fig.3: Different inventory levels of an item

5. Mathematical Formulation

Instantaneous state $q_i(t)$ of ith item is given by $\frac{dq_i}{dt} = -D_i$

Now for each item two scenarios may occur:- (i)$R_i+Q_i>Q_{is}$ and (ii)$R_i+Q_{is} \leq Q_{is}$. In each scenario two cases may arise- Case1: $R_i \geq Q_{i0}$, Case2: $R_i<Q_{i0}$.

$$\textbf{\textit{Scenario1-Case1:}}\ T_i = \frac{R_i+Q_i-Q_{is}}{A_i Q_{is}^{\gamma_i}} + \frac{Q_{is}^{1-\gamma_i} - R_i^{1-\gamma_i}}{(1-\gamma_i)A_i}$$

$$H_i = K_i p_i \quad where\ K_i = \frac{h_2(R_i+Q_i-Q_{is})^2}{2A_i Q_{is}^{\gamma_i}} + h_1\left\{\frac{R_i+Q_i-Q_{is}}{A_i Q_{is}^{\gamma_i-1}} + \frac{Q_{is}^{2-\gamma_i} - R_i^{2-\gamma_i}}{A_i(2-\gamma_i)}\right\}$$

$$\textbf{\textit{Scenario1-Case2:}}\ T_i = \frac{R_i+Q_i-Q_{is}}{A_i Q_{is}^{\gamma_i}} + \frac{Q_{is}^{1-\gamma_i} - Q_{i0}^{1-\gamma_i}}{(1-\gamma_i)A_i} + \frac{Q_{i0}-R_i}{A_i Q_{i0}^{\gamma_i}} \text{ and } H_i = K_i p_i$$

$$where\ K_i = \frac{h_2(R_i+Q_i-Q_{is})^2}{2A_i Q_{is}^{\gamma_i}} + h_1\left\{\frac{R_i+Q_i-Q_{is}}{A_i Q_{is}^{\gamma_i-1}} + \frac{Q_{is}^{2-\gamma_i} - Q_{i0}^{2-\gamma_i}}{A_i(2-\gamma_i)} + \frac{Q_{i0}^2 - R_i^2}{2A_i Q_{i0}^{\gamma_i}}\right\}$$

$$\textbf{\textit{Scenario2-case1:}}\ T_i = \frac{(Q_i+R_i)^{1-\gamma_i} - R_i^{1-\gamma_i}}{(1-\gamma_i)A_i},$$

$$H_i = K_i p_i \quad where\ K_i = h_1 \frac{(Q_i+R_i)^{2-\gamma_i} - R_i^{2-\gamma_i}}{A_i(2-\gamma_i)}$$

$$\textbf{\textit{Scenario2-Case2:}}\ T_i = \frac{(Q_i+R_i)^{1-\gamma_i} - Q_{i0}^{1-\gamma_i}}{(1-\gamma_i)A_i} + \frac{Q_{i0}-R_i}{A_i Q_{i0}^{\gamma_i}}$$

and $H_i = K_i p_i$ where $K_i = h_1\left\{\frac{(Q_i + R_i)^{2-\gamma_i} - Q_{i0}^{2-\gamma_i}}{A_i(2-\gamma_i)} + \frac{Q_{i0}^2 - R_i^2}{2A_i Q_{i0}^{\gamma_i}}\right\}$

So average profit $\tilde{Z} = \sum_{i=1}^{N} \tilde{Z}_i$, where $\tilde{Z}_i = \frac{\{Q_i(m_i - 1) - K_i\}\tilde{p}_i - c_{0i}}{T_i}$

So the problem reduces to the following fuzzy chance constrained programming problem:

$$\left.\begin{array}{l} Max\,\tilde{f} \\ Subject\ to\ constraints\ Pos(\tilde{Z} \geq \tilde{f}) \geq \alpha_1 \\ Pos\left\{\sum_{i=1}^{N}\{ar_i(Q_i + R_i)\} \leq (W_1 + \tilde{W}_2)\right\} \geq \alpha_2 \\ Pos\left\{\sum_{i=1}^{N}\tilde{p}_i Q_i \leq \tilde{R}\right\} \geq \alpha_3 \end{array}\right\} \quad (2)$$

6. Numerical Illustration

The model is illustrated for two items (N = 2) and for two examples as given below. Common parametric values assumed for both the examples are

c_{01}=40, c_{02}=30, W_1=15 units, ar_1=0.5 units, ar_2=0.45 units, $\alpha_1 = 0.9$, $\alpha_2 = 0.9$, $\alpha_3 = 0.9$, A_1=15, A_2=17.5, Q_{10}=5, Q_{20}=5, $\gamma_1 = 0.5$, $\gamma_2 = 0.45$, h_1=0.2, h_2= 0.1, m= 1.4.

Example 1: Let p_i=(p_{i1}, p_{i2}, p_{i3}, p_{i4}) be trapezoidal fuzzy number and R=(r_1, r_2), $\tilde{W}_2$=(w_{21}, w_{22}) be two fuzzy numbers whose membership functions are depicted in Fig-3. Here we assume p_{11}=10, p_{12}=12, p_{13}=13, p_{14}=15, p_{21}=8, p_{22}=11, p_{23}=12, p_{24}=14, r_1=\$1400, r_2=\$1500, w_{21}=50 units, w_{22}=70 units.

In this case using FCCPT and Lemma 1,2,3 the problem (2) reduces to following crisp problem.

$$\left.\begin{array}{l} Max\,f \\ Subject\ to\ \frac{f_4 - f}{f_4 - f_3} \geq \alpha_1,\ f_j = \sum_{i=1}^{N}\frac{Q_i(m-1)p_{ij} - K_i p_{ij} - c_{0i}}{T_i}\ for\ j = 3,4 \\ \frac{w_{22} - w_{20}}{w_{22} - w_{21}} \geq \alpha_2,\ w_{20} = \sum_{i=1}^{N} ar_i(Q_i + R_i) - W_1 \\ \frac{r_2 - P_1}{P_2 - P_1 + r_2 - r_1} \geq \alpha_3,\ P_j = \sum_{i=1}^{N} p_{ij}Q_i\ for\ j = 1,2 \end{array}\right\} \quad (3)$$

This problem is solved via GRG Method. In this example problem (2) is also solved using FSGA technique and results are presented in Table-1. It is observed that profits in both the cases are almost same. But GRG is applicable to only those problems whose possibility constraints can be reduced to equivalent crisp constraints.

Table-1: Results observed for Example1 and Example2

Example	Process	Q_1	Q_2	R_1	R_2	λ_1	λ_2	f
1	GRG	64.44	57.13	8.17	7.31	0.5528	0.4472	396.05
	FSGA	65.13	56.28	8.57	7.20	0.5686	0.4314	395.96
2	FSGA	55.44	48.16	6.50	11.02	0.4309	0.5691	494.76

Example-2: Let $p_i=(p_{i1}, p_{i2}, p_{i3})$ be parabolic fuzzy number with membership function:

$$\mu_{p_i}(x)=\begin{cases}1-\left(\dfrac{p_{i2}-x}{p_{i2}-p_{i1}}\right)^2, & \text{for } p_{i1}\le x\le p_{i2}\\ 1, & \text{for } x=p_{i2}\\ 1-\left(\dfrac{x-p_{i2}}{p_{i3}-p_{i2}}\right)^2, & \text{for } p_{i2}\le x\le p_{i3}\\ 0, & \text{otherwise}\end{cases}$$

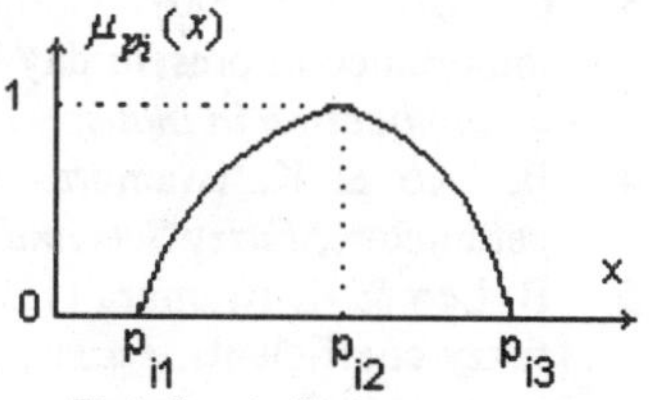

Fig4: Parabolic fuzzy number

$\widetilde{W}_2$ and R are same as example1. Here we assume p_{11}=10, p_{12}=13, p_{13}=15, p_{21}=14, p_{22}=16, p_{23}=19. In this case the problem (2) cannot be reduced to an equivalent crisp problem and so results are observed via FSGA and presented in Table-1.

6.1. ***Sensitivity Analysis*:** The results are observed for both the examples for different confidence levels of objective function (α_1) and presented in Table-2. It is observed that as confidence level increases profit decreases, which agrees with reality.

Table-2: Results observed for different confidence level of objective function

α_1	F		
	Example1		Example2
	FCCPT	FSGA	FSGA
0.91	395.32	395.30	493.23
0.92	394.59	394.57	491.60
0.93	393.86	393.85	489.87
0.94	393.12	393.10	488.01
0.95	392.40	392.39	485.99
0.96	391.67	391.66	483.75

7. Conclusion

For the first time a realistic inventory model is developed in fuzzy environment where constraints of the problem are represented as chance constraints and a generalized solution methodology is suggested which can be applied to other inventory problems in fuzzy environment.

Acknowledgement

The authors would like to thank AICTE, New Delhi for financial support of the project under which present research paper has been prepared.

References

1. K.V.S. Sarma (1987), A deterministic order-level inventory model for deteriorating items with two-storage facilities, *European Journal of Operation Research*, 29: 70-72.
2. A.K. Bhuina & M. Maiti, A two warehouse inventory model for a linear trend in demand, *Opsearch*, 31: 318-329.
3. C. Mohon (2000), Optimization in fuzzy-stochastic environment and its importance in present day industrial scenario, *Proceedings on Mathematics and its application in industry and business*, Narosa Publishing House, India, 2000.
4. B. Leo & K. Iwamura (1998), Chance constraint Programming with fuzzy parameters, *Fuzzy Sets and Systems*, 94: 227-237.
5. B. Leo & K. Iwamura (1998), A note on chance constrained programming with fuzzy coefficients, *Fuzzy Sets and Systems*, 100: 229-233.

Fuzzy Logic and its Application in Technology and Management
Editors: Debjani Chakraborty, S. Nanda and D. Dutta Majumder

A Reordering Strategy for Single-period Inventory Model with Imprecise Demand Information

Pankaj Dutta, Debjani Chakraborty and A.R. Roy

Department of Mathematics, Indian Institute of Technology, Kharagpur
Email: {pankaj, debjani, arroy}@maths.iitkgp.ernet.in

Abstract: In this paper, a reordering strategy for the products of a single period inventory system with fuzzy demand is developed. The entire period/season is divided into two slots and the reorder is to be made during the mid-season after the early-season demand has been observed. It is assumed that the demand rate during the period/season occurs at a relatively constant rate but not in a constant size. Replenishment is instantaneous and if there is any shortage during the middle of the period/season, then it is backordered. The objective is to find the expected optimal order quantity together with profit maximization. The optimization is achieved using fuzzy ordering of fuzzy numbers using their possibilistic mean values. The model is illustrated by a numerical example. It is shown that the explicit consideration of this reordering opportunity makes the decision maker very much richer.

Keywords: Single period inventory, Reorder, Fuzzy demand, Profit maximization

1. Introduction

The ordering strategy for a single period inventory model is the key to achieve success in the competitive business environment. Most of the single period inventory models (SPIM) describe a business strategy where an item is to be ordered only once to satisfy customer demand for a precise period [1]. However, the demand is uncertain, either random or vague, so for multi-product environments like seasonal or fashionable items, there needs a huge initial investment and enormous storage space capacity to avoid loss of good will. So the optimal ordering strategy has become a major issue to overcome the organization's limitations and restrictions as well as the customer satisfaction. This paper focus on a two ordering strategy throughout the whole season/period where the reorder is to be made during the mid-season after the early-season demand has been observed.

Practically, in real business environments, there exist lots of single period products (e.g., seasonal/fashionable items) that need replenishment later in the season / period either due to limited storage capacity or demand uncertainty. Lau and Lau [7, 8] are the only researchers who interpret the problem and developed a mathematical model with stochastic demand. They divided the whole season into two slots by the demand scale and allow shortages for both the slots. But, in competitive market, if shortages occur during the mid season of the period and customers are going back without fulfillment of their demand, it affects not only on the expected profit in the first slot but also on the demand rate of the next slot. Recently, in [6] we developed a model for single period products with reordering strategy under fuzzy demand. We divided the whole season into two slots in time scale and the demand rate during the season is considered as uniform. Hence forth, the demand rate during slot-1 is considered as θ% (60% say) of total season's demand. Shortages may occur during slot-1, but that is depending upon the corresponding profit function to be constructed by the decision maker with his credibility preference to the associated over stock or under stock profit. Here, in this paper, we seek to avoid the amount of shortages during slot-1 as much as possible. Replenishment rate is considered as instantaneous. Even though shortages occur before second order, if any, then it is backordered.

It is to be mentioned that, in literature, several inventory models have been developed under fuzzy environment [3-5], but in [2] Petrovic et al first developed a newsboy-type problem with discrete fuzzy demand. Later on, Li et al [9] and Kao and Hsu [10] also considered the SPIM with continuous fuzzy demand. The present work extends the SPIM together with the reordering strategy and fuzzy demand.

As the demands are linguistic in nature and the optimal order quantity in the second slot depends upon the demand that arises in the first slot, so the profit function as well as the decision variable during second slot are also fuzzy quantity. The purpose of this study is to determine a personal policy that will maximize the total resultant profit under the above state of affairs.

2. Modeling with Reordering Opportunity in Fuzzy Environment

2.1 Optimal policy

To develop a SPIM with two ordering strategy under imprecise demand information the following notations are used.

Q_1 order quantity at the beginning of the season/period

Q_2 order quantity at the beginning of the second slot

c unit purchase cost

p selling price per unit

h holding cost per unit for the next season

s shortage/under stock cost per unit

$\tilde{D}_1$ fuzzy demand for slot-1

$\tilde{D}_2$ fuzzy demand for slot-2

d_1 actual realization of demand during slot-1

d_2 actual realization of demand during slot-2

Suppose the shop-keeper has the opportunity for reorder during the middle of the season. Assuming that there is no option for substitution between the products we propose the reordering strategy for an individual item and develop the model for profit maximization. Note that, in [6] we divided the whole season into two slots in time scale and shortages may occur during slot-1, but that is depending upon the corresponding profit function to be constructed by the decision maker with his credibility preference to the associated over stock or under stock profit. Here, we seek to avoid the amount of shortages during slot-1 as much as possible.

It is assumed that the demand rate during the period/season occurs at a relatively constant rate but not in a constant size. Therefore, according to the expert's demand information, if $\widetilde{D}=(\underline{D}, D, \overline{D})$ be the total seasonal demand then the demand during slot-1 should be a certain percent of $\widetilde{D}$ (60% say). Consequently, $\widetilde{D}_1=(\underline{D}_1, D_1, \overline{D}_1)$ also becomes a fuzzy quantity. Without loss of generality it can be assumed that the maximum possible demand during slot-1 is $\overline{D}_1$ with lowest membership grade. Thus if the DM chooses $Q_1^*=\overline{D}_1$ then the possibility of shortages during slot-1 is nil rather there is obvious excess after the end of this slot. In this case the DM is not bothering about the certain excess of products during slot-1 rather he gives the priority to customer satisfaction during the middle of the season. In addition, as replenishment is instantaneous, shortages may be backordered, if any.

Therefore, if Q_1^+ be the number of leftover items after the end of slot-1, then Q_1^+ is defined as $Q_1^+=Q_1^*-d_1$ where d_1 is the actual occurrence of demand in slot-1. Let Q_2^* be the optimal quantity to be needed for maximizing the profit function in the second slot individually based on the fuzzy demand $\widetilde{D}_2$, if no leftover items are supplied from slot-1.

Then the optimal order quantity for the second slot to be purchased at the beginning of slot-2 is defined as follows:

$$Q_2^{\text{opt}}=\begin{cases} 0 & \text{for } Q_1^+ \geq Q_2^* \\ Q_2^*-Q_1^+ & \text{for } 0 \leq Q_1^+ < Q_2^* \end{cases} \tag{1}$$

Thus Q_2^* is the only decision variable left in our hand to be found such that the associated profit function for second slot attains its maximum individually.

2.2. Determination of Q_2^*

As demand $\widetilde{D}_2$ in slot-2 is imprecisely prescribed, it causes a fuzzy over stock profit ($O\widetilde{P}$) and fuzzy under stock profit ($U\widetilde{P}$) and hence the resultant profit function is also becomes a fuzzy quantity $\widetilde{P}(Q_2,\widetilde{D}_2)$ (say). If d_2 be the actual realization of fuzzy demand $\widetilde{D}_2$ in slot-2, then for this slot the profit function can be formulated as

$$P(Q_2,d_2)=\begin{cases} OP=(p+h)d_2-(c+h)Q_2 & \text{for } d_2 \leq Q_2 \\ UP=(p-c+s)Q_2-sd_2 & \text{for } d_2 > Q_2 \end{cases}$$

Now using several $\alpha-$level set and possibilistic mean value method [11], we can find out the required Q_2^* by maximizing the mean or expected value of this fuzzy profit function. Details of determining Q_2^* can be found in [6]. Below we give the following results.

If $P(Q_2,\alpha)$ be the $\alpha-$level set of $\widetilde{P}(Q_2,\widetilde{D}_2)$ then it can be derived as follows:

$$P(Q_2,\alpha)=[P_L(Q_2,\alpha),P_R(Q_2,\alpha)]$$
$$=[\min\{OP_L(Q_2,\alpha),UP_L(Q_2,\alpha)\},\max\{OP_R(Q_2,\alpha),UP_R(Q_2,\alpha)\}]$$

Thus the possibilistic mean value of fuzzy profit function $\widetilde{P}(Q_2,\widetilde{D}_2)$ is given by

$$\overline{M}(\widetilde{P})=\int_0^1 \alpha(P_L(\alpha)+P_R(\alpha))d\alpha \tag{2}$$

Case-2.2.a. When $s_2r_2-l_2(p+h)\le 0$ or equivalently

$D_{2,0}=\{(p+h)\underline{D}_2+s_2\overline{D}_2\}/(p+s_2+h)\le D_2$ then $Q_2^*\in[D_{2,0},D_2]$ and the possibilistic mean value $\overline{M}(\widetilde{P})$ is given by

$$\overline{M}(\widetilde{P})=(p-c+s_2)Q_2-\frac{1}{2}[(p+s_2+h)\alpha_2^2+s_2\{L(Q_2)\}^2]Q_2$$
$$+(p+h)\int_0^{\alpha_2}\alpha D_{2,L}(\alpha)d\alpha-s_2\int_{\alpha_2}^1\alpha D_{2,R}(\alpha)d\alpha-s_2\int_{L(Q_2)}^1\alpha D_{2,L}(\alpha)d\alpha \tag{3}$$

where $\alpha_2=[(p+s_2+h)Q_2-\{(p+h)\underline{D}_2+s_2\overline{D}_2\}]/\{(p+h)l_2-s_2r_2\}<L(Q_2)$; l_2 and r_2 are the left and right spreads of $\widetilde{D}_2$, respectively.

Therefore the optimal Q_2^* is derived by the equation

$$\frac{\partial\overline{M}}{\partial Q_2}(Q_2^*)\equiv(p-c+s_2)-\frac{1}{2}[(p+s_2+h)\alpha_2^2+s_2\{L(Q_2^*)\}^2]=0 \tag{4}$$

subject to the condition $p-2c-h\le 0$. Again if $p-2c-h>0$ then the optimal Q_2^* lies between D_2 and $\overline{D}_2$. In this case, $\overline{M}(\widetilde{P})$ is given by

$$\overline{M}(\widetilde{P})=-(c+h)Q_2+\frac{1}{2}[(p+h)\{R(Q_2)\}^2]Q_2$$
$$+(p+h)[\int_0^1\alpha D_{2,L}(\alpha)d\alpha+\int_{R(Q_2)}^1\alpha D_{2,R}(\alpha)d\alpha \tag{5}$$

along with $\dfrac{\partial\overline{M}}{\partial Q_2}(Q_2^*)\equiv-(c+h)+\dfrac{1}{2}[(p+h)\{R(Q_2^*)\}^2]=0$ (6)

Case-2.2.b. When $s_2r_2-l_2(p+h)>0$ or equivalently $D_{2,0}>D_2$ then $Q_2^*\in[D_2,D_{2,0}]$

In this case, $\overline{M}(\widetilde{P})$ is given by

$$\overline{M}(\widetilde{P}) = -(c+h)Q_2 + \frac{1}{2}[(p+s_2+h)\alpha_2^2 + (p+h)\{R(Q_2)\}^2]Q_2$$

$$-s_2 \int_0^{\alpha_2} \alpha D_{2,R}(\alpha)d\alpha + (p+h)[\int_{\alpha_2}^{1} \alpha D_{2,L}(\alpha)d\alpha + \int_{R(Q_2)}^{1} \alpha D_{2,R}(\alpha)d\alpha] \quad (7)$$

along with $\frac{\partial \overline{M}}{\partial Q_2}(Q_2^*) \equiv -(c+h) + \frac{1}{2}[(p+s_2+h)\alpha_2^2 + (p+h)\{R(Q_2^*)\}^2] = 0$ (8)

subject to the condition $-(c+h)+(p+h)\{(\overline{D}_2 - D_{2,0})/(\overline{D}_2 - D_2)\}^2 \le 0$. Otherwise, Q_2^* lies between $D_{2,0}$ and $\overline{D}_2$ where $\overline{M}(\widetilde{P})$ and Q_2^* can be computed as defined in equation (5) and (6), respectively.

2.3. Total optimal expected order quantity

In this subsection, we find out the total expected order quantity for the whole season. To calculate the expected Q_2^{opt}, the optimal policy defined in equation (1) can be written as

$$Q_2^{\text{opt}} = \begin{cases} 0 & \text{for} \quad \underline{D}_1 \le d_1 \le Q_1^* - Q_2^* \\ Q_2^* - Q_1^+ & \text{for} \quad \max\{\underline{D}_1, Q_1^* - Q_2^*\} < d_1 \le \overline{D}_1 \end{cases}$$

This is a fuzzy quantity as demand $\widetilde{D}_1$ is a fuzzy quantity, so uncertainty occurs in case of Q_2^{opt} also.

Thus, Expected $Q_2^{opt} = \overline{M}(\widetilde{Q}_2^{opt}) = \int_0^1 \alpha(Q_{2,L}^{opt}(\alpha) + Q_{2,R}^{opt}(\alpha))d\alpha$ (9)

Where $Q_2^{opt}(\alpha) = [Q_{2,L}^{opt}(\alpha), Q_{2,R}^{opt}(\alpha)]$ is given as follows:

Condition-2.3.a. When $Q_1^* - Q_2^* < \underline{D}_1$ then

$$Q_2^{opt}(\alpha) = [Q_2^* - Q_1^* + D_{1,L}(\alpha), Q_2^* - Q_1^* + D_{1,R}(\alpha)] \quad \text{for} \quad 0 \le \alpha \le 1$$

Condition-2.3.b. When $Q_1^* - Q_2^* \ge \underline{D}_1$ then if $Q_1^* - Q_2^* \le D_1$ then

$$Q_2^{opt}(\alpha) = \begin{cases} [0, Q_2^* - Q_1^* + D_{1,R}(\alpha)] & \text{for } \alpha \le L(Q_1^* - Q_2^*) \\ [Q_2^* - Q_1^* + D_{1,L}(\alpha), Q_2^* - Q_1^* + D_{1,R}(\alpha)] & \text{for } \alpha > L(Q_1^* - Q_2^*) \end{cases}$$

and if $Q_1^* - Q_2^* > D_1$ then

$$Q_2^{opt}(\alpha) = \begin{cases} [0, Q_2^* - Q_1^* + D_{1,R}(\alpha)] & \text{for } \alpha \le R(Q_1^* - Q_2^*) \\ [0,0] & \text{for } \alpha > R(Q_1^* - Q_2^*) \end{cases}$$

Thus total expected optimal order quantity combining slot-1 and slot-2 is given by

$$Q_1^* + \text{Expected } Q_2^{opt} = \overline{D}_1 + \overline{M}(\tilde{Q}_2^{opt}).$$

2.4. *Total expected resultant profit*

In order to find out the expected value of total resultant profit (*TRP*), we need to construct the individual resultant profit functions according to the optimal policy (1).

Case-2.4.a. When $Q_1^+ \geq Q_2^*$ then $Q_2^{opt} = 0$

$$TRP = \begin{cases} pd_1 - cQ_1^* + (p+h)d_2 - hQ_1^+ & \text{for } d_2 \leq Q_1^+ \\ pd_1 - cQ_1^* + (p+s)Q_1^+ - sd_2 & \text{for } d_2 > Q_1^+ \end{cases}$$

Case-2.4.b. When $Q_1^+ \geq 0$ and $Q_1^+ < Q_2^*$ then $Q_2^{opt} = Q_2^* - Q_1^+$

$$TRP = \begin{cases} (p-c)d_1 + (p+h)d_2 - (c+h)Q_2^*, & \text{for } d_2 \leq Q_2^* \\ (p-c)d_1 + (p-c+s)Q_2^* - sd_2, & \text{for } d_2 > Q_2^* \end{cases}$$

Obviously, this total resultant profit is a fuzzy quantity, $T\tilde{R}P$ (say). Combining above two individual resultant profits we can find out the expected or mean value of this fuzzy profit function. Thus, if $TRP(\alpha) = [TRP_L(\alpha), TRP_R(\alpha)]$ be the α – level set of this fuzzy resultant profit function, then its possibilistic mean value is given by

$$\overline{M}(T\tilde{R}P) = \int_0^1 \alpha(TRP_L(\alpha) + TRP_R(\alpha))d\alpha \tag{10}$$

When $Q_1^* - Q_2^* < \underline{D}_1$ then Case-2.4.a will never happen and we have

$$TRP(\alpha) = [(p-c)D_{1,L}(\alpha) + \overline{M}(\tilde{P}), (p-c)D_R(\alpha) + \overline{M}(\tilde{P})] \text{ for } \forall \alpha \in [0,1] \tag{11}$$

Again, when $Q_1^* - Q_2^* > \underline{D}_1$ then $TRP(\alpha)$ is given as follows:

If $Q_1^* - Q_2^* \leq D_1$ then

$$TRP(\alpha) = \begin{cases} [\min\{pD_{1,L}(\alpha) - cQ_1^* + \overline{M}(\tilde{P}_d), (p-c)(Q_1^* - Q_2^*) + \overline{M}(\tilde{P})\} \\ \quad , \max\{p(Q_1^* - Q_2^*) - cQ_1^* + \overline{M}(\tilde{P}_d), (p-c)D_{1,R}(\alpha) + \overline{M}(\tilde{P})\}] & \text{for } \alpha \leq L(Q_1^* - Q_2^*) \\ [(p-c)D_{1,L}(\alpha) + \overline{M}(\tilde{P}), (p-c)D_R(\alpha) + \overline{M}(\tilde{P})] & \text{for } \alpha > L(Q_1^* - Q_2^*) \end{cases} \tag{12}$$

and if $Q_1^* - Q_2^* > D_1$ then

$$TRP(\alpha) = \begin{cases} [\min\{pD_{1,L}(\alpha) - cQ_1^* + \overline{M}(\tilde{P}_d), (p-c)(Q_1^* - Q_2^*) + \overline{M}(\tilde{P})\} \\ \quad , \max\{p(Q_1^* - Q_2^*) - cQ_1^* + \overline{M}(\tilde{P}_d), (p-c)D_{1,R}(\alpha) + \overline{M}(\tilde{P})\}] & \text{for } \alpha \leq R(Q_1^* - Q_2^*) \\ [pD_{1,L}(\alpha) - cQ_1^* + \overline{M}(\tilde{P}_d), pD_{1,R}(\alpha) - cQ_1^* + \overline{M}(\tilde{P}_d)] & \text{for } \alpha > R(Q_1^* - Q_2^*) \end{cases} \tag{13}$$

where $\overline{M}(\tilde{P}_d)$ is the possibilistic mean of the dummy profit function $P_d(G(Q_1^+))$ [6].

Thus, the expected total resultant profit, $\overline{M}(\widetilde{TRP})$ can be determined from equation (10) using the α – cuts of $\widetilde{TRP}$ provided by either of the equation (11), (12) or (13) according to the position of $Q_1^* - Q_2^*$ where as $\overline{M}(\widetilde{P})$ is given by equation (3) or (5) or (7) according to the position of Q_2^*. In the following, we illustrate the model numerically. The benefit of product reorder instead of single order is also computed.

3. Numerical Example

In order to illustrate the above reordering strategy in single-period context, consider a spot seller stores woolen materials for winter season. Total seasonal time can be divided into two slots each carries two months. Suppose, imprecision occurs in the customer level like *'demand is about d'* and we transform this linguistic terms into triangular fuzzy numbers. Consider there are three items in stock and the system parameters for each item are presented in Table 1.

It is assumed that the demand rate during the period/season occurs at a relatively constant rate but not in a constant size. Therefore, according to the expert's demand information, if $\widetilde{D}$ be the total seasonal demand, then the demand during slot-1 should be a certain percent of $\widetilde{D}$ (approx. 60%, say). Results for different items are presented in Table 2.

The percentage increase in expected profit compares the benefit of ordering twice instead of one, and is computed as

$$\text{PIEP} = 100 \times (\overline{M}_{\text{re-order}} - \overline{M}_{\text{one-order}}) / \overline{M}_{\text{one-order}}$$

Table 1.
Parametric values for different items

Items	c	p	s	h	Demand information		
					during slot 1	during slot 2	for whole season
1	8	12	4	2	(1300,1500,1700)	(800,1000,1200)	(2100,2500,2900)
2	9	13	4	2	(800,1000,1200)	(500,700,900)	(1300,1700,2100)
3	10	14	4	2	(700,800,900)	(400,500,600)	(1100,1300,1500)

Table 2.
Comparison between one order case and two order case

Items	One order case		Two order case				
	$Q^*_{one-order}$	$\overline{M}(Q^*)$	Q_2^*	$\overline{M}(Q_2^{opt})$	total order quantity	expected profit	PIEP
1	2464	9118.73	982	782	2482	9559.36	4.83%
2	1658	5870.50	679	479	1679	6335.25	7.91%
3	1276	4712.34	488	388	1288	4956.18	5.17%

4. Conclusion

This paper considered a situation in which the single-period 'newsboy type' product may be ordered twice during a period. The reordering policy adopted here is quite relevant with real business environment. As the second order is being made after the realization of demand occurs in first slot, loss of goodwill and resource problem both can be saved. The strategy gives both decision maker's achievement and customer satisfaction.

Acknowledgement

The second author acknowledges the support given by the Department of Science & Technology, Govt. of India (DST/MS/157/01).

References

1. G. Hadley & T.M. Whiten (1963), Analysis of Inventory Systems, Prentice-Hall, Englewood Cliffs, NJ.
2. D. Petrovic, R. Petrovic & M. Vujosevic (1996), Fuzzy models for the newsboy problem, *Int. Journal Production Economics*, 45: 435-441.
3. H. Ishii & T. Konno (1998), A stochastic inventory problem with fuzzy shortage cost, *European Journal of Operational Research,* 106: 90-94.
4. C.H. Hsiesh (2002), Optimization of fuzzy production inventory models, *Inform. Sci.,* 146: 29-40.
5. J.-S. Yao & J. Chiang (2003), Inventory with out backorder with fuzzy total cost and fuzzy storing cost defuzzified by centroid and signed distance, *European Journal of Operational Research,* 148: 401-409.
6. P. Dutta, D. Chakraborty & A.R. Roy (2006), An inventory model for single period products with reordering strategy under fuzzy demand, *Computers and mathematics with applications,* Article in Press.
7. H. S. Lau & Amy H. L. Lau (1997), Reordering strategies for a newsboy-type product, *European Journal of Operational Research,* 103: 557-572.
8. Amy H. L. Lau & H. S. Lau (1998), Decision models for single-period products with two ordering opportunities, *Int. Journal Production Economics*, 55: 57-70.
9. L. Li, S.N. Kabadi & K.P.K. Nair (2002), Fuzzy models for single-period inventory problem, *Fuzzy Sets and Systems*, 132: 273-289.
10. C. Kao & W.K. Hsu (2002), A single period inventory model with fuzzy demand, *Computers and mathematics with applications*, 43: 841-848.
11. C. Carlsson & R. Full'er (2001), On possibilistic mean value and variance of fuzzy numbers, *Fuzzy Sets and Systems*, 122: 315-326.

Fuzzy Logic and its Application in Technology and Management
Editors: Debjani Chakraborty, S. Nanda and D. Dutta Majumder

Constrained Two-person Zero-sum Matrix Games with Fuzzy Goals

Vidyottama Vijay

Department of Mathematics
Indian Institute of Technology, Hauz Khas, New Delhi
Email: mar99005@ccsun50.iitd.ac.in

Abstract: A two-person zero-sum constrained matrix game with fuzzy goals is defined and shown to be equivalent to a primal-dual pair of fuzzy linear programming problems.

Keywords: Fuzzy matrix game, Fuzzy linear programming problem, Fuzzy duality.

1. Introduction

There are certain matrix game theoretic problems in real life where the strategies of the players are constrained to satisfy general linear inequalities. These decision problems give rise to constrained matrix games, which have initially been studied by Charnes [2] and then later in somewhat more generality by Kawaguchi & Maruyama [3].

Recently, Li [4] studied such constrained matrix games in fuzzy scenario where the payoff matrix is fuzzy. However, unlike the crisp scenario, [4] does not establish any equivalence with a pair of primal-dual fuzzy linear programming problem for the given constrained fuzzy matrix game.

The main aim of this paper is to conceptualize a two person zero sum constrained matrix game with fuzzy goals and show its equivalence to a primal-dual pair of fuzzy linear programming problems studied by Bector & Chandra [1].

The paper is organized as follows. Certain basic definitions and preliminaries with regard to crisp constrained matrix games and fuzzy linear programming duality are presented in Section 1. In Section 2, a two-person zero-sum constrained matrix game with fuzzy goals is introduced and its equivalence with an appropriate primal-dual pair of fuzzy linear programming problems is established.

2. Definitions and Preliminaries

Let R^n denote the n-dimensional Euclidean space and R^n_+ be its non-negative orthant. Let $A \in R^{m\times n}$ be an ($m \times n$) real matrix.

Let strategy spaces for Player I and Player II be S_1={ $x \in R^m_+$, $Bx \le c$} and S_2={ $y \in R^n_+$, $D^T y \le d$} respectively and expected payoff is given by

$K(x,y) = x^T A\, y$

where $K: S_1 \times S_2 \to R$ and $x \in R^m, y \in R^n, B \in R^{s\times m}$, $c \in R^s$, $D \in R^{n\times t}$, $d \in R^t$ and $B \in R^{s\times m}$. Then the constrained matrix games CG is defined by $CG = (S_1, S_2, A)$.

Definition 2.1. A point $(x^*, y^*) \in S_1 \times S_2$ is called a solution of the constrained matrix game CG if

$$x^T Ay^* \le x^{*T} Ay^* \le x^{*T} Ay, \forall x \in S_1, y \in S_2,$$

Here $x^{*T} A\, y^*$ is called the value of the constrained game CG.

Taking motivation from the conventional matrix game theory, Kawaguchi & Maruyama [3] established that a two person zero sum constrained matrix game CG is equivalent to the following primal-dual pair of linear programming problems

$$\begin{aligned} &\max \ d^T u \\ &\text{subject to,} \quad -A^T x + Du \le 0, \\ &\qquad Bx \le c, \\ &\qquad x, u \ge 0, \end{aligned} \tag{CLP}$$

and

$$\begin{aligned} &\min \ c^T v \\ &\text{subject to,} \quad -Ay + B^T v \ge 0, \\ &\qquad D^T y \le d, \\ &\qquad y, v \ge 0, \end{aligned} \tag{CLD}$$

Theorem 2.1. A constrained matrix game $CG = (S_1, S_2, A)$ has a solution $(x^*, y^*) \in S_1 \times S_2$ if and only if there exist $u \in R^t, v \in R^s$ such that (x, u) and (y, v) are optimal to the mutually dual pair of linear programming problems (CLP)-(CLD).

Another concept, which will be needed in this sequel, is the development of duality in linear programming under fuzzy environment. Here we present the recently constructed modified model of the fuzzy primal-dual pair of Bector & Chandra [1], which is motivated by Zimmermann [6] and Rodder & Zimmermann [5]. For this let us consider the fuzzy versions of the usual primal and dual problems as given below:

$$\begin{aligned} &\text{Find } z \in R^n \\ &\text{such that } c^T z \,\tilde{\succ}\, c^T z_0, \\ &\qquad Az \,\tilde{\prec}\, b, \\ &\qquad z \ge 0, \end{aligned} \tag{FP}$$

and,

$$
\begin{aligned}
&\text{Find } u \in R^m \\
&\text{such that } b^T u \,\tilde{\prec}\, b^T u_0, \\
&\qquad A^T u \,\tilde{\succ}\, c, \\
&\qquad u \geq 0.
\end{aligned} \tag{FD}
$$

Here '$\tilde{\succ}$' (respectively '$\tilde{\prec}$') is the fuzzified version of '$\geq$' (respectively '$\leq$'), termed as "essentially more than or equal to" (respectively, "essentially less than or equal to"), and is to be understood in the sense of Zimmermann [6] for a chosen membership function. Also $c^T z_0$ and $b^T u_0$ are the aspiration levels for the primal and dual objective functions which may be taken any scalars, say α and β respectively.

Let $p_i > 0$ $(i = 0,1\ldots m)$ and $q_j > 0$ $(j = 0,1\ldots n)$ be subjectively chosen constants of admissible violations for the objective function and constraints of the primal and dual problems respectively. Then choosing the linear membership functions and following Zimmermann's approach [6] the following pair of crisp linear programming problems is obtained for the fuzzy pair (*FP*) and (*FD*):

$$
\begin{aligned}
&\max \quad \lambda \\
&\text{subject to, } \lambda \leq 1 + \frac{c^T z - c^T z_0}{p_0}, \\
&\qquad \lambda \leq 1 + \frac{b_i - A_i z}{p_i}, \qquad (i = 1,2\ldots m), \\
&\qquad z \geq 0,\ \lambda \in [0,1],
\end{aligned} \tag{CP}
$$

and,

$$
\begin{aligned}
&\max \quad \eta \\
&\text{subject to, } \eta \leq 1 + \frac{b^T u_0 - b^T u}{q_0}, \\
&\qquad \eta \leq 1 + \frac{A_j^T u - c_j}{q_j}, \qquad (j = 1,2\ldots n), \\
&\qquad u \geq 0,\ \eta \in [0,1],
\end{aligned} \tag{CD}
$$

Here A_i (respectively A_j) denotes the ith row (respectively j^{th} column) of A. The pair (CP) and (CD) is termed as the modified fuzzy pair of primal-dual linear programming problems. Let $p = (p_1, p_2, \ldots, p_m)^T$ and $q = (q_1, q_2, \ldots, q_m)$. Then the following modified duality theorem is proved in Bector and Chandra [1]:

Theorem 2.2. Let (z, λ) be (*CP*)-feasible and (u, η) be (*CD*)-feasible. Then

$$(\lambda - 1) p^T u + (\eta - 1) q^T z \leq (b^T u - c^T z).$$

3. Two-Person Zero-Sum Constrained Matrix Game with Fuzzy Goals

Let S_1, S_2 and A be as introduced in Section 2. Let v_0, w_0 be scalars representing the aspiration levels of Player I and Player II respectively. Then a constrained matrix game with fuzzy goals denoted by *CFG*, is defined as

$$CFG=(S_1, S_2, A, v_0, \tilde{\succ}, p_0, w_0, \tilde{\prec}, q_0),$$

where, v_0 and w_0 are the scalars representing the aspiration levels of Player I and Player II respectively, p_0 and q_0 are the tolerance levels for Player I and Player II respectively, and '$\tilde{\succ}$' and '$\tilde{\prec}$' are fuzzified versions of '$\geq$' and '$\leq$' respectively. Now onwards we shall call a two person zero sum matrix game with fuzzy goals simply as fuzzy constrained matrix game.

Now we define the meaning of the "solution" of the fuzzy constrained matrix game *CFG*.

Definition 3.1. A point $(x^*,y^*) \in S_1 \times S_2$ is called a solution of the constrained fuzzy matrix game *CFG* if

$$x^{*T} Ay \tilde{\succ}_{p_0} v_0, \ \forall \ y \in S_2,$$

and,

$$x^T Ay^* \tilde{\prec}_{q_0} w_0, \ \forall \ x \in S_1.$$

By using the above definition, we now construct the following fuzzy linear programming problems:

$$(FP1) \ \widetilde{\min} \ (x^{*T} A)\, y, \ \text{subject to, } D^T y \geq d, \ y \geq 0,$$

and

$$(FD1) \ \widetilde{\max} \ x^T (A\, y^*), \ \text{subject to, } Bx \leq c, \ x \geq 0,$$

Now similar to Kawaguchi and Maruyama [3] for the crisp case, we introduce the following two fuzzy linear programming problems corresponding to the pair (*FP1*) and (*FD1*):

$$(\widetilde{CLP}) \ \text{Find } (u, x) \text{ such that } d^T u \tilde{\succ}_{p_0} v_0, \ -A^T x + Du \leq 0, \ Bx \leq c; x, u \geq 0,$$

and

$$(\widetilde{CLD}) \ \text{Find } (v, y) \text{ such that } c^T v \tilde{\prec}_{q_0} w_0, -Ay + B^T v \geq 0, D^T y \geq d; y, v \geq 0.$$

Now we define the membership function $\mu(d^T u)$, which gives the degree to, which u satisfies the fuzzy constraint $d^T u \tilde{\succ}_{p_0} v_0$, as follows

$$\mu(d^T u) = \begin{cases} 1 & , \ d^T u \geq v_0, \\ 1 + \dfrac{d^T u - v_0}{p_0} & , \ v_0 - p_0 < d^T u < v_0, \\ 0 & , \ d^T u \leq v_0 - p_0. \end{cases}$$

This choice of $\mu(d^T u)$ gives the crisp formulation of the fuzzy linear programming problem $(\widetilde{CLP})$ as

$$\begin{aligned}
&\max \quad \lambda,\\
&\text{subject to,} \quad \lambda \le 1 + \frac{d^T u - v_0}{p_0},\\
&\qquad -A^T x + Du \le 0, \qquad\qquad (FLP)\\
&\qquad Bx \le c,\\
&\qquad x, u \ge 0.
\end{aligned}$$

Similarly, the membership function $\nu(c^T v)$, which defines, the degree to which v satisfies the constraint $c^T v \,\tilde{\prec}_{q_0}\, w_0$ is taken as

$$\nu(c^T v) = \begin{cases} 1 & , c^T v \le w_0 ,\\ 1 + \dfrac{w_0 - c^T v}{q_0} & , w_0 < c^T v < w_0 + q_0,\\ 0 & , c^T v \ge w_0 + q_0, \end{cases}$$

$$\begin{aligned}
&\max \quad \eta\\
&\text{subject to,} \quad \eta \le 1 + \frac{w_0 - c^T v}{q_0},\\
&\qquad -Ay + B^T v \ge 0, \qquad\qquad (FLD)\\
&\qquad D^T y \ge d,\\
&\qquad y, v \ge 0.
\end{aligned}$$

From the above discussion we therefore observe that for solving the fuzzy constrained matrix game *CFG* we have to solve the crisp linear programming problems (*FLP)* and (*FLD*) for Player I and Player II respectively. Here it can be verified that (*FLP*) and (*FLD)* constitute a fuzzy primal-dual pair in the sense of [1].

The above can be summarized in the form of the following theorem:

***Theorem** 3.1.* The fuzzy constrained matrix game *CFG* described by CFG=$(S_1, S_2, A, v_0, \tilde{\succ}, p_0, w_0, \tilde{\prec}, q_0)$ is equivalent to two crisp linear programming problems (*FLP*) and (*FLD*) which constitute a primal-dual pair in the sense of duality for linear programming in a fuzzy environment.

References

1. C.R. Bector & S. Chandra (2000), On duality in linear programming under fuzzy environment, *Fuzzy Sets and Systems*, 125: 317-325.
2. A. Charnes (1953), Constrained games and linear programming, *Proc. Nat. Acad. Sci.*, 39: 639-641.
3. T. Kawaguchi & Y. Maruyama (1976), A note on minimax (maximin) programming, *Management Science*, 22: 670-676.
4. Dengfeng Li (1999), Fuzzy constrained matrix games with fuzzy payoffs, *The Journal of Fuzzy Mathematics*, 7: 873-880.
5. W. Rodder and H.-J. Zimmermann (1980), Duality in Fuzzy Linear Programming, In A.V. Fiacco and K.O. Kortanek (eds.), *External Methods and System Analysis*. Springer-Verleg, Heidelberg.
6. H.-J. Zimmermann (1978), Fuzzy Programming and Linear Programming with Several Objective Functions, *Fuzzy Sets and Systems*, 1: 45-55.

Fuzzy Logic and its Application in Technology and Management
Editors: Debjani Chakraborty, S. Nanda and D. Dutta Majumder

Solution of m×n Rectangular Fuzzy Games by Dominance

Prasun K. Nayak[1] *and Madhumangal Pal*[2]

[1]Department of Mathematics, Bankura Christian College, Bankura- 722101
[2]Department of Applied Mathematics with Oceanology and Computer Programming
Vidyasagar University, Midnapore-721 102, West Bengal
Email: madhumangal@lycos.com

Abstract: A solution of $m \times n$ rectangular fuzzy game with pay-off as imprecise numbers instead of crisp numbers namely interval and triangular fuzzy numbers are considered here. Solution of such fuzzy games with pure strategies by minmax–maxmin principle is discussed. The algebraic method to solve 2×2 fuzzy game without saddle point by using mixed strategies is also discussed. Here the $m \times n$ pay-off matrix is reduced to 2×2 pay of matrix by dominance method. One numerical example is provided to illustrate the method.

Keywords: Interval number, Triangular fuzzy number, Fuzzy game, Dominance.

1. Introduction

Solution of fuzzy games with pay-off as imprecise number was discussed in [1], by minmax–maxmin principle. Also the algebraic method of solving $m \times n$ rectangular fuzzy games with pay-off as interval numbers having no saddle point was introduced in [1].

Determination of 2×2 fuzzy games from a rectangular $m \times n$ fuzzy game without saddle point is a fundamental problem in 'fuzzy game theory'. In classical game theory the pay-off is crisp number. Practically, it may happen that the pay-off is not necessarily a fixed real number. In this paper, we take into consideration the pay-off as triangular fuzzy numbers and have tried to reduce the $m \times n$ matrix to 2×2 matrix.

2. Interval number and its arithmetic

An interval number is defined as $I = [X_L, X_R] = \{x : X_L \le x \le X_R, x \varepsilon \Re$, the set of all real numbers$\}$. Another way of representation of an interval number in terms of midpoint and width is $I = < m(i), w(i) >$ where $m(i) =$ midpoint of $I = (X_L + X_R)/2$ and $w(i) =$ half width of $I = (X_R - X_L)/2$. Addition of two interval-numbers $I = [X_L, X_R]$ and $J = [Y_L, Y_R]$ is $I + J = [X_L + Y_L, X_R + Y_R]$. Using mean width notations, if $I = < m_1, w_1 >$ and $J = < m_2, w_2 >$ then $I + J = < m_1 + w_1, m_2 + w_2 >$. Similarly, the other binary operations on interval numbers are defined.

Order relation among intervals [2]: If $I=[a, b]$, $J=[c, d]$, then $[a, b]<[c, d]$ if and only if $b<c$ and is denoted by $I<J$. I is contained in J if and only if $a \geq c$ and $b \leq d$ and this is denoted by $I \leq J$.

Definition 1. The 'dominated index' (DI) to the proposition I is dominated over J as

$$DI(I<J)=(m_2-m_1)/(w_1+w_1).$$

Using this DI we may define the following ranking order,

Definition 2. If $DI(I<J) \geq 1$, I is said to be 'totally dominating' over J in the sense of minimization and J is said to be 'totally dominating' over I in the sense of maximization, we denote this by $I<J$.

Definition 3. If $0<DI(I<J)<1$, then I is said to be 'partially dominating' over J in the sense of minimization and J is said to be 'partially dominating' over I in the sense of maximization. This is denoted by $I<J$.

When $DI(I<J)=0$, then $m_1=m_2$, we may emphasize on the width of the interval numbers I and J. If $w_1<w_2$, then the left end point of I is less than that of J and there is a chance that on finding a minimum distance, the distance may be on I. But at the same time, since the right end point of I is greater than that of J, if one prefers I over J in minimization then in worst case, he may be looser than one who prefer J over I.

Numerical Example: If $I=[110, 120]=\langle 115, 5\rangle$, $J=[150, 155]=\langle 152.5, 2.5\rangle$ $DI(I<J)=(152.5-115)/2>1$. So in minimization I is totally dominating over J.

A *triangular fuzzy number* can be written as (α, a, β), where α and β are left and right hand spreads respectively.

Definition 4. The dominated index (DI) of the proposition A = (α, a, β) is dominated over B = (γ, b, δ) is given by

$$DI(A<B)=(b-a)/(\beta+\gamma).$$

Using this DI index we may define the following ranking order.

Definition 5. If $DI(A<B) \geq 1$, A is said to be 'totally dominating' over B in the sense of minimization and B is said to be 'totally dominating' over A in the sense of maximization, we denote this by $A<B$.

Definition 6. If $0<DI(A<B)<1$ then A is said to be 'partially dominating' over B in the sense of minimization and B is said to be 'partially dominating' over A in the sense of maximization, we also denote it by $A<B$.

Lemma 1. If $DI(A<B)=0$ then A and B are said to be 'non comparable' and is denoted by $A \neq B$. In this case 'A is preferred over B' if (a) $\alpha=\gamma$ and $\beta<\delta$ or (b) $\alpha>\gamma$ and $\beta=\delta$, otherwise a pessimistic decision maker would prefer the number with smaller length of the support whereas an optimistic decision maker would do the converse.

Numerical example (1) If $A=<6, 0.2>$ and $B=<1, 0.2>$, $DI(A<B)=(6-1)/(0.2+0.2)>1$. Thus A is totally dominating over B in the sense of minimization, and B is said to be totally dominating over A in the sense of maximization.
(2) If $A=<0.4, 0.5>$ and $B=<0.3, 0.5>$, $DI(A<B)=(0.4-0.3)/(0.5+0.5)=0.1$. So A is said to be partially dominating over B.

3. Two-person Zero-sum Game

We shall consider the games relating to pure strategies (a particular executive) taken by the players. We further make two assumptions:

(*i*) The player *A* is in better position. The player *A* is called maximization player (or row player) and player *B* is called a minimizing player (or column player).

(*ii*) The total gain of one player is exactly equal to the total loss of other player.

In general, if the player *A* takes *m* pure strategies and *B* takes *n* pure strategies, then the game is called 'two person zero sum game' or $m \times n$ rectangular game.

4. Pay-off matrix

A pay-off matrix is always written for a maximizing player. Now consider the general $m \times n$ rectangular game, the pay-off matrix of *A* with *A*'s pure move A_1, A_2, ... A_m, and *B*'s pure moves B_1, B_2, ... B_n, is

$$\begin{array}{c} \\ A_1 \\ A_2 \\ \vdots \\ \vdots \\ A_m \end{array} \begin{array}{c} \begin{array}{ccccc} B_1 & B_2 & & & B_n \end{array} \\ \begin{bmatrix} <a_{11}, b_{11}> & <a_{12}, b_{12}> & \cdots & \cdots & <a_{1n}, b_{1n}> \\ <a_{21}, b_{21}> & <a_{22}, b_{22}> & \cdots & \cdots & <a_{2n}, b_{2n}> \\ \vdots & \vdots & \cdots & \cdots & \vdots \\ \vdots & \vdots & \vdots & \vdots & \vdots \\ <a_{m1}, b_{m1}> & <a_{m2}, b_{m2}> & \cdots & \cdots & <a_{mn}, b_{mn}> \end{bmatrix} \end{array}$$

The elements $<a_{ij}, b_{ij}>$ are symmetric triangular fuzzy numbers and for crisp game they may be positive, negative or zero. When a player *A* chooses strategy A_i and *B* selects B_j, it results in a pay-off of the triangular fuzzy number $<a_{ij}, b_{ij}>$ to the player *A*.

5. Solution of *2* × *2* games (mixed strategies)

Consider the fuzzy game whose pay-off is given by the adjacent matrix and for which there is no saddle point.

$$\begin{array}{c} \\ A_1 \\ A_2 \end{array} \begin{array}{c} \begin{array}{cc} B_1 & B_2 \end{array} \\ \begin{bmatrix} <a_{11}, b_{11}> & <a_{12}, b_{12}> \\ <a_{21}, b_{21}> & <a_{22}, b_{22}> \end{bmatrix} \end{array}$$

where the pay-off $<a_{ij}, b_{ij}>$ are symmetric triangular fuzzy numbers so that $b_{ij} = a_{ij} + \alpha$. If x_i and y_j be the probabilities with which *A* chooses his i^{th} strategy and *B* chooses his j^{th} strategy respectively, then

$x_1 = (a_{22} - a_{21})/(a_{11} + a_{22} - a_{12} - a_{21})$, $x_2 = (a_{11} - a_{12})/(a_{11} + a_{22} - a_{12} - a_{21})$,

$y_1 = (a_{22} - a_{21})/(a_{11} + a_{22} - a_{12} - a_{21})$, $y_2 = (a_{11} - a_{12})/(a_{11} + a_{22} - a_{12} - a_{21})$,

which are crisp numbers and the value of the game can be easily computed as *V* = <*a*, *b*> where, *b* = left and right hand spreads of triangular fuzzy numbers and

$$a = (a_{11}a_{22} - a_{12}a_{21})/(a_{11} + a_{22} - a_{12} - a_{21}),$$
$$b = (b_{11}b_{22} - b_{12}b_{21})/(b_{11} + b_{22} - b_{12} - b_{21}).$$

6. Game with no saddle point

The simplest case in a 2×2 fuzzy game with no saddle point is discussed in [1]. Here we consider an $m\times n$ fuzzy game. Now we discuss a particular method. In this method, the pay-off matrix can be reduced to 2×2 game so that it can be solved by using the method of [1]. The method of reduction of the pay-off matrix by this process is called the *dominance property* of the rows and columns of the pay-off matrix.

7. Concept of dominance

In game theory, if one pure strategy of one player be better for him then or as good as another, for all possible pure strategies of the opponent then the first is said to dominate the second. The dominated strategy can simply be discarded from the pay-off matrix, since it has no value. When this is done, the optimal strategies for the reduced matrix are also optimal for the original matrix with zero probability for the discarded strategies. When there is no saddle point in a pay-off matrix, then we can reduce the size of game by the 'dominance', before we set to solve the game problem.

Definition 7. If all the elements of the ith row of the pay-off matrix of an $m\times n$ rectangular game are dominating over r^{th} row in the sense of maximum, we discard r^{th} row and the deletion of the rth row from the matrix does not change the set of optimal strategies of the maximizing player.

Numerical example. Consider, for example, the fuzzy game with pay-off matrix

$$\begin{matrix} A_1 \\ A_2 \\ A_3 \end{matrix} \begin{bmatrix} <1,0.2> & <7,0.3> & <2,0.1> \\ <6,0.2> & <2,0.1> & <7,0.3> \\ <0,0.2> & <1,0.2> & <6,0.2> \end{bmatrix}$$

Here A is maximization player and B is minimization player. Now

$$DI(A_{31} < A_{21}) = (6-0)/(0.2+0.2) > 1,$$
$$DI(A_{32} < A_{22}) = (2-1)/(0.4+0.2) > 1,$$
$$DI(A_{33} < A_{23}) = (7-6)/(0.3+0.2) > 1.$$

So A_2 is dominating over A_3 in the sense of maximization. So we delete the row A_3. The reduced matrix is given by

$$\begin{bmatrix} <1,0.2> & <7,0.3> & <2,0.1> \\ <6,0.2> & <2,0.1> & <7,0.3> \end{bmatrix}.$$

Definition 8. If all the elements of the jth column are dominating over the sth column in the sense of minimization, we delete the sth column and the deletion of the sth column from the matrix does not change the set of optimal strategies of the minimizing player.

Numerical example. Consider the above pay-off matrix

$$DI(B_{11} < B_{13}) = (2-1)/(0.1+0.3) > 1,$$

$DI(B_{21} < B_{23}) = (7-6)/(0.2+0.3) > 1$

B_1 is totally dominating over B_3 in the sense of minimization. The resultant pay-off matrix is given by

$$\begin{array}{c} \\ A_1 \\ A_2 \end{array} \begin{array}{c} \begin{array}{cc} B_1 & B_2 \end{array} \\ \begin{bmatrix} <1, 0.2> & <7, 0.3> \\ <6, 0.1> & <2, 0.1> \end{bmatrix} \end{array}.$$

Definition 9. If the linear combination of pth and q^{th} rows dominates all the elements of the sth row in the sense of maximization, we discard s^{th} row and the deletion of the sth row from the matrix does not change the set of optimal strategies of the maximizing player.

Numerical example. Consider a particular pay-off matrix

$$\begin{array}{c} \\ A_1 \\ A_2 \\ A_3 \end{array} \begin{array}{c} \begin{array}{ccc} B_1 & B_2 & B_3 \end{array} \\ \begin{bmatrix} <1, 0.4> & <2, 0.1> & <-1, 0.1> \\ <3, 0.5> & <1, 0.3> & <2, 0.2> \\ <-1, 0.2> & <3, 0.4> & <2, 0.4> \end{bmatrix} \end{array}.$$

Using a convex combination of the second and third row we have $A_4 = \beta A_2 + (1-\beta)$ A_3; $0 \le \beta \le 1$. Taking $\beta = .5$ we get the elements of A_4 as < 1, .35 >, < 2, .35 >, < 2, .3 >.

Now A_4 is dominating A_1 in the sense of maximization, we discard the row A_1, the resulting pay-off matrix is given by

$$\begin{array}{c} \\ A_2 \\ A_3 \end{array} \begin{array}{c} \begin{array}{ccc} B_1 & B_2 & B_3 \end{array} \\ \begin{bmatrix} <3, 0.5> & <1, 0.3> & <2, 0.2> \\ <-1, 0.2> & <3, 0.4> & <2, 0.4> \end{bmatrix} \end{array}.$$

Definition 10. If the jth column is dominated by the convex combination of m and n^{th} column, we discard the jth column in the sense of minimization and the deletion of the jth column from the matrix does not change the optimal strategies of the minimizing player.

Numerical example. Consider the above pay-off matrix. The convex combination of B_1 and B_2 i.e., $B_4 = aB_1 + (1-a)B_2$; $0 \le \alpha \le 1$. Taking $\alpha = 0.5$, we get the elements of B_4 as

$$\begin{bmatrix} <2, 0.4> \\ <1, 0.3> \end{bmatrix}.$$

Now B_4 is totally dominating over B_3 and thus drop the third column, the resulting matrix is given by

$$\begin{array}{c} \\ A_2 \\ A_3 \end{array} \begin{array}{c} \begin{array}{cc} B_1 & B_2 \end{array} \\ \begin{bmatrix} <3, 0.5> & <3, 0.4> \\ <-1, 0.2> & <1, 0.3> \end{bmatrix} \end{array}.$$

Definition 11. When there is no saddle point and no course of action dominates any other, we compute the 'values' for different 2×2 sub games. As A is maximizing player he will definitely select that pair strategies which will give the best 'value' of 2×2 sub games and the corresponding sub matrix provides the optimal solution.

Similarly, as B is minimizing player he will definitely select that pair of courses, which will give the least 'value' of the 2×2 sub games, the corresponding sub matrix will provide the optimal solution to the given fuzzy game problem.

Numerical example. Consider the fuzzy game whose pay-off matrix is given by

$$\begin{bmatrix} <19, 0.2> & <15, 0.4> & <16, 0.1> \\ <0, 0.2> & <20, 0.4> & <5, 0.4> \end{bmatrix}.$$

There is no saddle point and no course of action dominates any other. We compute the values V_1, V_2, V_3 of the following three 2×2 sub games as obtained from the given matrix.

Sub game 1

$$\begin{bmatrix} <19, 0.2> & <15, 0.4> \\ <0, 0.2> & <20, 0.4> \end{bmatrix}.$$

The corresponding value $(V_1) = < 95/6,\ 7/5 >$.

Sub game 2

$$\begin{bmatrix} <19, 0.2> & <16, 0.1> \\ <0, 0.2> & <5, 0.4> \end{bmatrix}.$$

The saddle point solution value $(V_2) = <16, 0.1>$

Sub game 3

$$\begin{bmatrix} <15, 0.4> & <16, 0.1> \\ <20, 0.4> & <5, 0.4> \end{bmatrix}.$$

The corresponding value $(V_3) = < 245/16,\ 11/5 >$

Here min $\{V_1, V_2, V_3\} = V_3$ so the resulting pay-off matrix is given by

$$\begin{array}{cc} & \begin{array}{cc} B_2 & \quad B_3 \end{array} \\ \begin{array}{c} A_1 \\ A_2 \end{array} & \begin{bmatrix} <15, 0.4> & <16, 0.1> \\ <20, 0.4> & <5, 0.4> \end{bmatrix} \end{array}.$$

8. Illustrative example

To illustrate the 'dominance' method, we consider the following pay-off matrix

$$\begin{array}{cc} & \begin{array}{cccc} B_1 & \quad B_2 & \quad B_3 & \quad B_4 \end{array} \\ \begin{array}{c} A_1 \\ A_2 \\ A_3 \end{array} & \begin{bmatrix} <8, 0.3> & <15, 0.4> & <-4, 0.1> & <-2, 0.4> \\ <19, 0.1> & <15, 0.5> & <17, 0.4> & <16, 0.1> \\ <0, 0.3> & <20, 0.2> & <15, 0.5> & <5, 0.4> \end{bmatrix} \end{array}.$$

All $DI\ (A_1 < A_2\ ) \geq 1$, so A_2 is totally dominating over A_1 in the sense of maximization we delete the row A_1. The resulting pay-off matrix is

$$\begin{bmatrix} <19,0.1> & <15,0.5> & <17,0.4> & <16,0.1> \\ <0,0.3> & <20,0.2> & <15,0.5> & <5,0.4> \end{bmatrix}$$

Again all $DI(B_4 < B_3) \geq 1$ so B_4 is totally dominating over B_3 in the sense of minimization, so we delete the column B_3, the resulting pay-off matrix is

$$\begin{bmatrix} <19,0.1> & <15,0.5> & <16,0.1> \\ <0,0.3> & <20,0.2> & <5,0.4> \end{bmatrix}$$

This is no course of action dominates any other and there is no saddle point. We compute the values for different 2×2 pair of strategies. Since B is the minimizing player we take the minimum 'value' and the corresponding pay-off matrix will provide the optimal solution to the fuzzy problem. By the previous the least value of $\{V_1, V_2\ V_3\} = V_3$, so the optimal strategies of A are (A_2, A_3) and that of B are (B_2, B_3). The final pay-off matrix

$$\begin{array}{cc} & \begin{array}{cc} B_2 & B_3 \end{array} \\ \begin{array}{c} A_2 \\ A_3 \end{array} & \begin{bmatrix} <15,0.5> & <16,0.1> \\ <20,0.1> & <5,0.4> \end{bmatrix} \end{array}.$$

The probabilities are $x_1 = 0,\ x_2 = 15/16,\ x_3 = 1/16,\ y_1 = 0,\ y_2 = 11/16,\ y_3 = 0,\ y_4 = 5/16$, and the value of the game is $V = <245/16,\ 11/5>$. Hence the optimal solution of the complete game is $(0, x_1, x_2)$ for A and $(0, y_2, 0, y_4)$ for B, the value of the game being V.

References

1. A.L. Narayan, A.R. Meenakshi, & A.M.S. Ramasamy (2002), Fuzzy games, *The Journal of Fuzzy Mathematics,* 10: 817 – 829.
2. A. Sengupta & T.K. Pal (2000), On comparing interval numbers, *European Journal of Operational Research,* 127: 28 – 43.
3. D. Dubois and H. Prade (1980), *Fuzzy Sets and Systems: Theory and Applications,* Academic Press, New York, U.S.A.
4. G.J. Klir & B. Yuan (1995), *Fuzzy Sets and Fuzzy logic, Theory and Applications*, Prentice-Hall Inc., Englewood Chiff, N.J., U.S.A.

Part 4

Application of Fuzzy Logic in Technology & Management

Fuzzy Logic and its Application in Technology and Management
Editors: Debjani Chakraborty, S. Nanda and D. Dutta Majumder

Logic, Mathematics and Reality: A Cybernetics and Systems Approach

Dwijesh Dutta Majumder

Professor Emeritus, Indian Statistical Institute (ISI) &
Director-Secretary, Institute of Cybernetics Systems & Information Technology
(ICSIT) Kolkata – 700108

Extended Summary

After a brief introduction to different approaches to Logic and Mathematics, classical and modern, hard and soft, two-valued and many valued, Cybernetics and Synergetics, Physics and Philosophy, we intend to indicate the need to develop theories and methods to reduce the gap between the two so that the computational science and Information Technology can represent the reality or "relative realities" in science and technology, for decision making in technology and management.

Alan Turing while giving the proof of unsolvability of the Hilbert's Decision Problem, describes the limit of algorithmic problem solving and laid the foundation of modern computing science. But that created an incorrect belief that all current and future real computers are captured by Turing Machines (TM).

There exist many models like Cellular Automata, Neural Networks, Modern Internet Machines, $ Calculus, Quantum Computing, Evolutionary Computing etc. by which TM unsolvable problems can be tackled.

In the opinion of the present author the evolution of mathematics and logic is intrinsically linked with the evolution of qualitative/quantitative notions of consciousness and so also with the evolution of languages and scripts. The evolution of civilizations consisted of natural and bio-physico-social development of patterns and forms in the physical or external world and their conceptualization and comprehension by sense organs and their representation in the internal/mental world – that led to the evolution of languages, scripts, arts, culture, religion, philosophy, science etc.

Similarly, the evolution of mathematical systems (hard ones like: Arithmetic, Algebra, Geometry, Calculus, Trigonometry, etc., and soft ones like: Statistics, Probability, Fuzzy Set Theory, Rough Sets, Artificial Neural Network, Evolutionary Computing, Quantum Computing etc.) – stands forth as that which mediates between human mind and environment including nature and society, inner and outer world and our thought and perception.

The Cybernetics and Systems approach differs significantly from the conventional approach. Norbert Wiener established Cybernetics as a science of control and communication, H. Haken proposed synergetics as a science of self organization and D. Dutta Majumder in his formalism on "Cybernetics and General Systems Theory –

A unitary Science" (Norbert Wiener Award Winning paper) extended the approach in laying the foundation for unifying the observing systems, the observed systems and the act of observing – to include integrated complex human-machine-nature in a holistic systems. Next in the context of artificial intelligence and artificial consciousness research in relation to pattern recognition, computer vision and man-machine communication systems – we try to deal with evolution of consciousness, mind-body duality and a pluralistic view of 'Reality'. At this point I wish to propose a soft computing based emotion/intension/gesture recognition system for an intelligent man-machine interface for service robots like house keeping robots, entertainment robots, rehabilitation robots for disabled etc., which will have to have the ability to recognize a combination of bio-signals such as voice, gesture, ECG, EEG, EMG, facial expressions etc. The proposed soft computing toolbox may contain fuzzy logic (FL), artificial neural network (ANN), rough set (RS), evolutionary computation (EC), probabilistic reasoning (PR), belief network (BN), chaos theory (CT) – out of which a complementary combination will be chosen. In order to develop human-machine interfaces for service – robot systems that can be partners with humans in farms, factories, laboratories, domestic works – to help the users mentally and emotionally it has to understand the user's emotion and react depending on the monitored information. An understanding of the evolution of human consciousness – its functioning of internal and external world views about 'realities', is a necessary condition for such developments. The Unitary formalism of Cybernetics and General Systems Theory as proposed by D. Dutta Majumder mentioned earlier provides such a methodology, - which extending Progogine's theory of Non-Equilibrium Thermodynamics to Non-Equilibrium Psychodynamics a mathematical model has been proposed (Roy and Dutta Majumder 1998,2000) and applied to neurocognitive domain (Roy and Dutta Majumder 1999 and 2000).

We have analyzed how psychodynamics transformation is associated with changes in thinking, psycholinguistic, psychovisual and neurodynamics of brain functions.

We now describe how the psychodynamic transformation is associated with alteration of cognitive space and time and perceptual 'geometries' and 'logics' with some simple example. The plural term 'geometries' or 'logics' imply that there could be several varieties of internal representation and cognitive realities depending on psychodynamic transformation. For example in Fig.1, in reality of Riemanian (elliptical) geometry (one sees in a converging lens or car driver's mirror), a triangle displays this property in radian:

Riemanian world: $< A + < B + < C < 2\Pi$ (1a)

In Lobacheveskian (hyperbolic) reality, such as in a diverging lens (or microscope's mirror) the corresponding property is:

Lobacheveskian World $< A + < B + < C > 2\Pi$ (1b)

In Euclidian (linear or classical) reality, such as that observed in a plane mirror, the relevant characteristic is the basis of school geometry:

Euclidian World $< A + < B + < C = 2\Pi$ (1c)

We may encounter much more counterintuitive anomalous experience than that of non-Euclidian geometries of the Riemanian or Lobacheveskian type in non-standard logics like quantum logic, transfinite logic, fuzzy logic, and dream logic. It is well known that in these and other non-classical logics, the basic foundation of classical

Euclidean Space: Flat geometry, e.g., a plane (Three angles of a triangle = 180°)

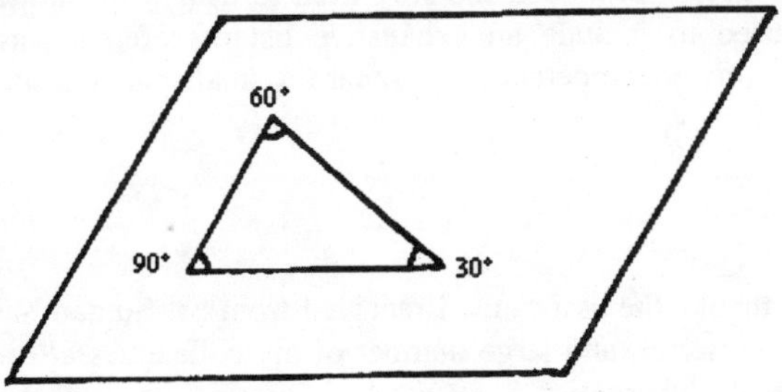

$\Delta = 180°$

$\kappa = 0$

Riemannian Space: Positively curved geometry, e.g. sphere
(Three angles of a triangle > 180°)

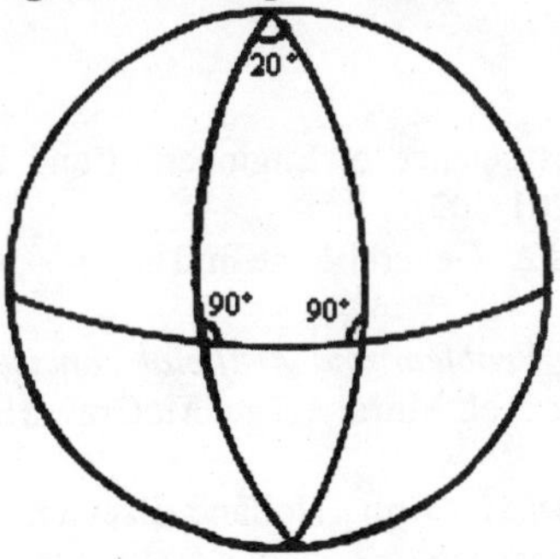

$\Delta = 200°$

κ = positive

Lobachevskian Space: Negatively curved space, e.g. saddle
(Three angles of a triangle < 180°)

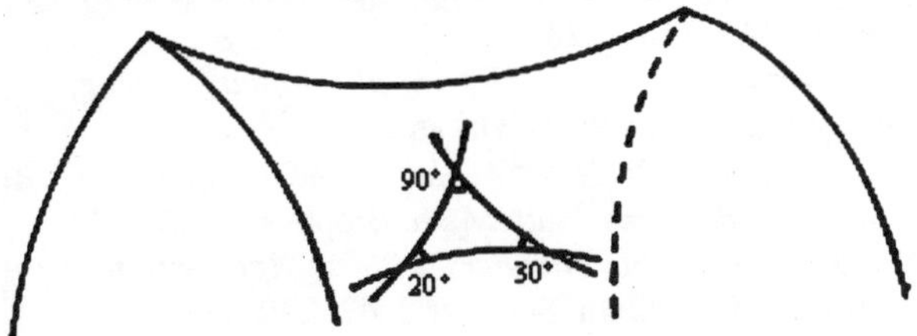

$\Delta = 140°$

κ = negative

κ denotes Curvature, i.e. degree of curvedness

Figure 1: Reality as moldable, mutable and malleable: Different realities with different laws, spaces and truth system

logic, i.e. the Principle of Exclude Middle Law or "Tertium non datur" may not hold – which implies that an object satisfies a property as well as a contradictory property.

The problem of perception and cognition of reality exists in epistemological contrapositions in science and philosophy, as evidenced in famous dialogues such as the Newton–Goethe dichotomy, Bohr–Einstein debate, Tagore–Einstein dialogues etc. In the second half of twentieth century several authors like, P.C. Mahalanobis, L.A. Zadeh, Raja Ramanna, Dutta Majumder and Roy tried to deal with Statistical and Fuzzy logics. I have tried to include an exhaustive list of reference as an unsatisfactory substitution of my incompetence to prepare a small paper on a big topic.

Acknowledgement

I wish to acknowledge with thanks the assistance I received from Sri Suman Sarkar of ICSIT in preparing the manuscript, and large number of my colleagues at Indian Statistical Institute, Institute of Cybernetics Systems And Information Technology and other academic bodies of India and abroad for the helpful discussion I had with them.

References

1. D. Dutta Majumder (1975), Cybernetics science of Engineering and Biology, Cybernetica, *Int. Assoc. Cybern.* 18(3): 201-207.
2. D. Dutta Majumder (1979), Cybernetics & General System Theory – Aunitary Science. *KYBERNETES*, 8: 7-15.
3. D. Dutta Majumder (1933), *Mind-body Problem and Artificial conciseness: A Cybernetic Approach*, Eds. A. Ghosal & P.N. Murthy, Tata McGraw Hill, New Delhi :337-345.
4. August stern (1992), *Matrix Logic And Mind*, North - Holland Elsevier.
5. N. Wiener (1961), *Cybernetics*, MIT Press, N.Y.
6. N. Wiener (1948), *Computing Machines and Nervous Systems, Cybernetics* MIT Press.
7. H. Von. Foerester & G. Zopf (1962), *Proc. Symp. On Principles of Self Organization*, Pergamon Press, N.Y., 181-228.
8. C. E. Shanon & W. Weaver (1949), *The Mathematical Theory of communication*, The university of Illionis Press, Urbana 3.
9. W.H. Pitts & W.S. McCulloch (1947) *How we know the universals – the Perception of Audiotory and Visual Forms*, Bull. Math. Biophys, 9: 124-147.
10. Allan M. Turing (1936), *On Computable Numbers with an Application to the Entscheidungs problem,* Proc. London Math. Soc. Ser.2, 42: 230-265.
11. Davis Martin (1958), *Computability and unsolvability*, McGraw Hill, N.Y.
12. John Von Neumann (1965), *Probabilistic Logics and the Synthesis of Reliable organisms from unreliable components*, Automata Studies (eds) C.E. Shannon & J. Mc Carthy, Princeton university Press, N.J., 43-98.
13. Jefferson Hane Weaver, *The World of Physics vol 2, The Einstein universe and the Bohr Atom, Simon and Schuster*, New York.
14. D. Dutta Majumder & B.B. Chowdhury (1980), Fuzzy Sets and Possibility Theory in Reliability Studies of Man-Machine Systems: in *Fuzzy Set Theory and Applications to Policy Analysis and Information systems*, Eds. P.P. Wang & S.K.Chang, Plenam Press, N.Y.

15. D. Dutta Majumder (1984), On the Relevance of Fuzzy Mathematical Systems, in: *Human Systems Management*, N.Y. vol 4, pp 301-324.
16. D. Dutta Majumder (1993), Fuzzy Mathematics and Uncertainty Management for Decision Making in Science Society, *Journal of Computer Science and Informatics*, 23(3): 1-31.
17. D. Dutta Majumder (1993), Mind-Body Problem And Artificial Consciousness for Computing Machines: A Cybernetic Approach, in: *Recent Advances in Cybernetics and systems*, (Eds) A. Ghosal & P.N. Murthy, Tata McGraw Hill, New Delhi – Jan. 18-23, pp 337-345.
18. D. Dutta Majumder (1933), *Mind-Body Duality: Its Impact on Pattern Recognition And Computer Vision Research*, Third APRDT, I.S.I. P.C. Mahalanobis Birth Centenary Volume, Dec., pp 3-17.
19. D. Dutta Majumder (1992), *Digital Computers Memory Technology*, Wiely Eastern (New Delhi) & John Wiley & Sons (New York, London), 3rd Edition.
20. D. Dutta Majumder (1985), *Fuzzy Sets in Pattern recognition, Image Analysis, Automatic Speech Recognition*, Aplikace Matematiku, Chekoslovak Academy of Sciences, Vol. 30: 237-254.
21. D. Dutta Majumder (1986), Pattern Recognition, Image Processing and Computer Vision in Fifth Generation Computer Systems, *SADHANA*, Indian Academy of Sciences, 9 (2): 139-156.
22. D. Dutta Majumder (1986), An Unified view of Pattern Recognition, Image Processing, Computer Vision and Artificial Intelligence, Proc. Nato Advanced Research Workshop on *Pyramidal Systems for Image Processing and Computer Vision* (Ed. S. Levialdi), Plenum Press, Maratea-Napples, Italy.
23. D. Dutta Majumder (1988), An Unified Approach to Artificial Intelligence, Pattern Recognition, Image Processing, and Computer Vision in Fifth generation Computer System, *Information Sciences*, 45: 391-431.
24. S. K. Paul & D. Dutta Majumder (1986), *Fuzzy Mathematical Approach to Pattern Recognition*, John Wiley & Sons. New York
25. D. Dutta Majumder (1993), Scientific contributions of Professor P. C. Mahalanobis, *Current Science*, 65(1).
26. P.C. Mahalanobis (1950), Why Statistics?, *Address of the General President, Indian Science Congress, Poona.*
27. P.C. Mahalanobis (1944), *Philos. Trans. R. Soc.*, London, B-231: 329-451.
28. N. Wiener (1961), *Cybernetics.* The MIT Press and John Wiley & Sons. Inc. New York, London 2nd edition.
29. L. Von Bertalnanfy (1950), An outline of general systems theory, *Brit. J. Philos, Scientist*, 1(2).
30. M.D. Mesarovic (1960), *Towards the development of general systems theory*, Neue Technik(1963) and *The Control of Multivariable Systems*, MIT Press and John Wiley and Sons, New York.
31. M.D. Mesarovic & D.P. Eckman (1961), On some basic concept of a general systems theory, Proc. *3rd Int. Conf. Cybernetics*, Namur.
32. D. Dutta Majumder (1975), Cybernetics, a science of engineering and biology, *Cybernetica* (Namur) XVIII, No.3.
33. D. Dutta Majumder (1979), Cybernetics and General Systems – A Unitary Science? *Kybernetes*, 8: 7-15.

34. D. Dutta Majumder (1984), Post Wiener Cybernetics, Proc *Sixth Int. Congress of Cybernetics & Systems*, Paris, AFCET.
35. J.R. Searle (1980), Minds brains and programs with open peer commentaries. *The Behavioral and Brain Sciences*, 3: 417-457.
36. J.R. Searle (1990), Is the brain's mind a computer program? *Scientific American*, 262: 26-31.
37. R. Penrose (1989), *The Emperors' New Mind*, Oxford University Press, Oxford, England.
38. S.A. Klein (1991), *The Duality of psychophysics, Representations of vision*, Ed. Andrel Gorea, Cambridge University Press, Cambridge, pp 231-249.
39. W.J. Freeman (1990), *On the Fallacy of assigning an origin to consciousness, Machinery of the Mind*, Ed. E.R. John, Birkhanser, Cambridge.
40. N. Bohr (1983), *Discussion with Einstein on epistomological problems in automatic physics, Quantum Theory and Measurement*, Ed. J.A. Wheeler and Zurek, Princeton University Press, Princeton.
41. J. Von Neumann (1983), *On the Fallacy of assigning an origin to consciousness, Machinery of the Mind*, Ed. E.R. John, Birkhanser, Cambridge.
42. Julian Huxley, *Evolution after Darwin*, Vol. III: 251-252 & 261-262.
43. Swami Vivekananda, Complete Works, Vol.II, Ninth Edition, pp 432.
44. Swami Vivekananda, 'Raja Yoga', Advita Ashrama, Calcutta.
45. P.N. Tandon (1993), *Consciousness: Clinical and Beyond*, INSA NEWS, New Delhi, Issue-112.
46. S.R. Ranganathan (1967), *Ramanujan - The Man and the Mathematician*, Asia Publishing House.
47. D. Dutta Majumder, Prasanta Chandra Mahalanobis (1993) – A Scientist with a Social Commitment, P.C. Mahalanobis Birth Centenary Volume, ISI Club, Calcutta.
48. P.C. Mahalanobis (1957), Dialectica, 8: 95-111 & 18:189-194.
49. L.A. Zadeh (1962), From Circuit Theory to Theory.
50. L.A. Zadeh (1965), Fuzzy Sets, *Inform. Control.*, 8: 338-353.
51. Raja Ramana (1993), Concept of discreteness, continuity and contour Continuum Theory as related to the life-time and masses of elementary particles, *Current Science*, 65 (6): 472-477.
52. Sisir Roy & D. Dutta Majumder, Indeterminacy Relations, Unsharp Observables and Wholeness of Individual Quantum Process, *Proc. The National Academy of Sciences*, India.
53. N. Bohr (1949), *A. Einstein: Philosopher – Scientist.* (The Library of Living Philosophers, La Salle, Illinois), Ed. P.A.Schilpp.
54. P.Busch & P.J. Lahti (1981): *Phis. Rev. D 29, 1634 (1984) P. Mittlested: Current issues in Quantum logic*, p.259, Plenum Press, NY.
55. Von Neumann (1953): *Mathematical Foundation of Quantum Theory.*
56. P. Busch (1985): *Int. J. Theoretical Physics*, 24, 63.
57. C.A. Murthy, S.K. Pal & D. Dutta Majumder (185), Correlation Between Two Membership Functions, *Fuzzy Sets and Systems*, 17: 23-28.
58. C.W Tyler (1991), *Some Tacit Assumptions in Visual Psychophysics, Representation of Vision*, Ed. Anderi Gorea, Cambridge Univ. Press, 252-278.
59. D. Dutta Majumder (1990), Man-Machine Interaction by Speech, Language and Vision, *Proc. KBCS Speech Workshop, TIFR, Bombay*, India.

60a. P.K. Roy, S. Banerjee & D. Dutta Majumder (1992), A study of the transformation of the Human Psyche through its different states in terms of Irreversible Thermodynamics, *SCIMA: Systems and Cybernetics* 21(2-3): 83-94.

60b. P.K. Roy, S. Banerjee & D. DuttaMajumder, Non-equilibrium Psychodynamics, *SCIMA: Systems and Cybernetics* 21(2-3): 103-112.

60c. P.K. Roy, S. Banerjee & D. Dutta Majumder (1992), Concept of Classical Thermodynamics as Applied to the States if the Human Psyche, *SCIMA* 21(2-3): 95-102.

61a. P.K. Roy & D. Dutta Majumder (2000), Sleep and Dream: A Biothermodynamics Analysis, *Journal of Intelligent Systems*, 10: 44-101.

61b. D. Dutta Majumder & P. K. Roy (2001), Evolution of Group Consciousness – A Cybernetic Approach (Basis of an anthropological dating technique, *Kybernetes – The Int. J. of Systems and Cybernetics*, 30 (9/10), MCB University Press, Bradford England.

61c. P. K. Roy, J. Miller & D. Dutta Majumder (2002), Neural Information Processing: Non-equilibrium information theory, *Springer Lecture Notes on AI*, 2275; 191-203.

61d. P. K. Roy & D. Dutta Majumder (2003), Cognitive Processing and Activation During Sleep and Dream: An Analysis using Stochastic Transition and Stochastic Resonance, *IETE Jr. of Research*, 49(2 & 3); 183-195.

62a. D. Dutta Majumder (1993), *Mind-Body Problem and Artificial Consciousness for Computing Machines: A Cybernetic Approach, Recent Advance in Cybernetics and Systems*, Eds. A. Ghosal & P.N. Murthy, Tata McGraw Hill, New Delhi, pp 337-345.

62b. D. Dutta Majumder (1993), Mind-Body Duality: Its Impact on Pattern Recognition and Computer Vision Research, *ICAPRDT-93*, P.C. Mahalanobis Birth Centenary Volume, ISI, ECSU, pp 1-16.

63. Eugene P. Wigner (1967), Remark on the Mind-Body Question (pp 171-184), and Two kinds of Reality (pp 185-199), *Symmetric and Reflections*, Indian University Press. .

64. P.K. Roy, R. Kozma & D. Dutta Majumder (2002), From Neurocomputation to Immunocomputation: A method and Algorithm for fluctuation Induced Instability in Biological Systems, *IEEE Trans. Evolutionary Computing*, 6(3): 292-305.

65. D. Dutta Majumder (1997), Pattern Recognition and Machine Learning: Current Status and Some Contributions, (Golden Jubilee Commemoration Lecture, 1994), *Review Bulletin of the Calcutta Math. Society*, 5(1): 1-66.

66. D Dutta Majumder & M Bhattacharya (1997), Uncertainity Management in Multimodal Image Fusion System for Diagnosis and Therapy Planning, *Intl. Conf. on Recent Advances in Statistics and Probability*, (Indian Statistical Institute & Bernoulli Society for Mathematics, Statistics and Probability) Calcutta, Dec. 29-Jan. 1.

67. D. Dutta Majumder & M. Bhattacharya (1997), A Shape Based Approach to Automated Screening of Malignancy in Tumors in Tomography and Other Related Images, *IETE NASELSOM*, New Delhi.

68. D. Dutta Majumder & M. Bhattacharya (1998), A Multimodal Medical Imaging Processing System for Diagnosis and Therapy Planning, Int. Conf. on *Computers and Devices for Communication* (*CODEC*), Calcutta, India, Jan. 14-17.

69. D. Dutta Majumder & M. Bhattacharya (1998), Multimodal Data Fusion for Alzheimer's Patients Using Dempster-Shafer Theory of Evidence, *3rd Asian Fuzzy Systems Symposium*, Korea, pp 713-719.
70. S. Banerjee & D. Dutta Majumder (1998), Some Recent Advances in Bio-medical Imaging - A Comparative Survey of Different Modalities, Proc. of the *National Academy of Sciences*, India, Vol. LX VIII, Sec. A, Part I: 1-21.
71. D. Dutta Majumder & M. Bhattacharya (1997), Registration and Fusion of CT, T1 and T2 Weighted MR Images of Brain of Alzheimer's Patient, *Journal of Computer Science and Informatics*, Computer Society of India, CSI, 4 (3-4): 12-22.
72. D. Dutta Majumder & M. Bhattacharya (1998), A Multimodal Medical Image Processing System for Therapeutic Planning, *Society of Management Science and Applied Cybernetics*, SCIMA, 25: 45-72.
73. D. Dutta Majumder & M. Bhattacharya (1998), Soft Computing Methods for Uncertainty Management in Multimodal Image Registration and Fusion Problems, Int. Conf. of *Information Sciences JCIS-98*. North Carolina, Duke University, USA.
74. D. Dutta Majumder & S. Banerjee (1998), A Multimodal Medical Image Processing System Using a Knowledge Based Approach, *Image Processing & Communications*, Poland, 4 (1 & 2): 147-180.
75. P. Roy, A Raj, D Dutta Majumder & R Shaw (1998), Discrete Perceptual Space-time and Non-Computability: From Experimental Findings to Theoretical Foundations, in: S Hametroff (Ed). *Proc. Third International Conf. Towards an Understanding of Human Consciousness*, University of Arizona, Tucson.
76. P. K. Roy, R. Kozma and D. Dutta Majumder (2000), *From Neurocomputing to Immunocomputing: Stability Analysis As New Anticancer Strategy*, Advances in PR and DT (Ed), Narosa Pub. House, New Delhi, pp 372 – 377.
77. D. DuttaMajumder, P. Roy, R. Kozma & J. Biswas (1999), A Cybernetic approach to Spontaneous Cancer Remission: Exploring a New Paradigm for Cancer Treatment, in R.Vallee & J.Rose (ed), *Recent Advances in Cybernetic and Systems*, ICCS Series. World Organization of Systems and Cybernetics, Paris.
78. D. Dutta Majumder & M. Bhattacharya (2000), Registration of Multimodal Images of Alzheimer's Patient: A shape Theoretic Approach, *Pattern Recognition Letters*, 21: 531 – 548, Elsevier Science.
79. D. Dutta Majumder & M. Bhattacharya (2000), Cybernetic Approach to Medical Technology: Application to Cancer Screening and Other Diagnostics, *Millennium Issue in the journal Kybernete*, U.K. (Frank George Research Award Paper). 29 (7-8): 871 – 895.
80. D. Dutta Majumder & P. Roy (2000), Cancer Self-remission and tumour instability – A Cybernetic analysis: Towards a fresh paradigm for Cancer treatment, *KYBERNETES*, Millennium Vol. 29 (7/8), Frank George Research Award Winning Paper II): 896 - 925.
81. M. Bhattacharya & D. Dutta Majumder (2000), *Multi Resolution Medical Image Registration using Mutual Information and Shape Theory, Advances in Pattern Recognition and Digital Techniques*, Narosa publ., New Delhi, pp 175 – 180.
82. D. Dutta Majumder (2001), *A Study on Mathematical Tools and Techniques for Multimodal Neuro-Imaging Framework for Registration and Fusion for Diagnostics and Therapy Planning*, India–USA Neuro-science Research

Collaboration Workshop, Virginia, USA, and National Institute of Health - NIH and National Brain Research Centre (NBRC), India.
83. D. Dutta Majumder (2003), Pattern Recognition and Machine Learning: Current Trends and some Applications, in J.C. Mishra(Ed): *Computational Mathematics, Modeling and Algorithms*, Narosa Publishing, New Delhi, pp 230-305.
84. S.R. Pal & S. Mitra (1999), *Neuro-Fuzzy Pattern Recognition: methods in soft computing,* Wiley Inter Science.
85. N.R. Pal (2003), *Pattern Recognition in Soft Computing Paradigm*, World Scientific.
86. H. Haken & R. Valle (1997), Synergetics and Cybernetics, *Encyclopedia of Applied Physics*, 20.
87. H. Haken (1987), *Advanced Synergetics*, 2nd ed., Berlin: Springer.
88. H. Haken (1988), *Information and Self-Organization*, Berlin: Springer.
89. H. Haken (1990), *Synergetic Computers and Cognition*, Berlin: Springer.
90. J. S. Hopfield (1982), *Proc. Natl. Acad. Sci.* U.S.A. 79, 2553-2558.
91. J.A.S. Kelso (1995), *Dynamic Patterns*, Cambridge, Mathematical: MIT Press.
92. T. Kohonen (1987), *Associative Memory and Self-Organization*, 2nd ed., Berlin: Springer.
93. B. Mandelbrot (1983), *The Fractal Geometry of Nature*, New York: Freeman.
94. G. Nicolis & I. Prigogine (1977), *Self-Organization in Non-equilibrium Systems*, New York, Wiley.
95. A. M. Turing (1952), *Philos. Trans. R. Soc.*, London B 237: 37-52.
96. R. Vallee (1995), *Cognition et Systeme, Essaid*, Epistemo-praxeologie, Lyon-Limonest: L'Interdisciplinaire.
97. H. Von Forster (1976), Objects: Tokens for Eigen-behaviors, *Cybernetics Forum* 8: 91-96.
98. L. Zadeh (1965), Fuzzy Sets and Systems, in: J. Fox (Ed), *Systems Theory*, Brooklyn: Polytechnic Press.
99. G. Klir (1969), *An Approach to General Systems Theory* (A general text in cybernetics), New York: Van Nostrand.
100. O. Lange (1962), *Introduction to Economic Cybernetics*, Oxford: Pergamon.
101. M. D. Mesarovic (1964), Foundation for a General Systems Theory, in: M.D. Mesarovic (Ed.), *Views on General Systems Theory*, New York, Wiley.
102. H. Haken, *Springer Series in Synergetics*, Vols. 1-67, Berlin: Springer.
103. Kaushik Majumder & D. Dutta Majumder (2003), *Pattern Generation and Pattern Recognition from a dynamical system Perspective: Towards a New Paradigm in Intelligent System Design*, ICSIT Technical Report, 2003.
104. S. Korner (1967), *Laws of Thought, Encyclopedia of Philosophy*, Vol. 4, McMillan, NY, pp 414-417.
105. C. Lejewski & J. Lukasijewcz (1967), *Encyclopedia of Philosophy*, McMillan, NY, Vol. 5, pp 104-107.

Fuzzy Logic and its Application in Technology and Management
Editors: Debjani Chakraborty, S. Nanda and D. Dutta Majumder

Sensor Selection in a Connectionist Framework

Nikhil R. Pal and Debrup Chakraborty

Electronics and Communication Sciences Unit
Indian Statistical Institute, Kolkata

Abstract: Suppose for a given classification or function approximation (FA) problem data are collected using l sensors. From the output of the ith sensor n_i features are extracted, thereby generating $p =$ features. So for the task we have $X \subset R^p$ as input data along with their corresponding outputs or class labels. Here we propose a modified Multilayered Perception (MLP) that can select the sensors online, which eliminates the effect of redundant or bad features and also do the required task, FA or classification. We call this network as group featured selecting multilayered perception (GFSMLP). Our simulations show that GFSMLP can do a good job of detecting the bad groups of features and eliminating their effect online ith the learning task.

1. Introduction

It is known that for a given problem all features that characterize a data point may not be equally important, even some features may be *bad* features. Thus, selection of a proper subset of features from the available set of features is important for designing efficient classifiers or other function approximation (FA) systems. Methodologies for selecting good features on the basis of the feature ranking etc. do exists [1, 4, 9], but most of these methods perform feature analysis in a separate phase, offline with the main learning process. In [4] and [7] it was noted that the goodness of f feature of a learning task depends on the learning algorithm itself. Hence, the feature subset, which is appropriate for one learning algorithm may not be the best for other learning algorithms. Therefore, the integration of the feature selection phase with the main learning task can yield better systems. There have been a few previously reported works where the feature selection phase is integrated with the main classification/FA algorithm to obtain better systems [2, 3, 5, 11, 10]. The present methodology follows the same philosophy.

This paper addresses the problem of feature selection in a different setting. Here we assume that the features available can be divided into a few groups. The motivation of the problem comes from the fact that now-a-days for a given problem w often data from multiple sensors. From a single sensory information one may generate/extract several features. Conventional feature selection methods select good features from all available features generated from all these sensors. But, our objective is to discard all the computed features from the signals obtained from redundant *sensors*, if any. The features computed from th output of a particular sensor can be taken as a group. This kind of grouping results in a natural partition of

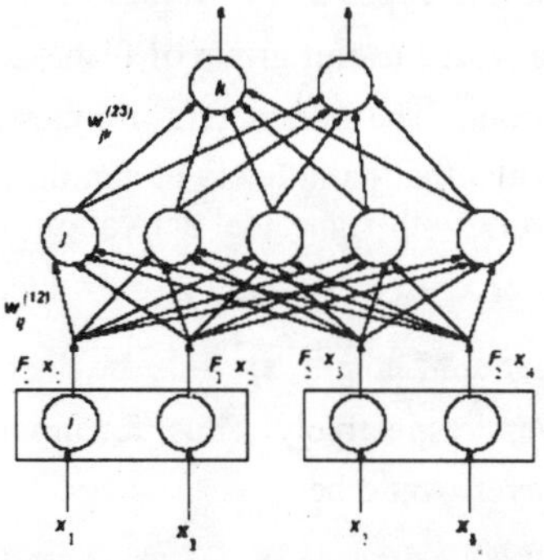

Figure 1: MLP with group selection feature

the total set of features according to their sensory origin. There may exist other natural groupings among features too. Here we address the problem of feature group selection, instead of individual feature selection. Of course, individual feature selection is a special case of this group feature selection methodology.

2. A Group Featured Selecting Multilayered Perceptron

A multilayered perceptron (MLP) can also be modified to do group feature selection. A feature selecting MLP was proposed by Pal and Chintalapudi [10], which used sigmoidal attenuation functions in the input layer. We generalize their idea to propose GFSMLP, a modified form of MLP for group feature selection.

A GFSMLP is similar to n ordinary MLP but it has a feature selection component inherent in it. Here we assume that all features present in the data are grouped according to the sensory origin. Each group of features (features related to the same sensor) has a *feature attenuator gate* associated with it. These gates restrict the entry of features from a bad sensor into the network. We start with all feature gates almost closed, and as training of the MLP proceeds, the gates related to the good or relevant features open to let them in, but remain closed for the bad features. The degree to which a gate remains open at the end of the training is a measure of the importance of the feature related to the gate.

Let F_l be an attenuator function attached to the lth group of feature. Each feature x_i of the lth group gets multiplied by the attenuator function F_l before it gets into the network. If $F_l = 0$, then no feature of the lth group will get into the network; while if $F_l = 1$, then every feature of the lth group enters the network unattunated. For intermediate values of F_l transformed values of the features entered into the network. Fig.1 shows architecture of the group feature selecting MLP (GFSMLP) with just one hidden layer and two groups of features with attenuation function F_1 and F_2. Each attenuation function F_l should be such that is has a tunable parameter β_l and $F_l \in [0, 1]$ to facilitate the learning of β_l, F_l should be differentiable. Moreover, F_l should be such that over the reasonably large interval (a, b), as β_l goes from a to b, F_l should either monotonically increase from 0 to 1 or monotonically decrease from 1 to 0. There can be any choices for F_l, we take $F_l = 1 - e^{-\beta^2}$. β_l is a parameter related to the lth group of features. If $\beta_l \to 0, F_l \to 1$ and if $\beta_l \to \pm \propto$, $F_l \to 0$. The objective here is to select the approximate values of $\beta_l s$ through training such that $F_l \to 1$, if l is associated with a useful group of features and $F_l \to 0$, if the lth group is a *bad* or *redundant* group. The parameters $\beta_l s$ can be learnt by the backpropagation algorithm along with other parameters of a multilayered perceptron.

Let us consider a network with sigmoidal activation functions and a single hidden layer. For each input vector $x = (x_1, x_2, ..., x_p)^T \in R^p$, let S_l denote the set of features which belongs to the lth group. Let $z_i^{(1)}, z_i^{(2)}, z_i^{(3)}$ be the outputs of the ith nodes of the input, hidden, output layers respectively. Thus, for inputs x, if $x_i \in S_l$, the output of the ith node in the input layer would be

$$z_i^{(1)} = x_i F_l \tag{1}$$

Let $w_{jk}^{(23)}$ be the weight of the link correcting the jth node of the hidden layer with the kth node of the output layer. Similarly, $w_{jk}^{(12)}$ denotes the weight correcting the ith node of the input layer with the jth node of the hidden layer. Thus,

$$z_i^{(2)} = \frac{1}{1+e^{-\sum_i z_i^{(1)} w_{ij}^{(12)}}} \tag{2}$$

and

$$z_i^{(3)} = \frac{1}{1+e^{-\sum_j z_j^{(2)} w_{jk}^{(23)}}} . \tag{3}$$

For an input x if the target is t, then we define an error E as

$$E = \frac{1}{2}\sum_k \left(z_k^{(3)} - t_k\right)^2 \tag{4}$$

We define

$$\delta_i^{(q)} = \frac{\partial E}{\partial z_i^{(q)}}, \qquad q = 1, 2, 3 . \tag{5}$$

Thus, the update equations for the two sets of weights can be derived as

$$w_{jk}^{(23)}(t+1) = w_{jk}^{(23)}(t) - \eta \delta_k^{(3)} z_j^{(2)} z_k^{(3)} (1 - z_k^{(3)}) \tag{6}$$

and

$$w_{ij}^{(12)}(t+1) = w_{ij}^{(12)}(t) - \eta z_j^{(2)} (1 - z_k^{(2)}) z_i^{(1)} \delta_j . \tag{7}$$

In the above equations, η is a predefined learning constant. The update equation for the feature attenuator of the lth group, β_l can be derived as

$$\beta_l(t+1) = \beta_l(t) + \mu \sum_{i \in S_l} 2\delta_i^{(1)} z_i^{(1)} \beta_l(t) . \tag{8}$$

Here μ is also a predefined learning constant. The $\beta_l s$ are so initialized that at the beginning of training no feature group is important (i.e., no feature gets into the network- all gates are almost closed). As training continues, the gradient descent changes those $\beta_l s$ more which can reduce the error faster. Thus, with iteration, for each important group $F_l \to 1$.

Although, we have shown the derivation for an MLP with only one hidden layer, its extension to MLPs with more than one hidden layer is straightforward. Next we discuss some simulation results.

3. Simulation Results

We provide here experimental results on two data sets: Chem, RS14-Data.

The Chem data [12] set is used to test the function approximation capability of the network. Chem contains the data for operator's control of a chemical plan for producing a polymer by polymerization of some monomers. There are five inputs and one output. The input variables are monomer concentration (u_1), change of monomer concentration (u_2), monomer flow rate (u_3) and the two local temperatures inside the plant (u_4 and u_5). The only input (y) is the set of points for monomer flow rate. In [12] there is a set of 70 data points obtained from an actual plan operation. It is known that u_4 and u_5 *do not significantly contribute* to the output [12]. For convenience we used a normalized version of this data, where each feature and the output is normalized to lie between 0 and 1.

RS-data set [6] is a 256 level seven-channel satellite image data of size 512×512 pixels obtained from Landsat-TM3. The 512×512 ground truth data provide the actual distribution of classes of objects captured in the image. From these images we produce the labeled data set with each pixel represented by a 7-dimensional feature vector and a class label. Each dimension of a feature vector comes from one channel. This data have 8 classes representing different landcover types. We have enhanced this data for our simulations. We generated an additional feature from the each of the 7 channels of the image. For each pixel p in an image we considered its 8-neighbourhood over a 3×3 window and computed the standard deviation of the 9 pixels (the neighbourhood of p and the pixel itself), we call it d_p. In the new data set, for each channel we take the gray value of p and d_p, as features. So we have 14 features divided into 7 groups. We call this as the RS14 data.

As started earlier, the attenuation parameters for the network are initialized with the aim that at the onset of training the network considers all feature groups to be unimportant. Thus we set $\beta_j = 3$ thus making $F_j = 0.0001, \forall j$.

***Chem*:** In this example we demonstrate the feature selection capability of our network for the function approximation task. As stated earlier, the data set consists of 5 input features. The five input features can be easily divided into 3 groups with respect to the type of features. The monomer concentration (u_1) and change of monomer concentration (u_2) can constitute one group, the monomer flow rate (u_3) as the second group and the temperature parameters (u_4 and u_5) can form the third group. The Chem data set was used by Lin and Cunningham in [8].

Table-1: PI and number of features selected by GFSMLP for Chem data (considering 3 groups of features)

Hidden Nodes	PI		No. of groups selected	
	Mean	Std. Dev.	Mean	Std. Dev.
5	0.002360	0.0000	2.0	0.0
7	0.002295	0.0000	2.0	0.0
10	0.002258	0.0000	2.0	0.0
15	0.002022	0.0000	2.0	0.0

Table 2: Misclassifications and no. of groups of features selected for RS14 data with GFSMLP

No. of hidden nodes	Training Error %		Test Error %		No. of groups selected	
	Mean	Std. Dev.	Mean	Std. Dev.	Mean	Std. Dev.
10	26.52	4.25	20.88	2.53	3.4	0.54
20	21.82	2.67	20.76	3.12	3.8	0.44
25	18.02	2.26	17.96	1.38	4.2	0.44
30	17.63	2.74	18.86	1.65	4.0	0.70

To evaluate a system they used a performance index PI defined as

$$PI = \frac{\sqrt{\sum_{k=1}^{n}(t_k - o_k)^2}}{\sum_{k=1}^{n}|t_k|} \qquad (9)$$

Here t_k and o_k are desired and actual outputs respectively. They got a PI of 0.002245 on Chem by using features u_1, u_2, u_3, which they reported as the only useful ones.

The performance of GFSMLP on Chem data for different hidden nodes are sgown in Table 1. Here too the average performance in terms of PI and the number of groups selected is shown for 5 independent runs for each network configuration. Table 1 show that the GFSMLP selects two groups for all runs. Also the two groups selected are the first and second groups. The PI value suggests that it can also do the function approximation job with a good accuracy.

***RS14 Data*:** Table 2 shows the percentage misclassification and the number of groups of features selected for different number of hidden nodes for GFSMLP. For each number of hidden nodes average value of misclassification and the number of feature selected for 5 independent runs are reported in Table 2. For all cases the result is better than in [6], where a misclassification of 21.8% is reported on the test data. Also, using only a few features attains this performance.

4. Conclusions and Discussion

Many real life applications use data from several sensors for decision-making. Intelligent systems for automatic inspection and controlling of welding, medical diagnosis, controlling of range safety for missile testing are some such examples. Typically, each sensor output is converted into a set of features. For example, X-ray radiograph may be used to compute a set of features for weld inspection. More sensors mean more cost, more processing time, and sometimes more hazardous too (X-rays). So one would like to work with the minimal number of sensors where each sensor could be responsible for a set of features. This is a more difficult problem than selecting individual features. In this paper we have proposed a novel scheme for the same using the MLP. Using real-life data we have demonstrated that the proposed system can do an excellent job of sensor selection.

References

1. L.M. Beleue & K.W. Bauer (1995), Determining input features for multilayered perceptron, *Neurocomputing*, 7 (2): 111-121.
2. D. Chakraborty & N.R. Pal (2001), Integrated fature analysis and fuzzy rule-based system identification in a neuro-fuzzy paradigm, *IEEE Trans. on Systems Man Cybernetics B*, 31 (3): 391-400.
3. D. Chakraborty & N.R. Pal (2004), A neuro-fuzzy scheme for simulations features selection and fuzzy rule-based classification, *IEEE Trans. Neural Networks*, 15 (1): 110-123.
4. R. De, N.R. Pal, S.K. Pal, Feature analysis: neural network and fuzzy set theoretic approaches, *Pattern Recognition*, 30 (10): 1579-1590.
5. C.D. Huang, C. T. Lin, N.R. Pal (2003), Hierarchical learning architecture with automatic feature selection for multiclass protein fold classification, *IEEE Trans. on Nanobioscience*, 2 (4): 221-232.
6. A.S. Kumar, S. Chowdhury & K.L. Mazumder (1999), Combination of neural and statistical approaches for classifying space-borne multispectrul data, *Proc. of ICAPRDT99*, Calcutta, India: 87-91.
7. F. Kohavi & G. John (1997), Wrappers for feature subset selection, *Artificial intelligence*, 97 (1): 273-342.
8. Y. Lin & G.A. Canningham III (1995), A new approach to fuzzy-neural system modeling, *IEEE Trans. Fuzzy Systems*, 3 (2).
9. N.R. Pal (1999), Soft computing for feature analysis, *Fuzzy Sets and Systems*, 103: 201-221.
10. N.R. Pal & K.K. Chintalapudi (1997), A connectionist system for feature selection, *Neural Parallel and Scientific Computations*, 5 (3): 359-381.
11. N.R. Pal, T.C. Cahoon, J.C. Bezdek, K. Pal, A new approach to target recognition for LADAR data, *IEEE Trans. on Fuzzy Systems*, 9 (1): 44-52
12. M. Sugeno & T. Yasukawa (1993), A Fuzzy-Logic based approach to qualitative modeling, *IEEE Trans. on Fuzzy Systems*, 1 (1): 7-31.

Fuzzy Logic and its Application in Technology and Management
Editors: Debjani Chakraborty, S. Nanda and D. Dutta Majumder

PROMETHEE: A Decision Support Tool for e-Business

B.K. Mohanty

Indian Institute of Management, Off Sitapur Road, Lucknow-226 013
Email: mohanty@iiml.ac.in

Abstract: In this paper we have introduced a procedure to obtain the customer's preference ranking of the products in on-line business. This is based on the concepts of Fuzzy sets, Data Mining and the PROMETHEE methods of Multiple Criteria Analysis. This methodology takes into account the customers' flexibility behavior (or indifference attitude) to articulate the above preference ranking. Customers' data on business transactions are considered here as an input in order to enunciate various association rules amongst the product attributes. This is possible by taking the customers' transactions as the fuzzy subset of the product attributes. The membership values of the attributes corresponding to a transaction depends on the users preference level of the product attributes and the number of products are contained in that transaction. The concept of PROMETHEE is further used here to determine the net association amongst the attributes. This procedure leads to obtain the preference ranking of the products as per the customers' choice.

1. Introduction

Over the last decade many companies have focused attention on the use of Internet to commercialize their products and services. The development of information technology, on-line business strategies, and Internet tools has established more convenient ways to commercialize their business transactions efficiently. Many organizations consider this as a major paradigm shift from the traditional concepts of across the table marketing to the world of easy and effortless business processing. More and more organizations from all sectors are attracted by the tremendous benefits of Internet applications. To name a few these include cost and time saving, increase of productivity, flexibility, consistency, reduced workload, timeliness and facilitator for global business across the country and removal of regional boundaries. Internet business is considered as a technological driving force in the twenty first century.

Although many successful stories involving the use of Internet have been reported, numerous Internet investors are disappointed and impatient as they wait for benefits that may never come. This is mainly because, in the Internet Business, the underlying technology should be more focused to the current business profile and help in satisfying the customer's needs. Unfortunately, there has been a little study of quality measurement in the Internet business applications with respect to the customers needs. In this context, quality is customer focused, and hence the relative importance of quality characteristics of Internet business applications should be determined for different types of users or business environments. This specific nature of Internet technology helps to evaluate the trustworthiness of the product in eyes of the consumers especially in the Internet market. The consumer's inability to discriminate a fraudulent product from a legitimate one is a serious problem for the sustained viability of the Internet commerce. The Internet might become the next "lemon market": in an environment where it is difficult to tell the difference between good and bad products, the bad ones poison the market and drive away the good products and eventually the consumers. In addition to this, the lack of product definition in the Internet market may lead to increase in cost of doing business on-line and decrease the market's competitiveness.

Mainly a customer wants to avoid the risk while purchasing the product in the Internet. Risk refers to a consumer's perceptions of uncertainty and adverse consequences of engaging in an activity. When risk is present trust is needed before a buyer is willing to transact with a seller. Perceptions of risk have been found to be greater especially when a consumer purchases the products through the Internet market rather than in store channels. This is because the costumers loose the ability to engage in direct observation of the product attributes, which he/she normally experiences and communicates with the seller in consumer stores. The possibility that a seller withholds information about the quality of a product, or its attributes (which are in multiple numbers), or about the performance of a company, is a source of risk for potential buyers. Also the uncertainty due to risk increases when the Internet tools do not provide the required information about the product attributes in a multiple fashion. In this case, the market is considered as not reliable and hence the customer focused. Moreover, in this case the buyer cannot scrutinize the desired attributes of the product (which are in general multiple in numbers) and prefers to face-to-face interaction with the seller in the traditional market in compare to the Internet market.

2. Multiple Attribute Concepts in Internet Business

The above problems of the Internet business can be well addressed by taking into account the customers need in a multi-dimensional view. That is a customer would like to examine and analyze the multiple aspects of the product before making the final purchasing decision. For example in a car purchasing problem the multiple attributes may be cost, fuel consumption, re-sale value, engine type, design, color etc. In this particular case, the customer has to select a car that offers the best match, in terms of its attributes to his/her requirements. Concepts of Multiple Attribute Decision Making help here to handle this problem. The multi-attribute approach is one of the many models that have been developed to deal with decision-making situations where several factors have to be considered simultaneously [1, 5]. In practice the decision maker searches for a best satisfying compromise among several conflicting non-commensurable and fuzzily defined attributes.

3. Fuzzy Logic in Internet Business

In any business environment (on-line or traditional) a customer with his / her product knowledge or product information goes to the shop, compares different products and selects a product of best choice. Normally a sales assistant helps and suggests a product(s) that satisfy the customer's product requirements. So far most on-line business approaches are based on precision data, non-ambiguity knowledge and exact information. We have not yet come across any model, which incorporates the customer's realistic behavior of fuzzily defined product attribute requirements. Non-fuzzy data assumptions in the Internet business will have difficulty in implementation because of the fact that impreciseness is inherent in the customers mind and also in almost all business problem domains. Ignoring or approximating the customer's fuzzy requirements and using the existing precise models will not only lead to an unpredictable result but also will cause to a failure because of the misrepresentation of the customer's real opinion in terms of his/her knowledge content. For example in buying a shirt a customer, normally do not consider precisely the amount of percentage of cotton, the style and softness of the fabric and the durability of the shirt. In another example of car purchasing the attributes of the product CAR may be Cost, Resale Value, Mileage, Comfort, Maintenance Cost etc. In the customer's view these attributes are defined as, the cost of the car should around 200 thousand, Resale vale should be very high, More or less the mileage should be around 20 Kms, the car should be comfortable and the maintenance cost must be relatively low. The above representations of the customer viewpoints are realistic and a day to day language, but are vague in computational sense. Fuzzy logic quantifies these linguistically defined statements and represents them in the form of fuzzy numbers [3]. This work aims to address the customer's main issue of fuzzy requirements and to formulate a model to meet the customer's desired level of the product attributes.

4. Hierarchical Product Classification

Another problem in the marketing, a customer very often faces, especially in Internet business, because of the availability of similar alternative products in the network. This creates some sort of ambiguity in the customer's mind while making the final choice of a product. The ambiguity is mainly because of two reasons. Firstly how to make a final product choice to purchase, and secondly on what basis the other products will be rejected. In order to answer the above questions the customer may like to classify the products in different preference levels, preferably through some numerical strength of preference. Achievement of this classification will serve as a decision aid to the customer in the sense that, while purchasing a product he/she will come to know the information that at what level the product is chosen and what extent the other available products in the network are either superior or inferior to the chosen product. This procedure will remove in the ambiguities in the customer's mind about the other products, and inherently will create a measurable instrument the light that how far the chosen product is below or above to the other available products. In addition to this the customer will also know, how far he/she is compromising to the best available product (zero compromise, in case he/she purchases the best available product). The main problem here is how to classify the products into different preference hierarchy levels. The concept of attribute flexibility [4, 8] is defined here to classify the products from best to worst as per the customer's requirements. The multiple attributes of the product in the form of fuzzy sets are considered to rank the products in the Internet market.

4.1. Example

If two cars cost (2, 0.8) and (2.5, 0.6) and the attribute cost has the flexibility 0.2, the costs 2 and 2.5 are considered at par and kept in the same preference class. The numerical quantities 0.8 and 0.6 represent the membership values of the attribute cost.

In the Internet business, we have not come across any literature, which incorporates the multiple aspects of the product attributes along with the customer's opinion of fuzziness, in each attribute, for assessment of the product. This work handles the above-mentioned problems and after classifying the available products arrives at a viable solution which best satisfies the customer's requirements. Briefly the procedure can be explained in the following steps.

1. Quantify each attribute of the product in the form of Fuzzy Numbers.
2. Choose a specific attribute and measure the extent to which this attribute's level is satisfied to that of the customer's desired level. Note that for K number of products we will have K such attribute levels.
3. Determine the attribute flexibilities across the product attributes.
4. Ryu's algorithm is used to classify the products into different hierarchical levels.

5. Data Mining Concepts in the Internet Business

In the reference [4] it has been assumed the customer's preference order of product attributes subjectively. However, in reality the customer might not prefer the product attributes in that order or he/she may change his/her preference order depending on the business scenario. Moreover, in the business interest, particularly on on-line business situations, a vendor may take interest the actual attribute preference order to which really the customers prefer. This necessitates articulating a preference order of the product attributes as per the customers choice. It is a difficult task and becomes a real problem particularly in the on-line business system. In order to achieve this present work takes into consideration a large set of data of actual customer transactions. The data mining techniques further assist us by exploring the past business transactions and finding the explicit as well as implicit preference order of the product attributes used by the customers.

In practice the actual business transactions record the products and not the product attributes. However, our requirements are the preference order of the product attributes as per the customers choice. Articulating the customer choice from the available information of the product transaction records is a difficult problem. This problem becomes more difficult and complex when the product attributes are imprecise, ambiguous, vague, confusing or fuzzily defined. We have handled this situation and arrived at a desirable solution of preference order by using the concepts of association rules of the Data Mining [7]. Briefly this procedure is as given below.

Initially customer's transactions are taken and each transaction is defined as a fuzzy set of the product attributes. The membership functions of the above fuzzy set corresponding to a transaction depend on the user's preference level of the product attributes and the number of products is contained in that particular transaction. The product transactions and hence the fuzzy sets of the attributes corresponding to each transaction establishes various associations amongst the attributes. In the context of data mining these derived associations leads to some association rules and each rule has a numerical confidence level. These association rules take into account the customer's minimum requirements of the attribute levels. The customer's minimum level requirement on each attribute is considered here by taking into account the customers flexibility level that he/she can compromise on different attributes. That is the flexibility values of the attributes play an important role in identifying the minimum attribute level requirements. Details about the flexibility values are found in [4]

6. PROMETHEE Multiple Criteria Analysis Concepts in the Internet Business

Use of Data Mining concepts to rank the product attributes in the Internet business are explained in [6]. The association rules amongst the attributes of the products help us to determine a particular order of the attributes as per the customer's choice. This is done by giving an interpretation to the association rules in the following way. Let there are 3 attributes, (say, A_1, A_2 and A_3) and three association rules with degree of confidence as 0.52, 0.75 and 0.5 respectively.

$$A_2 \xrightarrow{0.52} A_1, A_3 \xrightarrow{0.75} A_1, A_2, A_3 \xrightarrow{0.5} A_1$$

In [6] the association rules are interpreted as, as per the customer's viewpoint the presence of the attribute A_2 enhances the preference level of A_1 additionally 52%. This can be explained by the following example.

6.1. Example

Suppose the price (attribute A_1)) of a car is very costly and satisfies the customer to a satisfaction level of 60 % only. Take another attribute of the same car say mileage (attribute A_2) and it is very satisfactory to the customer. That is even though the cost attribute is of average satisfaction; the customer is tempted to purchase the car because of the attribute mileage. That is by taking the above association rule the satisfaction level of the attribute cost (A_1) got the support of the attribute mileage (A_2) with a confidence level of 0.52. That is the attribute A_2 is similar to that of A_1 to a degree of 52% level as per the customers lookout. Similarly we can have the interpretation for all other association rules.

The real difficulties of using the concept of data mining or for that matter the association rules are of accounting the numerical support (confidence level) in one direction. For example if A_2 supports A_1 with a confidence level 0.52 ($A_2 \xrightarrow{0.52} A_1$) it is also necessary to account how far A_1 supports A_2, in order to calculate the net support gained by A_1. This because as per [6], the association rule $A_2 \xrightarrow{0.52} A_1$ represents as if the attribute A_2 as an addition of 52% of more of A_1. With out consideration of the association rule in the opposite direction we are adding the similarities of A_1 and A_2 twice, once while taking $A_2 \longrightarrow A_1$ to A_1 and the other in the direction $A_1 \longrightarrow A_2$ to A_2. In fact we need the net support of an attribute. This can be obtained by using the PROMETHEE Multi Criteria Analysis method [2]. This is explained through an example. From the above association rule the support gained by the attribute A_1 is

$$\phi^+(A_1) = 1/3(0.52 + 0.75 + 0.5) = 0.6$$

As given in [6], the other association rules the support to the attribute A_2 from A_1 is

$$A_1 \xrightarrow{0.25} A_2, A_1 A_3 \xrightarrow{0.75} A_2$$

That is we have $\phi^-(A_1) = 1/2(0.25 + 0.75) = 0.5$

The net support gained by the attribute A_1 is $\phi(A_1) = 0.6 - 0.5 = 0.1$

Similarly we can have the net support of the other attributes numerically. This gives an ordering of the product attributes. As this ordering information is based on the data of customer's product transactions, we have our desired order of product attributes as per the customer's choice.

References

1. B. Aouni & O. Kettani (2001), Goal Programming model: A Glorious history and promising future, *European J of Operational Research*, 133: 225 – 231.
2. J.P. Brans, B. Mareschal & Ph. Vincke (1984), *PROMETHEE: A New Family of Outranking Methods in Multicriteria Analysis*, in *Operational Research*, JP Brans(editor), Elsevier Science Publishers , North Holland, IFORS:477 – 490.
3. D. Dubois & H. Prade (1978), Operations on Fuzzy Numbers, *Int. J Systems Science*, 9: 225 – 231.
4. B.K Mohanty (2001), Product Classification in Internet Business, in: *Proc. of the Int. Conference on Information Technology*, IMIT, Cuttack, Nov. 2001.
5. B.K. Mohanty & TAS Vijayaraghavan (1995), A multiobjective programming problem and its equivalent Goal Programming problem with appropriate priorities and aspiration levels- A fuzzy approach, *Computers and Operations Research*, 22: 529 – 541
6. B.K. Mohanty (2002), *Fuzzy Applications in the Internet Business*, Proceedings in International conference on *Analysis and Discrete Structure*, to be held at IIT Kharagpur during December 22-24.
7. A.K. Pujari (2001), *Data Mining Techniques*, University Press (India) Ltd.
8. Y.U. Ryu (1999), A hierarchical constraint satisfaction approach to product selection for electronic shopping support, *IEEE Trans in Sys. Man. & Cybernetics* –A, 29: 525– 532.

Fuzzy Logic and its Application in Technology and Management
Editors: Debjani Chakraborty, S. Nanda and D. Dutta Majumder

Fuzzy Multiobjective Decision Making in an e-University System

Umesh Deshpande, Arobinda Gupta and Anupam Basu

Department of Computer Science and Engineering
Indian Institute of Technology Kharagpur
Email: {*uad, agupta, anupam*}@cse.iitkgp.ernet.in

Abstract: In this paper, a model of an e-University system is presented where different universities collaborate to offer courses to the students. Multiple criteria and preference ratings could be associated for the curriculum (referred to as a task) by a student. Task assignment in this system is important since a good assignment would make it feasible to satisfy the objectives. Fuzzy multiobjective decision making process has been adapted for task assignment in a dynamic distributed system. The results of the simulation studied suggest that the fuzzy technique has significant performance gains over other commonly used heuristic techniques for a large range of task arrival rates.

1. Introduction

With the advent of the information revolution, today there is an increase in the need for specialized learning. Since there is a rapid change in the technology and practices, the industry continually needs skilled manpower for customized applications. It is difficult for the single learning institution to cater to these challenges. Collaboration among different universities and colleges is important so that they can share their resources, like the faculty members, laboratories etc. This is the primary objective of an *e-University* system [1]. In an e-University system, different universities or colleges collaborate to form a consortium and offer courses to the students (we will henceforth use the term a *node* for a university or college in the e-University system). Once a student registers for a degree or diploma, the complete curriculum is chalked out depending on the background of the student and pre-requisites necessary. The student may have a deadline before which he/she has to finish the curriculum. He/she is initially assigned to a node and the curriculum (the curriculum would henceforth be termed as a *task*) of the student is completed in cooperation with other nodes.

Each course has duration, a cost, and a quality associated with it. The student may have multiple conflicting criteria (or objectives) that need to be achieved for their tasks depending on their specific requirements. The relative importance for each of those objectives may be different. For example, a student may not be too much bothered about the cost but may want a fast response with a high quality. So, the relative importance of the response time and the quality are much higher than the cost. Decisions in such a scenario must be made in an environment where a multiobjective function governs constraints on the problem, and the relative value of each of these objectives is different. Task assignment, which is a multiobjective decision making problem, is important since a wrong assignment may make it difficult, if not impossible, to satisfy the objectives of the task.

At the time of the task submission, a student normally has only an intuitive and imprecise notion of the preference ratings to be associated with the objectives. It is very convenient for him/her to express one rating as a ratio with the ratings for each of the other criteria using linguistic terms or subjective ordinal values. For example, the quality may have a strong important over cost but the timely response may have absolute importance over the other two. Such a procedure, suggested by Saaty [2], of comparing two elements at a time using ordinal values is considerably simpler than trying to work with all the objectives at once, specially if the number of objectives is quite large. In this paper, multiobjective decision making process proposed by Yager [3] using fuzzy sets [4] is adapted for the task assignment problem. The proposed technique is evaluated against commonly used heuristic approaches for task assignment. It is observed that the fuzzy technique outperforms all of them for a large range of task arrival rates.

The paper is organized as follows. In Section 2, we present the formal e-University model. The multiobjective decision making process using fuzzy sets is reviewed in Section 3. The task assignment is formulated as a muliobjective decision making process in Section 4. The performance evaluation of the task assignment is discussed in Section 5. Section 6 concludes the paper.

2. The System Model

There are n nodes, N_1, N_2... N_n in a distributed system, which communicate through messages. Each node has a collection of agents controlling the resources present at that node. It is assumed that the communication delay for message passing between any two nodes N_i and N_j can be estimated and is constant Ψ_{ij}. A special unit called Liaison unit (LU), with an agent in it, is identified at every node that is responsible for coordination across nodes for resource sharing. The LU of a university migrates a task to a remote node if the task requires a resource that is not present locally. The detailed coordination mechanism, used by the LU, is presented in [5].

3. The Liaison Unit: Data Structures and Algorithms

The LU on each node N_i maintains the following two tables.

Local Resource Table (LRT) – The fields are discussed below.

Name	Duration	Surplus

Name is the name of the resource present locally at the node N_i. *Duration* is the time required for the resource, e.g. number hours required to complete a course or a laboratory. *Surplus* is defined as the number of seats that do not get filled each time a course was defined offered in a past window of length WL_k. WL_k is chosen depending on the duration of the resource r_k. The surplus is an indication of the load on a resource. The surplus information is sent to all the remote nodes periodically.

Remote Resource Table (RRT) - At the node N_i, the RRT keeps information about the resource present at the remote nodes. There is an entry for every resource type. The entries for all the nodes that have a resource type are kept one after another as described below. The fields are as follows.

Name	Id	Duration	Surplus

Name is the name of resource type. *Id* is the identifier of the node where the resource type is present. *Duration* is time of the resource of that node. *Surplus* is the one that was most recently obtained from the remote node.

3. Fuzzy Multiobjective Decision Making

The typical multiobjective decision problem involves the selection of one alternative, a_i, from a universe of alternatives containing n elements, $A = \{a_1, a_2...a_n\}$ given a set of p elements $O = \{O_1, O_2...O_p\}$ of criteria or objectives that are important to the user. In [6], each objective has been represented as a fuzzy set over the set of alternatives $\{A\}$. Thus, the grade of membership, $\mu_{O_i}(a_i)$ of the alternative a in the ith objective O_i, denotes the degree to which a satisfies the criteria specified by O_i. For a collection of objectives, the decision function $D = \wedge_{i=1}^{p}(O_i)$. The optimum solution a^* will be the alternative that maximizes D.

In [3], the above approach is extended to allow the introduction of importance ratings with the objectives. With each objective O_i, a non-negative number α_i is associated that indicates the power or importance of the objective in the decision. A higher number indicates that the corresponding objective is more important. Yager suggests that each fuzzy objective should be raised set to its power, with the exponent being the preference rating, before combining to form the decision function D. Thus, the sets $O_1^{\alpha_1}, O_2^{\alpha_2}, ..., O_p^{\alpha_p}$ are formed where

$$\mu_{O_i^{\alpha_i}}(a) = [\mu_{O_i}(a)]^{\alpha_i}$$

Therefore the grade of membership that the new decision function, D has for each alternative a is given by

$$\mu_D(a) = \min[\{\mu_{O_1}(a)\}^{\alpha_1}, \{\mu_{O_2}(a)\}^{\alpha_2}, ..., \{\mu_{O_p}(a)\}^{\alpha_p}]. \quad (1)$$

The optimum decision, a^*, will then be the alternative that satisfies

$$\mu_D(a^*) = Max_{a \in A}(\mu_D(a)). \quad (2)$$

3.1. Procedure for Determining the Power of Importance

There are p objectives and a scale should be constructed, rating these objectives as to their importance with respect to the decision. Saaty [2] described a method for obtaining a ratio scale of importance based on paired comparison. The user is asked to compare the objectives in paired comparisons. If objective O_i is being compared with objective O_j, values r_{ij} and r_{ji} are assigned as follows.

Table1 Pairwise Importance Ratings

Value of r_{ij}	***Definition***
1	Equal importance
3	Weak importance of one over other
5	Strong importance of one over other
7	Demonstrated importance of one over other
9	Absolute importance of one over other
2, 4, 6, 8	Intermediate values between two adjacent judgments

1. $r_{ij} = 1/r_{ji}$ and $r_{ii} = 1$
2. If O_i (O_j) is more important than O_j (O_i), then r_{ij} (r_{ji}) gets assigned a number according to Table 1.

Having obtained the above judgments from the user, a *pxp* matrix, *R* is constructed with the r_{ij} values. Saaty [2] has show that an eigenvector corresponding to the maximum eigenvalue, $\lambda_{\max}$, is a cardinal ratio scale for the elements compared. In this case, the eigenvector whose total is *p* should be found out, since this makes the average importance 1. Thus

$$W = [\alpha_1, \alpha_2, ..., \alpha_p] \text{ where } RW = \lambda_{\max} \text{ and } \sum_{i=1}^{p} \lambda_i = p$$

W is the vector corresponding to the decision maker's subjective assessment of the importance of each objective in the decision. It is to be noted that the procedure of comparing two elements at a time considerably simpler than trying to work with all the objectives at once.

4. The Task Assignment

In a distributed system, task assignment has to be done when the global system state is not available at any single node. Task assignment could be done by any of the nodes of the system or by a special node used solely for that purpose. For example, a task for a student may first migrate to one of the nodes of the system where the task assignment decision has to be made. The agent would return after the decision is made and the task is assigned to the target node. The node where the decision is taken has to collect the global state first. There is imprecision in the state collection itself because of the dynamic nature of the system. Since the resource utilization changes continuously, the system state might have changed in real time from the one that was collected.

For task assignment, the alternatives, *A*, are the different nodes of the system, i.e., $A = \{N_1, N_2 ... N_n\}$. There are four objectives defined in this setup – *Execution Time (ET), Quality (Q), Cost (C), and Migration Overhead (MO)*. The *Execution Time, Quality, and Cost objectives* have the obvious meanings but the *Migration Overhead* is not so obvious. When a task *T* is input to a node of the system, all the resources required by *T* may not be present at that node. Certain cost and time is spent for migrating those subtasks of *T* to remote nodes where the required resources are present. This is referred to as the *Migration Overhead*. Along with an input task *T* and the matrix *R* associated with the pair-wise preferences among different objectives are input by the user.

4.1. Task Assignment as Multiobjective Decision Making

The objectives O = {*ET, Q, C, MO*} are defined using fuzzy sets for every node of system. From the pair-wise preference matrix R, the preference ratings of the individual objectives specified by the vector $W = \{\alpha_1, \alpha_2, \alpha_3, \alpha_4\}$ are computed. The four objectives O = {*ET, Q, C, MO*} are defined using fuzzy sets as follows.

$$ET = \left\{\frac{h_{11}}{N_1}, \frac{h_{21}}{N_2}, ..., \frac{h_{n1}}{N_n}\right\} \quad Q = \left\{\frac{h_{12}}{N_1}, \frac{h_{22}}{N_2}, ..., \frac{h_{n2}}{N_n}\right\}$$

$$C = \left\{\frac{h_{13}}{N_1}, \frac{h_{23}}{N_2}, ..., \frac{h_{n3}}{N_n}\right\} \quad MO = \left\{\frac{h_{14}}{N_1}, \frac{h_{24}}{N_2}, ..., \frac{h_{n4}}{N_n}\right\}$$

In the sets above, h_{ij}, is the membership value of an alternative N_i in the fuzzy set for the objectives O_j. In this work, all the values of h_{ij}, $\forall i$: 1..*n* and $\forall j$:1...4 are so computed that they belong to the unit interval [0, 1]. The fuzzy sets of the objectives are computed when a task T is input taking into account the current available state of the system. The idea of computing the membership values of the fuzzy setes is illustrated with an example of a fuzzy set for *Execution Time* (*ET*).

First the time required, TR_i for the task T at every node N_i is estimated. Then the estimated values are normalized as follows. The minimum of the TR_i values (referred to TR_{min}) is found out. TR_{min} is divided by the individual TR_i values. With this normalization, the node with less estimated time required gets a value 1 and all others will get values between [0, 1]. Moreover, a node having lower estimated time required value will have a higher normalized value.

Second, a representative fuzzy set called *Time Criticality* (*TC*) is defined whose universe of discourse in the interval [0, 1]. The values of the membership function of *TC*, μ_{TC}, are also in the range [0, 1]. The fuzzy set *TC* captures a user's perception about how each node satisfies the *ET* objective given the estimated time required. It maps the normalized *TR* values to the membership values of *TC* such that a higher normalized values gets a higher membership value. h_{i1} which is the membership value of a node N_i in the fuzzy set *ET* is computed as follows:

$$h_{i1} = \mu_{TC}(TR_{\min} / TR_i) \tag{3}$$

The estimation of the time required for a task cannot be precise because of the dynamic nature of the system. For example, the completion time for a task T at a node N_i can be known only after T has finished execution. It is known at the point when T is submitted. Also, those subtasks of T that require a resource not present at N_i would have to be migrated. The destination of the node migration is found using the coordination mechanism mentioned in Section 2. It is also not known when T is submitted. Hence in all the following computations, the destination node is assumed to be the best node N_b. The best node is the one where the required resource is present and that has the maximum surplus.

The fuzzy sets for the other objectives are also computed in similar way. Representative fuzzy sets are called *Good Quality (GQ), Less Cost (LC), and Less Migration (LM)* are defined and used in the same way as the set *Time Criticality (TC)*. These may be provided by the user or default fuzzy sets may be defined at the system initialization time.

After the computation of the fuzzy sets and the preference ratings from the pairwise comparison matrix R, the multiobjective decision process discussed in Section 3 is executed and T is assigned to the node, N_i identified by the process. In the nest few paragraphs the computation of fuzzy sets is explained.

4.1.1. Computation of the Execution Time Fuzzy Set

The values h_{i1} for all i $(1 \leq i \leq n)$ indicate the goodness of a node N_i with respect to the execution time factor. As explained earlier, an estimate of the time required for T at N_i, TR_i is computed. To obtain this, first consider resource r present at N_i (found from the LRT of N_i). This gives an approximate measure of time required for r. When the surplus is very low, a bound on this measure is made. Summation Sum_1 of this computation for all such resources present at N_i is done.

Next, consider a resource r^1 required by T but not present at N_i. In this case, the duration of r^1 at the best node N_b is scaled by the surplus of r^1 at N_b. In addition, the time for migrating subsets of T to N_b has to be considered. Hence the term $2^* \psi_{ib}$ is added. The multiplication by 2 is there since two communications between N_i and N_b would be required - one for sending the task by N_i and other for sending the results by N_b. Summation Sum_2 of this computation for all such resources not present at N_i is done. The time required by T at N_i, TR_i is estimated to be the addition of the two summations Sum_1 and Sum_2.

In the following expression for TR_i, the term $\forall r \in T \wedge r \in N_i$ denotes a resource r that is required by T and is present at N_i and the term $\forall r \in T \wedge r \notin N_i$ denotes a resource r that is required by T but is not present at N_i. TR_i for all values of i $(1 \leq i \leq n)$ is calculated as follows.

$$TR_i = \sum_{\forall r \in T \wedge r \in N_i} \frac{\Delta_i^r}{S_i^r} + \sum_{\forall r \in T \wedge r \notin N_i} \left(\frac{\Delta_b^r}{S_b^r} 2 * \psi_{ib} \right) \tag{4}$$

Where Δ_i^r is the duration of r at N_i and Δ_b^r is the duration of r at the best node N_b. Similarly, S_i^r is the surplus at the node N_i and S_b^r is the surplus at the best node N_b. Let TR_{min} be the minimum of TR values, i.e., $TR_{min} = MIN_i(TR_i)$. Then $h_{i1} = \mu_{TC}(TR_{\min} / TR_i)$.

4.1.2. Computation of the Quality Fuzzy Set

The values h_{i2} for all i $(1 \leq i \leq n)$ are computed as follows. First the following values of Q_i are found $\forall i : 1 \ldots n$,

$$Q_i = \sum_{\forall r \in T \wedge r \in N_i} Q_i^r + \sum_{\forall r \in T \wedge r \notin N_i} Q_b^r \tag{5}$$

where Q_i^r is the quality of r at N_i and Q_b^r is the quality of r at the best node N_b. For resources that are not present at N_i, the quality of the best node is chosen. Let Q_{max} be the maximum of the Q values, i.e., $Q_{max} = MAX(Q_i)$. Then $h_{i2} = \mu_{GQ}(Q_i / Q_{\max})$.

4.1.3. Computation of the Cost Fuzzy Set

The computation of h_{i3} $(1 \leq i \leq n)$ is similar to that of h_{i2}. First the following values are found $\forall i : 1 \ldots n$,

$$C_i = \sum_{\forall r \in T \wedge r \in N_i} C_i^r + \sum_{\forall r \in T \wedge r \notin N_i} C_b^r \tag{6}$$

where C_i^r is the quality of r at N_i and C_b^r is the quality of r at the best node N_b. Let $C_{min} = MIN_i(C_i)$. Then $h_{i3} = \mu_{LC}(C_{\min} / C_i)$.

4.1.4. Computation of the Migration Overhead Fuzzy Set

The values h_{i4} for all values of i ($1 \le i \le n$) are computed as follows. First the estimation of the migration overhead, MO_i for task T input at a node N_i is carried out. The time required for the actual migration and the cost due to the migration are included in the calculation of MO_i as explained below

$$MO_i = \sum_{\forall r \in T \wedge r \in N_i} \left((2 * \psi_{ib}) + \left(MAX_{\forall j \wedge r \in N_j}\left(C_b^r - C_j^r\right)\right)\right) \tag{7}$$

The first term in the summation, $2* \psi_{ib}$ represents the actual migration time of the task. The second term indicates the cost overhead. C_j^r is the cost of the resource r at a node N_j and C_b^r is the cost or r at the best node N_b. The maximum of the difference is chosen as a pessimistic estimate. MO_i is computed for all values of i ($1 \le i \le n$). Let $MO_i = MIN_i(MO_i)$. Then $h_{i4} = \mu_{LM}(MO_{\min} / MO_i)$.

5. Performance Evaluation

A network of six nodes is simulated for carrying out the experiments. A total of 43 resources are distributed randomly across the nodes. The duration of the usage of the resources is randomly chosen with the average being 30 days of simulated time. The task arrival is modeled as a Poisson process with different arrival rates at each node. Let X denote the *system wide arrival rate*. For the simulation study, a single fuzzy set is used to represent all the following fuzzy sets – *Time Criticality* (*TC*), *Good Quality (GQ), Less Cost (LC), and Migration (LM)*.

The elements in the matrix R are generated as follows. Consider two objectives O_i and O_j. One of the two objectives is randomly selected for having a higher importance and its value in R is then chosen randomly from the 9-point ordinal scale of Table 1 proposed by Saaty. When the execution of T is over, a *performance measure* – which is the weighted average of the actual duration required, the quality accumulated, the cost, and the migration overhead incurred for the execution of T is calculated. The preference ratings (α values) for the corresponding objectives are the weights. All the objectives need to be minimized except the quality objective. Hence the value of the quality accumulated is multiplied by -1 in calculating the performance measure. In this way, the performance measures for all the tasks are summed up and the *Average Performance Measure P* = (*performance measures for all tasks / total number of tasks*) is calculated.

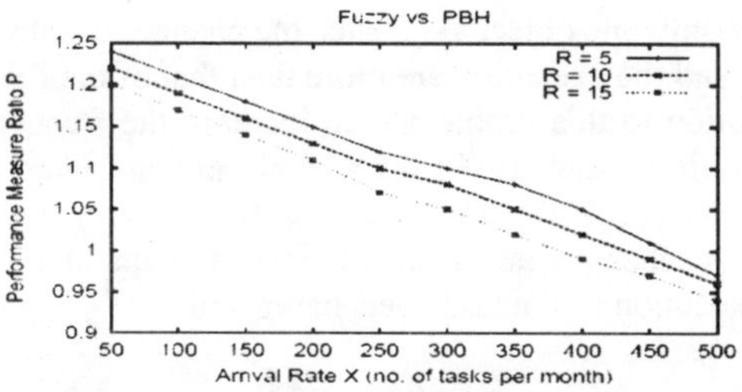

Figure 1: Comparison of Fuzzy and PBH Task Assignment Method

We have performed simulation studies comparing the fuzzy technique with different task assignment algorithm like the *Random,* the *Round-Robin,* and the *Preference-Based Heuristic (PBH)* algorithms. It has been observed that the fuzzy technique out forms all of them for a large range of task arrival rates. In the forthcoming subsection, the results of the comparison of the fuzzy technique with the *PBH* algorithm only are shown due to lack of space.

5.1. Comparison with the PBH Algorithm

In the *Preference-Based Heuristic (PBH)* task assignment algorithm, an input task is assigned to a node of the system that is considered to be the best for the objective for which the user has given the highest preference rating for that task. For example, if the user has the highest preference for the Execution Time objective, then equation (4) is used and the node giving the least value for that is chosen. All other objectives are ignored.

Let P_p be the *Average Performance Measure* of the *PBH* algorithm. $P = P_{fuz} / P_p$ which is the ratio of the *Average Performance Measure* of the fuzzy against the *PBH* technique. Graph for P is plotted against the values of the arrival rate X varying from 50 tasks/month to 500 tasks/month. Each such experiment is performed for three different values of R, the number of resources required by a task; the values being R=5, R=10, and R=15. Figure 1 shows these plots. The plots have to be interpreted as follows. Any point on a curve is a ratio comparing the performance measure of the fuzzy technique versus that of the *PBH* technique. For example, if the point is 1.2 then it means that the performance of the fuzzy technique is 20 percent better than the *PBH* algorithm. If the point is 0.9 then it means that the performance of the fuzzy technique is 10 percent worse than that of the *PBH* technique. From the observations the following conclusions can be drawn.

For a large number of the arrival rates (till 450 tasks/month for R = 5, till 400 tasks/month for R =10, and till 350 tasks/month for R = 15), the fuzzy approach gives consistently high gains than the *PBH* technique. The maximum gain is 24% for R=5, 22% for R = 10, and 20% for R = 15. Thus the fuzzy technique that uses the combination of all the objectives using Yager's model does perform better than the PBH technique that considers only a single objective.

For rates higher than these, the performance of the fuzzy technique is worse than that of the PBH technique. At this point, the system is *overhead.* It is to be noted that the system is constantly evolving where the surplus values keep varying depending on the arrival rate (of the system load). The surplus values are updated infrequently as compared to the very high arrival rates. The fuzzy technique uses the surplus values for the computation of the fuzzy sets for all the objectives while the PBH technique uses it for only one objective. Hence the chances of introduction of error in the fuzzy technique and the execution are more than that of the PBH technique.

One possible solution to this problem is to increase the frequency of updating of the surplus values with the increase in the arrival rate. However, if this is done, the system throughout plunges. If the updating frequency is very high, there are a huge number of message exchanges and a lot of time is spent in just processing those messages and the execution of the tasks gets hampered.

6. Conclusions

In this paper, we presented a model of an e-University system and addressed the important problem of assigning a student to a node of the system. Multiple criteria and preference ratings could be associated for a curriculum (a task) of a student. Fussy multiobjective decision-making process has been adapted for task assignment is a dynamic distributed system. The simulation results suggest that the fuzzy technique has significance performance gained over the commonly used heuristic techniques for a large range of task arrival rates.

References

1. e-University Task Force (2000), The University of Texas at Austin e-University Initiative, *http://www.utexas.edu/e-Univerity*.
2. T. Saaty (1980), The Analytic Hierarchy Process, *Mc-Graw Hill.*
3. R. Yager (1978), Fuzzy Decision Making including Unequal Objectives, *Fuzzy Sets and Systems*, 1: 87-95.
4. L. Zadeh (1965), Fuzzy Sets, *Information and Control*, 8: 338-353.
5. U. Deshpande, A. Gupta & A. Basu (2004), Coordinated Problem Solving through Resource Sharing in a Distributed Environment, *IEEE Trans. on Systems, Man and Cybernetics – Part B – Cybernetics*, 2: 1299-1304.
6. R. Bellman & L. Zadeh (1970), Decision Making in a Fuzzy Environment, *Management Science,* 17: 141-164.

Fuzzy Logic and its Application in Technology and Management
Editors: Debjani Chakraborty, S. Nanda and D. Dutta Majumder

Image Texture Segmentation Using 2-D Fuzzy Gabor Filters

Arindam Samanta, A.K. Ray
Sangram Ganguly and Shuvabrata Nandi

Department Electronics & Communication Engineering
Indian Institute of Technology Kharagpur.

Abstract: In this paper we propose the "2-D Fuzzy Gabor filters" i.e., the fuzzyfication of the tunable orientation parameter (θ) of the filter. This is due to the following reasons: firstly, the fuzzyfication of orientation (θ) helps in achieving a predefined, continuous angular spread about any principal orientation/angle (θ), thus, approximating continuous sampling of the orientation parameter. Secondly, this achieves a smoothening effect on the filtering process, thus, producing very well delineated texture fields. Moreover, the filter bank has been designed so as to adapt its frequency parameter to the input image so as not to arbitrarily choose frequencies for processing purpose This achieves image based frequency sampling, and thus, further enhances the overall results of filtering as would be expected. A comparative study of the fuzzy and non-fuzzy filters has also been presented.

Keywords: Gabor Filters, Fuzzy Theory, Membership Function, Textured Image Segmentation, Image Entropy and Energy.

1. Gabor Filters and Mathematical Representation

The 2-D Gabor functions are complex sinusoid (carrier) gratings modulated by 2-d Gaussian functions in the space domain, and shifted Gaussians in the spatial frequency domain which means that they are complex valued functions on R^2. The Gabor filters can be configured to have various shapes, bandwidths, center frequencies, and orientations by the adjustments of parameters. Regardless of the region of frequencies passed, the 2-d Gabor functions uniquely minimize the 2-d space-frequency uncertainty principle for complex-valued functions on R^2. Hence Gabor functions can be interpreted as the product of a modulating amplitude envelope with a complex carrier function whose argument is a modulating phase envelope, both of which can be computed and analyzed separately.

A. Gabor Channel Filters:

The convolution version of the complex 2-d Gabor functions has the general form

$$(j=\sqrt{-1}), \qquad h(x,y)=g(x',y')\exp[2\pi j(Ux+Vy)] \qquad (1)$$

where $(x',y')=(x\cos\varphi+y\sin\varphi,-x\sin\varphi+y\cos\varphi)$ are rotated co-ordinates and

$$g(x,y) = \left(\frac{1}{2\pi\lambda\sigma^2}\right)\exp\left[-\frac{(x/\lambda)^2 + y^2}{2\sigma^2}\right] \tag{2}$$

thus *h(x, y)* is a complex sinusoidal grating modulated by a 2-d Gaussian function with aspect ratio λ, scale parameter σ, and major axis oriented at an angle φ from the x-axis. If $\lambda=1$, then φ need not be specified because *g(x, y)* is circularly symmetric. The spatial frequency response of the Gabor Function (1) is

$$H(u,v) = \exp\left\{-2\pi^2\sigma^2\left[(u'-U')^2\lambda^2 + (v'-V')^2\right]\right\} \tag{3}$$

where primed values denote the rotated co-ordinate values in φ. Thus *H(u, v)* is a band-pass Gaussian with minor axis oriented at an angle φ from the *u*-axis, aspect ratio $1/\lambda$, radial centre frequency $F = \sqrt{U^2 + V^2}$ (measured in cycles/image) and orientation $\theta = tan^{-1}(V/U)$ in radians. Usually it is most convenient to consider those filters whose modulating Gaussians have the same orientation as the complex sine grating (i.e. $\theta=\varphi$) when (1) and (3) reduces to,

$$h(x,y) = g(x',y')\exp[2\pi jFx'] \tag{4}$$

and

$$H(u,v) = \exp\left\{-2\pi^2\sigma^2\left[(u'-F)^2\lambda^2 + (v')^2\right]\right\} \tag{5}$$

B. Gabor Filters Properties

It is intuitive to specify Gabor filters in terms of their spatial frequency and orientation bandwidths. But the bandwidths having the form (1) is not clearly meaningful, as the range of radial frequencies passed varies with the orientation of the frequencies. The frequency (octave) and the orientation (radian) half-peak bandwidths of the daisy-petal Gabor Filter (F4) are $B = \log_2\left[\frac{\pi F\lambda\sigma + \alpha}{\pi F\lambda\sigma - \alpha}\right]$ and $\Omega = 2\tan^{-1}\left[\frac{\alpha}{\pi F\sigma}\right]$ respectively where $\alpha = \sqrt{\frac{\ln 2}{2}}$. By varying the free parameters B, Ω, F and θ, a Gabor filter of arbitrary orientation and bandwidth characteristics can be generated. Now it has been proved that shifting the frequency response of a Gaussian function does not change its localization in either domain. Daughman [ref] proved that 2-d Gabor functions *h(x, y)* to be the only function mapping $R^2 \rightarrow C$ simultaneously achieving the lower bounds of the uncertainty relations $\Delta x.\Delta u \geq \frac{1}{4\pi}$ and $\Delta xy.\Delta v \geq \frac{1}{4\pi}$. Thus 2-d Gabor filters optimally and uniquely achieve simultaneous localization in space and spatial frequency, which implicates that they can be defined to have very narrow frequency and orientation responses without disturbing spatial localization.

Now the Gabor function(4) has real and imaginary components given respectively by

$$h_c(x,y) = g(x',y')\cos[2\pi Fx']$$

and

$$h_s(x,y) = g(x',y')\sin[2\pi Fx'] \tag{6}$$

The results of convolving $h_c(x, y)$ and $h_s(x, y)$ with any 2-d function are approx. equal other than a difference of $\pi/2$ in the phase spectra along the direction of θ, $h_c(x,y)$ and $h_s(x,y)$ are in phase quadrature. The 2-d Gabor filter $h(x,y)$ is subsequently the analytic function associated with $h_c(x,y)$ and - $h_s(x,y)$ is the quadrature filter associated with $h_c(x,y)$ to within a good approximation. Since the convolution of an analytic function with another function is also analytic filtering an image with a 2-d Gabor filter results in a complex-valued analytic image containing a limited range of frequencies and orientations. The complex magnitude of the filtered image will be maximized over those regions of the original image characterized by the particular spatial frequency attributes to which the filter is tuned.

Figure 1: Quadrature components of a 2-d Gabor filter (a) real (symmetric) and (b) imaginary (asymmetric) components.

2. Fuzzy Theory and Fuzzyfication of the Orientation Parameter (θ) of the 2-D Gabor Filter

The fuzzy theory of sets was proposed by Zadeh [4] in 1965. Let U be the universe of discourse, $U = \{(u_1, f_A(u_1)), (u_2, f_A(u_2)), \ldots\ldots, (u_n, f_A(u_n))$ and let A be a fuzzy set of U. Then the fuzzy set A can be represented as

$$A = \{(u_1, f_A(u_1)), (u_2, f_A(u_2)), \ldots\ldots\ldots\ldots, (u_n, f_A(u_n))\} \qquad (7)$$

where f_A, f_A: $U \rightarrow [0, 1]$, is the membership function of the fuzzy set A; $f_A(u_i)$ indicates the degree of membership of u_i in A. If the universe of discourse U is an finite set, then the fuzzy set A can be expressed as follows:

$$A = \sum_{i=1}^{n} f_A(u_i)/u_i, \quad u_i \in U \qquad (8)$$

If the universe of discourse is an infinite set, then the fuzzy set A can be expressed as follows:

$$A = \int_U f_A(u_i)/u_i, \quad u_i \in U \qquad (9)$$

Membership functions can be of various shapes depending on choice and usage like sigmoid, Gaussian, trapezoidal, triangular etc, see figure 2.

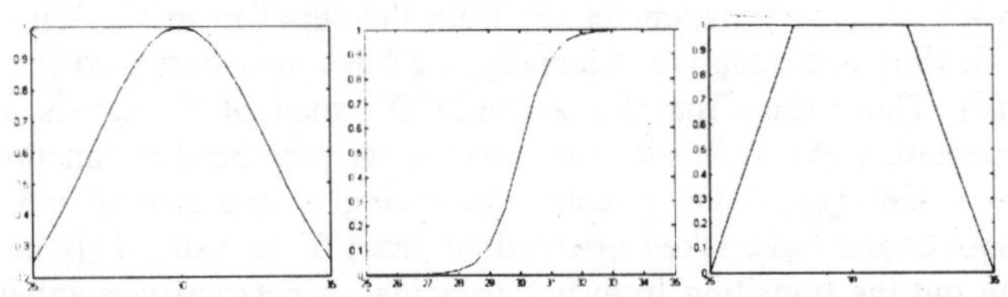

Figure 2: (a) Gaussian (b) sigmoid (c) trapezoidal membership functions.

The 2-d Gabor Channel filters yields excellent results for edge-detection in images with low noise level. In case the images are corrupted with high noise level, the performance of the rose-petal Gabor filters is not that effective. Here we have proposed fuzzyfication of the orientation θ in the 2-d Gabor Channel Filters. The new fuzzified Gabor filters take the following form:

$$h(x,y) = g(x',y')\exp(2\pi jFx') \qquad (10)$$

with the orientation θ being fuzzy i.e. $\theta = [\theta_i, \theta_f]$ where θ_i= starting value of orientation angle and θ_f = final value of orientation angle. Here we have used a triangular membership function to represent the fuzziness in the image, as shown in Figure 3, where $\mu(\theta_{crisp}) = 1$ and the starting and final values are so chosen that, $\theta_{crisp} = (\theta_i + \theta_f)/2$. A suitable interval of $(\theta_f - \theta_i)$ has been chosen here as 10°.

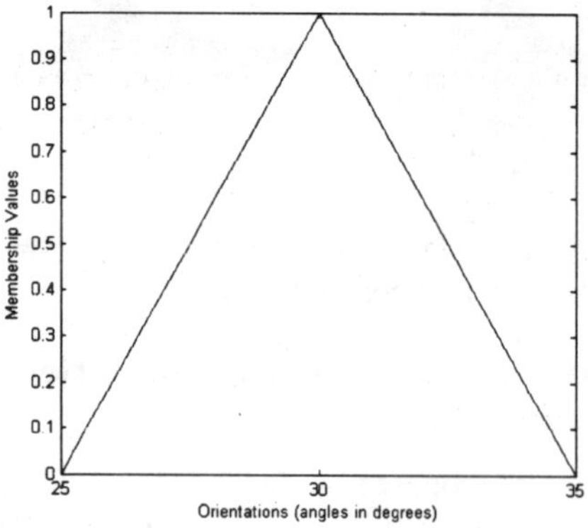

Figure 3: Membership function for orientation angle with x-axis showing θ-θ_{crisp} & y-axis the membership value.

3. Results and Discussions

We have used 256×256 monochromatic (multiple) texture images from the Brodatz texture image library. Also a set of remotely sensed images have been chosen in our experiment. The design of the Gabor filter bank has two broad aspects i.e. frequency (F) and orientation (θ). For selecting the frequencies, we have used the 2-D Fourier transform of the input image i.e. the 2-D FFT of the image has been used to supply the centre frequencies of the filter bank by way of an efficient algorithm. This is in keeping with the fact that each texture field has it's own characteristic centre spatial frequency and a narrow band of frequencies about it. This has helped in avoiding the arbitrary selection of centre frequencies and, thus, has helped to achieve better filtering results because the filter bank has been designed to adapt to the input image i.e. pick up it's frequency parameter (F) from the supplied image. This makes the filter more flexible and adaptive. Secondly, we have also fuzzyfied the orientation parameter (θ). This means that for a bandwidth span of 5 degrees about every principal orientation (θ) we have used a triangular membership function to assign membership values (μ) to the orientations having a step size of 0.5 degrees as discussed above, and have taken orientations from 0° to 180°. This has helped in smoothening out the transition from one principal orientation to another (say from

20° to 30°) unlike in the case of crisp logic. This is very true of the nature of fuzzy logic, it helps in the inclusion of imprecise and vague data in the process of analysis which would otherwise be rejected by classical or crisp logic i.e. a statement like "orientation about 60°", now, finds meaning in the filtering process which enhances the robustness and performance (segmentation ability) of the Gabor filter bank. This is because it takes care of all the major and minor orientations about a principal orientation. Furthermore, we have used the membership value (μ) of each orientation to weigh the filter response for that particular orientation and then, we have taken the weighted average of the filter responses for the bandwidth about each principal orientation (θ). In this way the statement "about 60°" finds meaning in the filtering process. We give a comparison between images filtered by a non-fuzzy Gabor filter bank and those filtered using a fuzzy Gabor filter bank based on a performance index which measures the average entropy of the images, using Shannon's definition as well as Pal's definition of entropy. We call this the Average Entropy Index (AEI) of the images. Furthermore, we also measure and plot the average energy of the images. As is evident from figures 4, 5 and 6 the images processed using the 2-D Fuzzy Gabor filters show better segmentation than the ones processed using the conventional non-fuzzy Gabor filters. The features of the fuzzy filtered images (atleast 10 out of 15 samples) are closer to those of the original image than the non-fuzzy filtered images implying better segmentation.

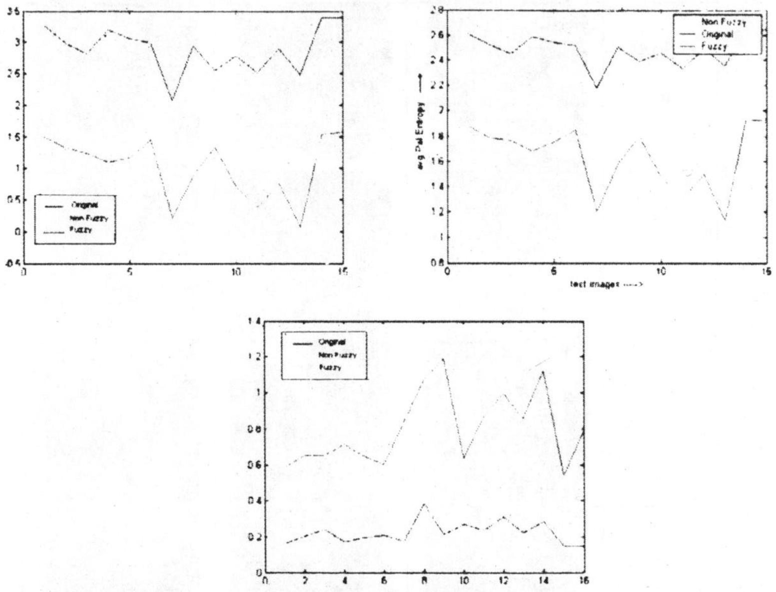

Figure 4: (a) Average Entropy Index (AEI) (Shannon), (b) Average Entropy Index (AEI) (Pal et al) and (c) Average energy of the original and filtered images

4. Concluding Remarks

A novel technique of texture recognition has been presented here employing fuzzy Gabor Transform, which yields excellent results on a large set of natural and artificial textured scenes. By fuzzifying the orientation parameter of the filter, the various texture orientations have been appropriately delineated.

Results

Sample No.	Original image	Image processed by fuzzy Gabor filter	Image processed by non-fuzzy Gabor filter
1).			
2).	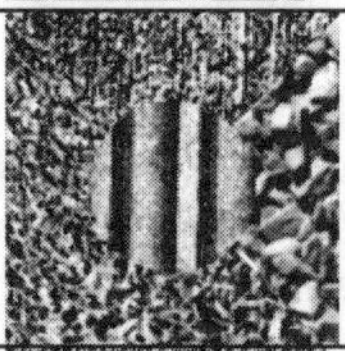	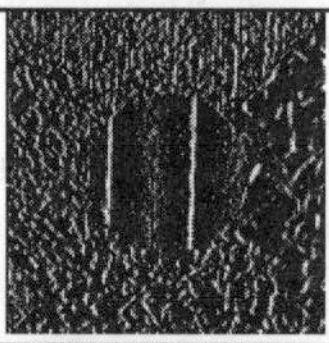	
3).		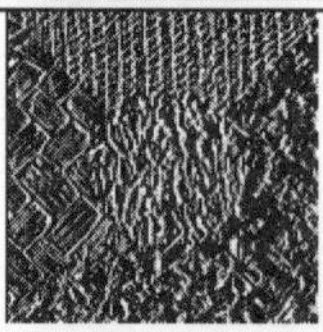	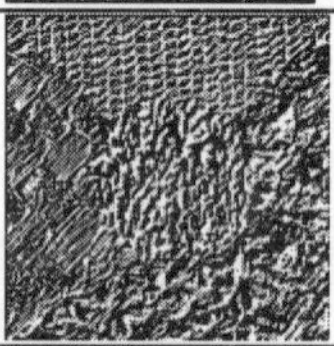
4).		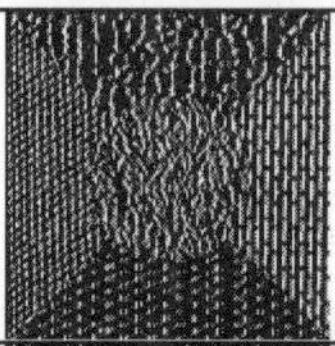	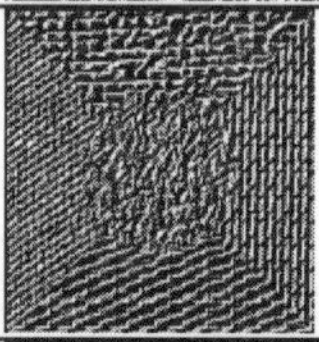
5).	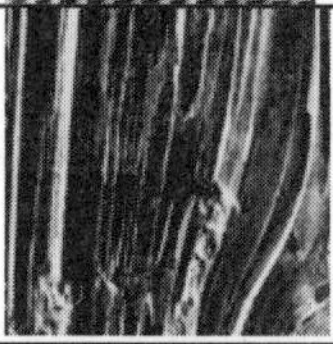	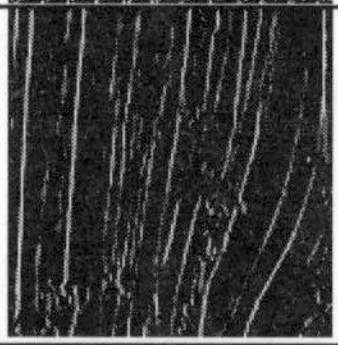	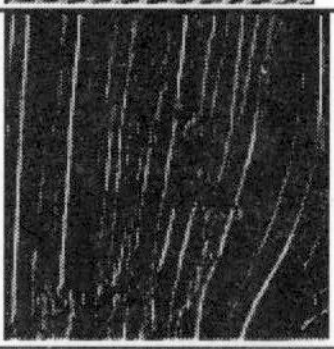
6).	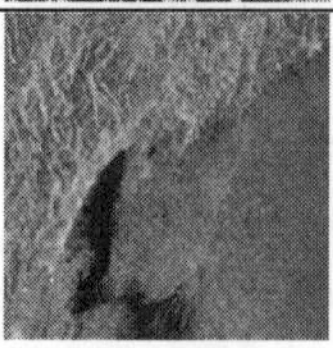	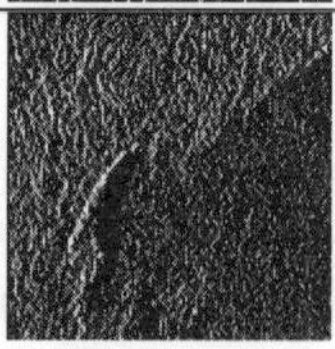	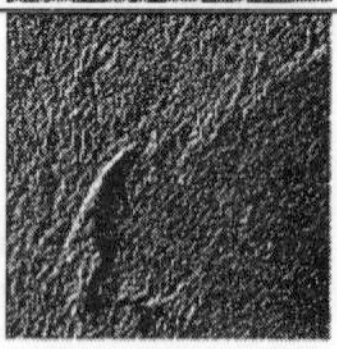

References

1. A.C Bovik, M. Clark & W. S. Geisler (1990), Multi-channel texture analysis using localized spatial filters, *IEEE Trans Pattern Anal. Machine Intell,* l PAMI-12: 55-73.
2. A.C. Bovik (1991), Analysis of multi-channel narrow-band filters for image texture segmentation, *IEEE Trans. Signal Processing,* 39: 2025-2043.
3. W.K. Pratt, *Digital Image Processing,* New York, Wiley 3rd edn..
4. M.R. Turner (1986), Texture discrimination by Gabor functions, *Biol. Cybern.*, 55: 71-82.
5. M. Porat & Y.Y. Zeevi (1988), The generalized Gabor scheme for image representation in biological and machine vision, *IEEE, Trans. Pattern Anal. Machine Intell,* 10: 452-568.
6. L.A. Zadeh (1965), Fuzzy sets, *Information and Control,* 8: 338-353.
7. L.A. Zadeh (1978), Fuzzy sets as a basis for a theory of probability, *Fuzzy sets and Systems*, 1: 3-28.
8. S.K. Pal & N.R. Pal (1993), A review on image segmentation techniques, *Pattern Recognition,* 26: 1277-1294.
9. P. Brodatz (1966), *Textures*: *A photographic album for artists and designers.* New York: Dover.
10. H.C. Nothdurft (1985), Orientation sensitivity and texture segmentation, *Vision Res.*,25: 551-560.
11. C.V. Jawahar & A.K. Ray (1996), Fuzzy statistics of digital images, *IEEE Signal Processing Letters,* 3: 225-227.
12. S.K. Pal (1998), A note on the quantitative measure of image enhancement through fuzziness, IEEE *Trans Pattern Anal. Machine Intell,* PAMI-4 (2): 204-206.

Fuzzy Logic and its Application in Technology and Management
Editors: Debjani Chakraborty, S. Nanda and D. Dutta Majumder

Fuzzy Logic Applications in Civil Engineering Management Problems

S.V. Barai

Department of Civil Engineering, IIT Kharagpur 721 302
Email: skbarai@civil.iitkgp.ernet.in

Abstract: Soft Computing tools such as fuzzy logic, neural networks, genetic algorithm and probabilistic reasoning have been successfully applied in civil engineering domain. The paper will address two such applications in the field of civil engineering management problems. Author and his research group members have developed the management problems applications using fuzzy logic and neural networks. They are - bridge evaluation management and construction management problems.

1. Introduction

Zadeh (1965) developed the fuzzy set theory. He believes that our ability to make precise and significant statements concerning a given system diminishes with increasing complexity of the system. The statement is very true in the context of civil engineering management. The civil engineering management domain often involves handling of uncertain and vague information. The complexity arises from the use of subjective opinion and numerical data. Fuzzy logic is now part of the movement towards developing effective intelligent systems for civil engineering management problems solving and decision making (Folger and Klir, 1991). Such systems can deal with complex and ill-structured situations. During development phase of intelligent systems, there are various choices of computational models (e.g. approximate reasoning, neural networks, genetic algorithms etc.) combining fuzzy logic, which can make application more realistic. The objective of the paper is to demonstrate two interesting applications of Fuzzy Logic in Civil Engineering domain. They are given below.

Maintaining records of serviceability of bridges and consequently retaining their level of reliability during their lifetime deserves high priority from techno-economic considerations. The aspects of bridge *inspection data* utilization and the assessment are of key importance and deserve mathematical treatment. The key issue to be addressed in such cases is to get an effective method of interpretation of the qualitative data and then mathematically manipulating it to yield reliable results. Quantitative evaluation however accurate is too cumbersome, hence the procedure of inspection needs to be standardized and the analysis needs to be done qualitatively. Hence there is a need to develop system, which can be well-organized bridge management system using fuzzy logic. Further, application on Internet can provide versatility.

Tasks like design of construction methods, concrete mixing and placement, constructability analysis, project planning, scheduling and control, construction quality control, etc. are receiving attention as application domains. Among them, developing Decision Support System (DSS) for constructability analysis is a challenging task. Computerizing of constructability analysis steps is a key to effective construction process automation. However, it is impossible to develop DSS for such analysis without acquiring formal constructability knowledge for various structural systems. This constructability knowledge cannot be manually acquired because of the complexity of problems involved. Also, the domain experts find it extremely difficult, if not impossible, to articulate complicated relationship among many design decisions regarding structural system and its constructability. The objective is to describe a novel application of NeuroFuzzy model (Kosko, 1996) for constructability analysis based on the use of machine learning (ML) and to demonstrate their feasibility.

Following sections explain the development of above discussed two applications in detail.

2. Bridge Evaluation Management Application Using Fuzzy Logic

***General*:** Bridges form an integral part of any country's transportation scheme, be it railway network or highway network. Maintaining records of serviceability of bridges and consequently retaining their level of reliability during their lifetime deserves high priority from techno-economic considerations. This aspect, unfortunately, has not hitherto been given the importance it deserves. For the optimal allocation of resources, a clear cut procedure, guidelines and maintaining records is desirable which is the sole motivation underlining the development of a Bridge Management System (BMS). The key aspects addressed in a BMS are: (a) Inspection (b) Maintenance (c) Repair and Rehabilitation.

***Scheme Identification For Bridge Structure Discretization*:** As per the guidelines given by IRC SP-35 (1990) a bridge can be divided into 13 components for visual inspection. These 13 components can then be subdivided into subcomponents to make the evaluation process easy. The overall scheme has these components: (1) Approaches (2) Protective Works (3) Waterway (4) Foundation (5) Substructure (6) Bearings (7) Superstructure (8) Expansion Joints (9) Wearing Coat (10) Drainage Spouts and Vent Holes (11) Handrails and Parapets (12) Footpaths (13) Utilities. These components further have been divided into subcomponents. These components are judged on the basis of some parameters relevant to them, and the details of these parameters are given elsewhere (Yadav, 2002).

***Identification Of Linguistic Terms*:** There are two parts of evaluation of components and subcomponents of bridges: (a) Evaluate the importance / weightage of subcomponent based on its requirement in the overall functioning of the component. (b) Evaluating its condition on the basis of its functioning. The main categories in which the evaluation type is classified are: (i) Evaluating Importance (ii) Evaluating condition. (iii) Evaluating damage (iv) Checking whether functioning is satisfactory or not. (v) Checking for abnormalities. The terms for various requirements are enumerated in Table 1:

Fuzzy Weighted Average Approach: The governing expressions such as fuzzy Weighted Average which has been adopted for assessing the degree of global damage in bridge structures in its simplest form is given below for $i = 1,2,n$ number of members.

$$y = \sum_{i=1}^{n} w_i x_i \Big/ \sum_{i=1}^{n} w_i \quad (1)$$

Table 1: Fuzzy variable descriptions

Fuzzy Variable	Fuzzy Terms
Condition Level	Extremely Good, Very good, Good, Medium, Poor, Very poor
Damage Level	Highly damaged, Significantly damaged, Moderately, Less, Slightly, Not damaged at all
Abnormalities Level	Very high, High, Reasonable, Little, Very little, None
Functioning level	Completely satisfactory, Quite satisfactory, Reasonably satisfactory, Barely satisfactory, Unsatisfactory, Highly unsatisfactory
Importance Level	Very Important, Important, Moderately Important, Less Important, Very less Important, Not Important

Where, x_i denotes the fuzzy condition rating of the ith element and w_i denotes the fuzzy importance factor for the same. The $[W_{ia,} W_{ib}]$ and $[X_{ia,} X_{ib}]$ are α-cut values of the fuzzy variable W_i and Xi corresponding to the fuzzy set μ_{Wia} and μ_{Xia}. Once the fuzzy function y (above expression) has been converted to the interval function Y, the solution can be obtained using the Vertex method (Barai, 2000).

***System Development Process*:** The process involves (a) Implementation of the forms for various components and subcomponents of bridge structure. (b) Implementation of the vertex method (Barai, 2000). (c) Integration of all the forms and the vertex method to get the unified system.

The Forms here is the basic input device that we have. A form consists the information as to how a component has to be inspected, it provides user with an exhaustive option list from which he/she can select his opinion, and finally forms are a means to store information for further processing. A typical form has following features: (i) Graphical User Interface. (ii) The guidelines to inspect various components. (iii) An exhaustive list of options according to the kind of assessment. (iv) Means of acquiring and storing the data. All these features are implemented using Java (Naughton and Schildt, 2000) environment. The typical screen shots of the system are shown in Figure 1.

The choice of the *Internet* Based Bridge Management System stems from the wide accessibility that the Internet is capable of providing. Another reason to make this particular system web based is owing to the fact that the systems, which are not available on Internet, are totally closed ended from the point of view of the user. The system has been developed using fuzzy weighted approach. With advancement of Information Technology, there is a scope to develop the system on wireless communication based Personal Digital Assistants (PDAs), Laptops and other mobile computers.

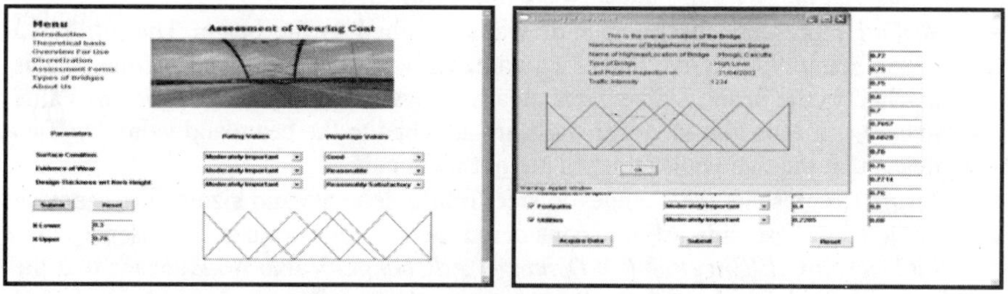

Figure 1: Typical screen shots of system

3. Construction Management Application using NeuroFuzzy Model

General: Construction as a sub-discipline of civil engineering is perhaps at the forefront of testing the applicability of Decision Support Systems (DSS) technology. There is a potential in developing DSS for constructability evaluation (Tatum 1987, Eldin 1988). In developing DSS, an important issue of knowledge acquisition and implementation comes into picture. Artificial Neural Networks (ANN) can be an excellent learning model for DSS (Haykin 2000). For exploring the possibility of ANN application to constructability analysis, the objectives of the current study can be divided into four major issues. They are as follows: (a) Qualitative constructability analysis data modeling using NeuroFuzzy models (b) Comparison of NeuroFuzzy Model with ML generated Rules (c) Performance reliability of the NeuroFuzzy model (d) Missing data handling in NeuroFuzzy model for constructability analysis

Data collection: The relevant data for the current study is obtained from literature (Skibniewski et al., 1997). In this collection of 31 data examples, each example represents a structural design concept evaluated from the point of view of its constructability. This data is used to train the ANN models. The data set considers the constructability of a beam in a 12-storied building. There are seven independent attributes describing the beam design. The dependent attribute is constructability evaluation (ConEva). It is a measure of the constructability of a given design and has been attributed three values: *Poor, Good and Excellent.* The independent attributes are described below:

- ***ReRa*** represents reinforcement ratio of the beam, and three values are attributed: *Low, Average and High.*
- ***CoBeRa1*** represents the first beam to column connection with the values attributed as: Low, *Average and High.*
- ***CoBeRa2*** represents the second beam to column connection with values attributed as: Low, *Average and High.*

- ***NoSla*** represents the number of slabs attached to the beam. The attributed values being *None, One, SameTwo and DiffTwo*. Value *One* means one slab is attached to the beam. Value *None* means no slab is attached to the beam. Value *SameTwo* means two identical slabs are attached to the beam and value *DiffTwo* means that the two slabs attached are not identical.
- ***NoWall*** represents the number of walls attached to the beam. The attributed values being *None, One, SameTwo* and *DiffTwo*. Value *One* means one wall is attached to the beam. Value *None* means no wall is attached to the beam. Value *SameTwo* means two identical walls are attached to the beam and value *DiffTwo* means that the two walls attached are not identical.
- ***BcCha1*** represents the changes in steel reinforcements and size of the beam on the left or first side of the considered beam. Five values are used: *None, SliChgereinf, AllChgeReinf, WDchange, AllChange*. Value *None* means that the beam on the left side of the column has exactly the same shapes and size as that of the beam on the right side. Value *SliChgereinf* means that the beam on the left side has same size but slightly different reinforcement as the beam on the right hand side. Value *AllChgeReinf* means that the beam on the left side has same size but different reinforcement as the beam on the right hand side. Value *WDchange* means that the width or depth of the beam on the left side has changed but the reinforcement has not. Value *AllChange* means that the two beams are entirely different in size and reinforcement.
- ***BeCha2*** represents the changes to the beam on the right side as in the previous attribute regarding the beam on the left.

Data Modeling: Fuzzy Information Presentation- the independent and dependent attributes were modeled as trapezoidal fuzzy functions and also discrete binary values. Various previous works has shown that trapezoidal fuzzy functions can closely imitate the modeled parameters. The modeling of dependent and independent attributes for the typical data is given in (Nair, 2000).

Incomplete/Missing Data Modeling- The data in which the 'or' estimations were given was modeled by taking the average of the attributed values. The empty fields in Tables of (Skiebniewski et al., 1997) will be treated as a missing data for ANN modeling purpose. For the missing data, it is considered as the average of all the attribute values associated with that attribute data used for training the network. Missing data values are modeled as trapezoidal fuzzy functions and the final representation of the missing values are given in Skiebniewski et al. (1997).

NeuroFuzzy Modeling: Backpropagation based multilayer perceptron was selected. The neural network architecture had two hidden layers with 48 and 48 hidden units having sigmoid activation function in each layer. The program was implemented using improved backpropagation in C programming Language in Turbo C Environment. After several exercises, keeping compromise between the accuracy and the computational time, sum square error was selected as 0.0001, learning rate as 0.7 and momentum rate as 0.9. Note that no optimization of the architecture or training parameters was performed. In contrast, common architecture and parameters were selected so that the time consuming exercise will take reasonable time.

Neural Networks Evaluation: The evaluation methods used to estimate the performance of ML models in general as well as ANN models in particular are:

Resubstitution (R), Hold out (H), cross-validation such as leave – one – out (L) or k-fold cross - validation (K). Degree of response is defined as the ratio of examples correctly predicted to number of examples tested.

- Resubstitution (R): The resubstitution exercise gave a degree of response equal to 1. All the examples were mapped perfectly as expected. This is a highly optimistic method.
- Leave-one-out (L) & K-fold (K): Cross-validation exercise was carried out 10 times on randomly generated data. Both these exercise gave a mean degree of response equal to .68. The deviation of the results was from .74 to .61. The uniformity of the results of these two exercises indicates that the performance of the model is reliable.
- Holdout (H): Holdout exercise was carried out 10 times, over randomly generated data. This exercise gave a mean degree of response equal to 0.64. The results deviated from .59 to .69. This exercise gave a low degree of response, as this is a highly pessimistic approach to check the performance of the model.

The efficiency of a model can be defined as the degree of response to the missing data. The ML approach generated 22 rules (Skiebniewski et al., 1997). Out of the 22 (10+12) tested examples, 18 examples mapped correctly. The degree of response was 0.82, i.e., 82% of the total data set had been correctly mapped.

4. Summary

In this paper fuzzy logic based applications in the domain of civil engineering domain were demonstrated. The first application was fuzzy logic combining with Internet application for bridge management system. The second application was fuzzy logic combining with neural networks for constructability analysis. The demonstrated application has shown greater potential of fuzzy logic in civil engineering management problems.

Acknowledgment

Author would like to acknowledge his former graduate students Mr. Rajeev Nair and Deepesh Yadav for carrying out developmental work and numerical simulations.

References

1. S.V. Barai (2000), Fuzzy Inference Based Procedure for Visual Inspection of Steel Bridges, *IRC Journal*, 61(3): 525-545
2. N.N. Eldin (1988), Constructability improvement of project design. *J. Constr. Engineering and Management, ASCE* 114(4): 631-640.
3. T.A. Folger & G.J. Klir (1991), *Fuzzy Sets, Uncertainty and Information*, Prentice Hall of India Private Limited, New Delhi – 110001.
4. IRC-SP-35 (1990), *Guidelines for Inspection and Maintenance of bridges*, Indian Roads Congress Special Publication 35, New Delhi
5. S. Haykin (2000), *Neural networks - A comprehensive foundation*, Pearson Education Asia, Second edition, India
6. B. Kosko (1996), *Neural networks and fuzzy systems*, Prentice-Hall International, Englewood Cliffs, N.J., U.S.A., 3 rd edition.

7. R.S. Nair (2000), *Neuro-fuzzy Learning Models for Constructability Analysis*, M.Tech dissertation Department of Civil Engineering, IIT, Kharagpur, India.
8. P. Naughton & H. Schildt (2000), *Java 2: The Complete reference*, Third edition, Tata McGraw-Hill Publishing Company Limited, New Delhi.
9. M. Skibniewski, T. Arciszewski & K. Lueprasert (1997), Constructability analysis: Machine learning approach, *J. of Computing in civil Engineering ASCE* 10(3): 6-16.
10. C.B. Tatum (1987). Improving constructability during conceptual planning. *J. Constr. Engineering And Management, ASCE*, 113(2): 191-207.
11. D. Yadav (2002), *Development of Internet Based Fuzzy Inference Driven Bridge Management System*, B.Tech Thesis Report, Department of Civil Engineering, IIT Kharagpur, India.
12. L.A. Zadeh (1965) Fuzzy Sets, *Information and Control*, 8: 338-353.

Fuzzy Logic and its Application in Technology and Management
Editors: Debjani Chakraborty, S. Nanda and D. Dutta Majumder

Personalized Mobile Fuzzy Fractal Ring Tone Generator

Suthikshn Kumar

Communication and Embedded Systems Division, Larsen and Toubro Infotech Ltd
#2 Church Street, Bangalore 560 001
Email: suthikshn.kumar@lntinfotech.com

Abstract: Fractals are very popular for generating images and music. We propose to use them for generating mobile ring tones. These fractal ring tones are both entertaining and melodious. As the computations required for generating melodious fractal tones are simple iterations, the ring tone generator can be integrated with the mobile handset. The Fuzzy fractals are proposed for extending the usefulness of the ring tone generator. This ring tone generator is personalized by using the audiogram. People with hearing impairment will benefit by the personalized ring tone generator. A PC based fractal ring tone generator demo is being developed based on the Nokia series 60 SDK for Symbian OS mobile handsets.

1. Introduction

Ever since the Nokia introduced the downloadable ring tones way back in 1990s, ringtones have gained popularity and are becoming the success factors for new handsets being introduced in the market. The current day handsets can play polyphonic ringtones, music clips, ring-back tones, voice-mail greeting messages and music messaging. Growth in the ringtone market has been exponential, almost doubling every quarter [17].

The current day market for mobile phones is buzzing with customer craze for new features. The Mobile handset makers are packing more and more features while bringing down the prices. In this scenario, the mobile ring tones have gained importance. The most recent mobile handsets have features such as polyphonic ring tones. The available PC based software or the tune composers on the handset itself can be used for generating new melody ring tones from music and can be downloaded into the handsets. The new generation mobile phones have additional features such as assigning of multiple ring tones for different callers, time based ring tone selection etc.

The customers of mobile handsets have a plethora of choice for ring tones. The ring tones can be downloaded from Internet, transmitted through messages, generated offline etc. While there exist ring tones generated from classical, pop, rock, jazz, etc, there is always a need for generating melodious ring tones, which are both catchy and easy to generate.

There are tools such as Composer for RingTone [22], which convert MIDI files, RTTTL ring or melodies composed by users to mobile telephone ring tones. These tools support almost all mobile phone brands available in market, including Nokia,

Ericsson, Motorola, Sony, Simens, Panasonic, Sumsung, Alcatel, LG etc. It can export 2 types of ring tone formats: key-press sequence ring tone and RTTTL ring tone. These tools display MIDI files or user's melodies in numbered musical notations, which is more convenient for editing. Users can freely modify the imported scores (cut, paste & change intercepts as part of score....) and listen to what they have created before exporting it to ring tone format.

This paper proposes the use of fractals such as Mandelbrot sets [2,3] for generating mobile ring tones. The Mandelbrot sets have earlier been used for generating music [4]. As the mandelbrot equations are computationally inexpensive, the mandelbrot ring tone generator application can be loaded onto the mobile handset and the ring tones are generated in real time.

The personalization of ring tones enhances the usefulness of the mobile handsets. Ring tone generator to generate the personalized ring tones uses the handset user's hearing profile. The ring tone frequencies are adapted to match the hearing profile and thus enhance their audibility to the user. Based on the concept proposed in this paper, an application has been developed using Nokia Series 60 SDK for symbian OS mobile phones.

This paper is structured as follows: In section 2 we discuss about Mandelbrot equations, fractals and mandelbrot music. The concept of Fractal Ring Tone generator for Mobile phones is presented in section 3 along with the proposal for Fuzzy Mandelbrot sets. In section 4, the implementation details along with the personalization are presented. Summary and conclusions are given in the section 5.

2. Fractals

Fractals are objects displaying self-similarity at different scales. The fractals have dimensions greater than their topological dimensions. The following figure (fig 1) illustrates the generation of fractals called Koch curve using simple iteration. The method is to replace a line segment with a triangle and repeat it forever. The first little iteration is shown in the figure and result is an interesting snowflake like curve.

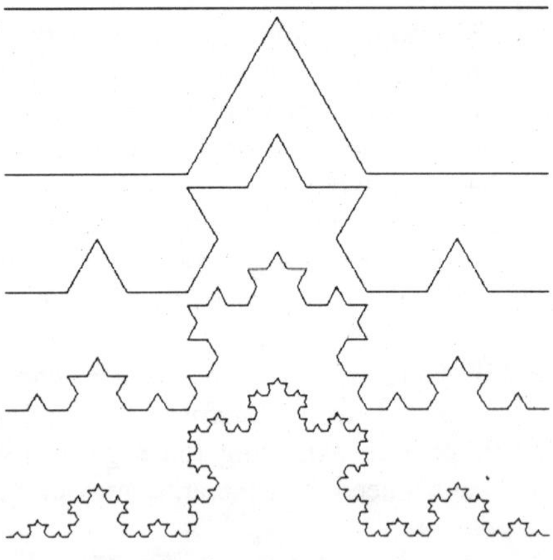

Figure 1: Fractal generation

The length of the fractal is infinite and size of it cannot be measured. An iterative algorithm unlike a Euclidean shape, which is given by a simple formula, describes it. There are several types of fractals and Mandelbrot sets are very popular.

2.1. Mandelbrot Sets

Mandelbrot set is a fractal named after its creator [3]. Unlike the other fractals, even though the Mandelbrot set is self-similar at magnified scales, the small-scale details are not identical to the whole, i.e., the Mandelbrot set is infinitely complex. But the process of generating it is based on an extremely simple equation.

The Mandelbrot set M is a collection of complex numbers. The numbers Z which belong to M are computed by iteratively testing the Mandelbrot equation:

$$z = z^2 + c$$

C is a constant. If the equation converges for chosen Z, then Z belongs to M. The following image (fig 2) shows the Mandelbrot set plotted on a complex plane.

The number of iterations n required to find whether the Mandelbrot equation diverges for a particular value of c is useful. While drawing color images of Mandelbrot sets, different colors can be assigned for different ranges of iterations to get a very colorful plot. Fig 2 shows the color plot of Mandelbrot set using the GNU XAOS software[27].

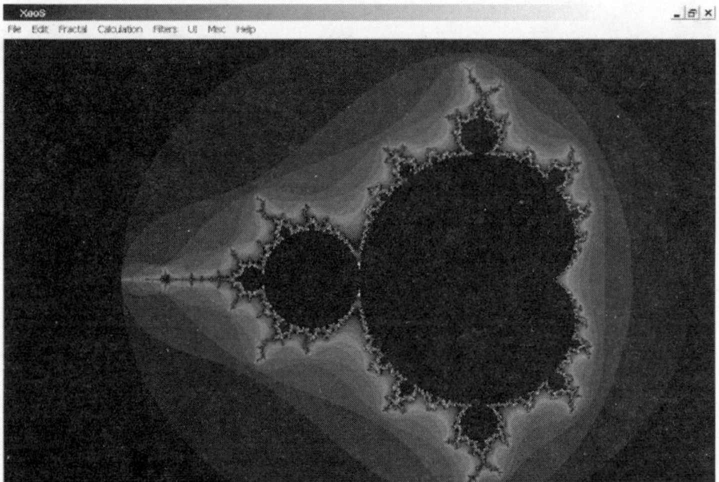

Figure 2: Color Mandelbrot Set Plot

2.2. Mandelbrot Music

There are earlier works, which have proposed schemes for synthesis of music using fractals [14] and mandelbrot sets [4]. Yo Kubota [4] has described in detail the steps involved in generating music from Mandelbrot sets [4]: The music parameters such Transition on time, Tone Pitch, Tone length, Tone volume, polyphony are calculated based on the Mandelbrot sets.

2.3. Fractals: Geometry of Music
In several fields such as economics, life sciences, cloud formation, the fractal geometry have been observed [16]. Self-similarity has also been observed in music [14, 25]. This shows that melodious music can be synthesized from fractals.

3. Fractal Ring Tone Generator
The mandelbrot ring tone generator proposed in this paper generates polyphonic ring tones in real time based on the mandelbrot music. When the user of the mobile receives a call, his/her mobile handset will invoke the mandelbrot ring tone generator application. This application in turn will compute the mandelbrot set in real time. This is used for generating the music using the user's preferences and map of mandelbrot set to ring tones. Whenever a user gets the mobile call, the ring tone will be different and thus will be both entertaining and melodious. The user selects the regions of interest in the mandelbrot plot and the ring tones are based on the computation of mandelbrot equations in that region.

3.1. Fuzzy Mandelbrot Sets
Fuzzy Mandelbrot sets are based on the fusion of concepts of Fuzzy sets [21] and Mandelbrot Sets. Fuzzy sets differ from crisp sets in that an element in a fuzzy set can belong to it with a degree of membership. The degree of membership can vary from 0 to 1.

The Mandelbrot equations in section 2.1 are used for computing the iteration values n. The values of n are plotted on a complex plane based on a fuzzy set as shown in the fig 3. Fuzzy Mandelbrot Set will have a smoother color transition when plotted on complex plane.

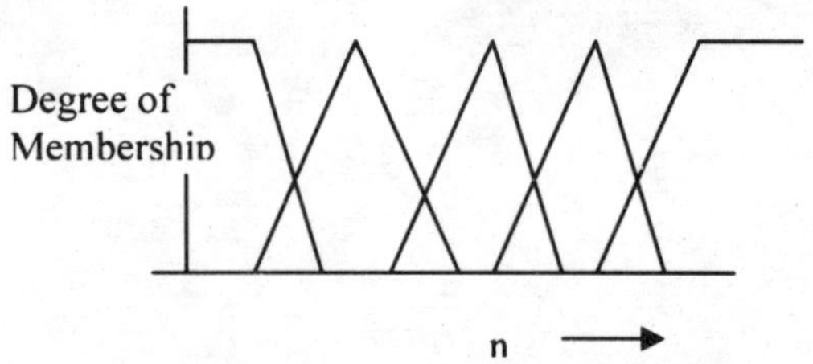

Figure 3: Fuzzy Mandelbrot Sets

4. Implementation

4.1. Nokia Series 60 Platforms
Nokia Series 60 platform for Symbian OS based mobile phones is a reference platform providing ample opportunities for developing new applications [7]. Its rich API library and the powerful SDK make it attractive for mobile application developers. A number of useful tools, libraries and the support for programming using languages such as C++ and Java prompted us to develop the Mandelbrot Ring Tone Generator using the Series 60 SDK.

4.2. Ring Tones Generator

The Ring Tone generator is being implemented using the Nokia Series 60 SDK. When the application is launched, it shows the color plot of the Mandelbrot set on the mobile Handset screen. The ranges n is mapped onto rainbow colors VIBGYOR, which in turn are mapped onto musical notes ABCDEFG. Different musical notes can thus be assigned to the different iteration ranges n. The initial value of C and incremental value are specified. The ring tone generator will use the Mandelbrot equation to compute the iteration values for range of values of C and generate corresponding musical tones. The resulting musical ring tones are stored in a scalable polyphone SPMIDI format [10].

4.3. Personalization

Personalization enhances the user's experience and effectiveness of the mobile phone. It serves the user by customizing the various services based on the individual's preferences and characteristics [13]. Earlier acoustic personalization of mobile phones has been proposed for improving the speech quality [1].

Hearing loss in individuals can be gradual and the quality of hearing can vary from person to person. Hearing loss can also result in difficulty for individuals in recognizing the mobile ring tones in the presence of background noise. Hence the Ring tone generator needs to be personalized depending on the individual's hearing requirements. For a person with hearing loss, the ring tone generator uses the hearing profile based on the audiogram

An Audiogram of a person shows the amount of hearing loss in each of the frequency bands [12]. Hearing is measured in unit of loudness called the DECIBEL. The decibel (dB) is a logarithmic scale. Physically, every 6 dB increase represents a doubling of sound pressure level. Every 10 dB increase sounds twice as loud. A person with hearing loss will perceive different frequencies at different levels [11]. An audiogram application has been developed for Symbian OS based phones using Nokia Series 60 SDK.

The fig 4 shows the screen shot of the Mandelbrot Ring Tone generator on the Nokia Series 60 Epoc Emulator. On selecting the Ring Tone generator application, the user is presented with options to select the regions of interest on Mandelbrot plot.

Figure 4: Screenshot of Mandelbrot Ring Tone Generator Emulated on Nokia 60 SDK

A real time fractal zoomer is provided similar to XAOS [27]. The user can import the hearing profile file generated by the audiogram application. On selecting the option to play will start the generated ring tone to be played. The user is also presented with the options for selection of notes for different colors (iteration values), duration of the notes, ringtone volume etc. The output ring tones generated are in Scalabale Polyphony MIDI (SP-MIDI) format [10]. The SP-MIDI can be played on mobiles, which have differing polyphony such as 8-note polyphony or 32-note polyphony.

5. Conclusion

Mandelbrot Ring Tone generators for Mobile handsets will have several advantages. It produces non-repeating melodious ring tones, which keep the user interested and entertained. Its ease of implementation as it is computationally inexpensive makes it attractive for developers. It is personalized to suit user's hearing requirement. Additionally, user can program to choose a vast range of tones.

This paper has introduced Mandelbrot sets and showed how it can be used for generating mobile ring tones. The fractal nature of music motivates us to use Mandelbrot sets. Implementation and Personalization aspects of the ring tone generator have been described with respect to Nokia series 60 Platform based mobile handsets.

References

1. Suthikshn Kumar (2003), *Acoustic Personalization of Mobile Phones*, 114th AES Convention, Amsterdam, The Netherlands, March 2003.
2. B. Mandelbrot (1977), *Fractals: Form, Chance and Dimension. San Francisco*, CA: W. H. Freeman and Company, 1977, xviii+265 pp.
3. B. Mandelbrot (2004), *Fractals in Chaos: The Mandelbrot Set and Beyond*, New York: Springer. 2004, xii + 230p.
4. Yo Kubota, *How to Compose Mandelbrot Music*, Available online: http://www.fin.ne.jp/~yokubota
5. J. Gleick (1998), *Chaos – Making a new Science*, Vintage Publishers, 1998.
6. M.J. Jipping (2002), *Symbian OS Communications Programmming*, John Wiley and Sons, 2002.
7. *Series 60 Application Framework Handbook*, Nokia, 2003.
8. *White Paper on Polyphonic Ring Signals, T300*, Sony-Ericsson, August 2002.
9. J.P. Mackenzie & M.B. Sandler, *Fractal Interpolation Functions for Sound Synthesis*, AES 90th Convention, 1991-92.
10. *Scalable Polyphony MIDI Specification (SP-MIDI)*, Midi Manufacturer's Association, Available online: www.midi.org
11. S. Launer (1995), *Loudness Perception in Listeners with Sensorineural Hearing Impairment*, PhD Thesis, University of Oldenburg, 1995.
12. E. Zwicker & H. Fastl (1999), *Psychoacoustics – Facts and Models*, Springer-Verlag, 2nd Edition.
13. G.C. van den Eijkel, M.M. Lankhorst, H. van Kranenburg & M.E. Bijlsma, *Personalization of Next-Generation Mobile Services*, Wireless World Forum Kick-off Meeting. Telematica Instituut, Enschede, The Netherlands

14. S. Yadegari (1991), Using Self –Similarity for Sound/Music Synthesis, *Proc of ICMC, Montreal*, pp. 423-424, ICMA, San Francisco.
15. J. Heckroth (1995), *Tutorial on MIDI and Music Synthesis*, MIDI Manufacturers Association.
16. B. Mandelbrot (1977), *The Fractal Geometry of Nature*, Freeman Publishers, New York.
17. T.Mohan (2004), Ring in the Moolah, Voice & Data, Feb 2004.
18. WinFract (FractInt for Windows), ver 18.21, Stone Soup Group.
19. R.T.Stevens (1991), *Fractal Programming and Ray Tracing with C++*, M&T Books, Calif.
20. D. Sen (2004), Understanding Digital Music and Sound Forge 6, BPB Publishers.
21. B. Kosko (1997), *Fuzzy Engineering*, Prentice Hall.
22. *Composer for Ringtones Pro V1.0*, www.ringtone-composer.com
23. *LEMUR: League of Electronic Musical Urban Robots*, Available online: http://www.lemurbots.org/
24. Jonathan Foote & Matthew Cooper (2001), Visualizing Musical Structure and Rhythm via Self-Similarity, *Proc. International Conference on Computer Music (ICMC), Habana, Cuba, September 2001.*, September 12.
25. S.D. Yadegari (1992), *Self-Similar Synthesis: On the border between Sound and Music*, Masters Thesis, MIT, 1992.
26. D.R. Hofstadter (1999), *Godel, Escher, Bach: An Eternal Golden Braid*, Harper Collins.
27. *XAOS: A Real time fractal zoomer*, Free Software Foundation, http://xaos.sf.net
28. DNA Music Central, http://www.dnamusiccentral.com

Fuzzy Logic and its Application in Technology and Management
Editors: Debjani Chakraborty, S. Nanda and D. Dutta Majumder

A Parallel Self-Organizing Neural Network (SONN) based Graded Color Object Extractor using a MUSIG Activation Function with CONSENT

S. Bhattacharyya[1], P. Dutta[2] and D. Dutta Majumder[3]

[1]Department of Information Technology, Kalyani Government Engineering College, WB
Email: siddhartha@kucse.wb.nic.in
[2]Department of Computer Science & Technology, Kalyani Govt. Engg. College, WB
Email: param@kucse.wb.nic.in
[3]ECSU, Indian Statistical Institute, Kolkata
Email: ddm@isical.ac.in

Abstract: In this article, a parallel self-organizing neural network (SONN) employing a MUltilevel SIGmoidal (MUSIG) activation function with CONtext SENsitive Thresholding (CONSENT) for graded color object extraction is presented. The hostility of an image pixel with its neighbors is indicative of the neighborhood context of a pixel and its contribution in the neighborhood. The threshold values for the activation function are derived from these hostility indices. A parallel SONN, comprising three component SONNs, with error compensation based on linear indices of fuzziness, is then employed to extract graded color objects from a color image scene. An application of the proposed approach is demonstrated with a noisy color image and a real life color image. The standard correlation coefficient between the original and the extracted images is used as the figure of merit of the proposed approach.

Keywords: SONN, Index of fuzziness, MUSIG, CONSENT, Hostility index

1. Introduction

Object extraction and detection is a major problem in the fields of defense, surveillance and target tracking. The problem of object extraction becomes more challenging when the entire color gamut comes to the fray. Neural networks can be used to deal effectively with the extraction of binary, multiscale and color (both pure and graded) objects [1-6]. Attempts in this direction mainly rest on evolving novel network architectures to accommodate the different gray scale and color responses arising from the image content [4]. However, other approaches are centered on resorting to the conventional architectures, while modifying the characteristic transfer functions of the networks [5]. Hence, conventional networks are able to extract multiscale objects. A parallel version of these networks is efficient in handling graded color objects [6]. Multiscaling capabilities are induced in these networks by means of a multilevel sigmoidal activation function (MUSIG). But, the MUSIG activation function generally uses fixed point thresholding assuming homogeneous gray scale and color distribution in the image scene. Real life images, however, bear heterogeneity in color information.

In this article, we have proposed an approach of exploiting the heterogeneity of color information in a color image scene to decide on the thresholding values of the MUSIG activation function. This image context sensitive thresholding (CONSENT) is determined by computing the hostility of an image pixel with its neighbors. A parallel SONN architecture comprising three component SONNs for the three primary component colors, employs this MUSIG activation function for the extraction of graded color objects. An application of the approach is demonstrated with two color images.

2. Mathematical Concepts

A brief idea about the fuzzy set theory and the hostility index is presented.

2.1. Fuzzy sets and subsets

A fuzzy set [7, 8] contains elements $\{x_1, x_2, x_3, \ldots\}$ with a certain degree of membership, $\mu_A(x)$. $\mu_A(x)$ lies in [0,1]. The higher the value of $\mu_A(x)$, the higher is the membership value of an element x in the fuzzy set A. The height (*hgt*) of a fuzzy set is the maximum membership value of the elements of a fuzzy set. If, for a fuzzy set, *hgt*=1, then it is a normal fuzzy set, otherwise, it is referred to as a subnormal fuzzy set. A normal fuzzy set is a superset of several nonempty subnormal fuzzy subsets. A subnormal fuzzy subset A_s can be normalized to a normal form by means of the normalization operator given by,

$$NormA_s(x) = \frac{A_s(x)}{hgt(A_s)} \tag{1}$$

For a subnormal fuzzy subset with support $[L, U]$, the normalization operator is expressed as,

$$NormA_s(x) = \frac{A_s(x) - L}{U - L}. \tag{2}$$

The corresponding denormalization can be achieved by

$$DeNormA_s(x) = L + (U - L)NormA_s(x) \tag{3}$$

The index of fuzziness [3] of a fuzzy set A having n supporting points is defined as

$$\nu(A) = \frac{2}{n^k} d(A, \underline{A}) \tag{4}$$

where $d(A, \underline{A})$ denotes the distance between fuzzy set A and its nearest ordinary set $\underline{A}$. An ordinary set $\underline{A}$ nearest to the fuzzy set A is defined as

$$\mu_{\overline{A}}(x) = \begin{cases} 0, & if \ \mu_A(x) \le 0.5 \\ 1, & if \ \mu_A(x) > 0.5 \end{cases} \tag{5}$$

The value of k depends on the type of distance metric used. For Hamming distance metric, k=1. The corresponding index of fuzziness is called the linear index of fuzziness $\nu_l(A)$. It is given by,

$$\nu_l(A) = \frac{2}{n}\sum_{i=1}^{n} |\mu_A(x_i) - \mu_{\underline{A}}(x_i)| = \frac{2}{n}\sum_{i=1}^{n}[\min\{\mu_A(x_i), (1 - \mu_A(x_i))\}] \tag{6}$$

2.2. Hostility Index

A $(M \times N)$ image can be represented by a neighborhood system, where a pixel (i, j) in the image is related to its neighbors by a dth order neighborhood (N_{ij}^{d}). For a second order neighborhood, d=2 and there are eight neighbors to a pixel. The level of heterogeneity in an image is reflected by the distribution of intensity levels in the neighborhoods and hence on the affinity of a pixel with its neighbors in terms of the intensity values. The closer the intensity values of a pixel and its neighbors, the higher is the homogeneity in the neighborhood and less is the pixel hostile to its neighbors. On the contrary, a heterogeneous neighborhood implies greater deviation in the intensity values of the neighborhood pixels and more hostile is a pixel to its neighbors. The degree of heterogeneity of the neighborhoods is indicative of the color distribution in the image scene. To account for this heterogeneity in a second order neighborhood, a hostility index can be defined as,

$$\zeta = \frac{1}{8}\sum_{i=1}^{8}\frac{|p-q_i|}{|p+1|+|q_i+1|} \tag{7}$$

where p is the gray value of the candidate pixel and q_i, i=1, 2, 3, ...,8 are the gray values of its neighbors. ζ lies in [0,1]. A higher value of ζ implies lower pixel level homogeneity and a lower value of ζ implies higher pixel level homogeneity.

3. SONN architecture for Graded Color Object Extraction using a MUSIG with CONSENT

A single multilayer SONN [3] can be used to extract binary objects. The same network architecture using a MUSIG activation function [5], can be used to extract multiscale objects from a multiscale image scene. A parallel version of the network comprising three such networks and a source layer and a sink layer [6] is efficient in the extraction of graded color objects from a color image scene. The neurons in each of the different layers of the different networks in the parallel architecture correspond to pixels in the image scene. There are no connections between neurons in the same layer of a network. Each neuron in a layer of a network is connected to the corresponding neuron and to its neighbors in the previous layer of that particular network following a neighborhood-based topology. Neurons in the output layer of a network are connected to the corresponding neurons in the input layer of the same network, for feedback of outputs. The inputs to each of the three parallel component networks are supplied from the source layer and are proportional to the component color values of the image scene and lie in [0,1] and the interconnection weights are initially set to 1.

The total input (I_i) to any neuron in any layer is given by

$$I_i = \sum_j w_{ij} o_j \tag{8}$$

where, o_j is the output of the jth neuron in the previous layer and w_{ij} is the connection weight between the ith node of one layer and the jth node of the previous layer. The output of a node i is obtained as,

$$o_i = f(I_i) \tag{9}$$

where f is the MUSIG activation function. The MUSIG activation function uses fixed point thresholding. This is valid if and only if the pixel neighborhood regions are homogeneous. However, in real life images, there is always some heterogeneity in a neighborhood region. This heterogenity of the image context influences the threshold values for the different levels of the MUSIG activation function. Thus the context sensitive threshold (CONSENT) value for a particular level is a function of the pixel hostility and is given by,

$$\theta_c = \zeta(\frac{1}{\alpha_i} - \frac{1}{\alpha_{i-1}}) \qquad (10)$$

where α_i and α_{i-1} are the class responses of the ith and the $(i\text{-}1)^{th}$ levels. So, the CONSENT based generalized sigmoidal activation function takes the form,

$$f(x) = \frac{1}{\alpha + e^{-(x-\theta_c)\lambda}} \qquad (11)$$

where, α controls the multilevel class responses, θ_c is the CONSENT value and λ is the steepness factor of the function. The MUSIG activation function is derived from this generalized form as,

$$f(x) \leftarrow f(x) + (\gamma\text{-}1)f(c), (\gamma\text{-}1)c \leq x < \gamma c \qquad (12)$$

where, γ represents the gray scale index and $1 \leq \gamma \leq K$, the number of color scale objects. Here, c represents the color contribution (assumed to be equal for all grades). Different responses for different grades of color are obtained by varying the α factor. The inputs of a network are propagated in the network from one layer to the next. The output layer outputs of a component network can be regarded as a fuzzy superset of several subnormal fuzzy subsets of color intensity. Since the component networks operate in a self-supervised mode, the system error is determined by computing the linear index of fuzziness in intensity levels of each of these fuzzy subsets. Back propagation of errors is used to adjust the interconnection weights between the neurons of different layers of the component networks. The outputs of these component networks are then fed back to the input layers of the same network. This is carried on until the system error gets reduced to zero or to some tolerable limit. Finally, the input image gets segregated into different regions representing the different grades of color, which is obtained at the sink layer. Figure1a shows MUSIG with fixed point thresholding and Figure1b shows MUSIG with CONSENT.

4. Proposed Object Extraction Methodology

The proposed technique for graded color object extraction using a parallel SONN architecture is carried out in the following two phases.

4.1. Determination of pixel hostility for computing CONSENT values

The pixel level hostilities, which are the measures of the level of homogeneity in the neighborhood regions are computed using equation (7). These hostility values are then used to determine the CONSENT values of the MUSIG activation function. This activation function is employed by each of the component SONNs in the parallel architecture for mapping the input color information into different grades of color.

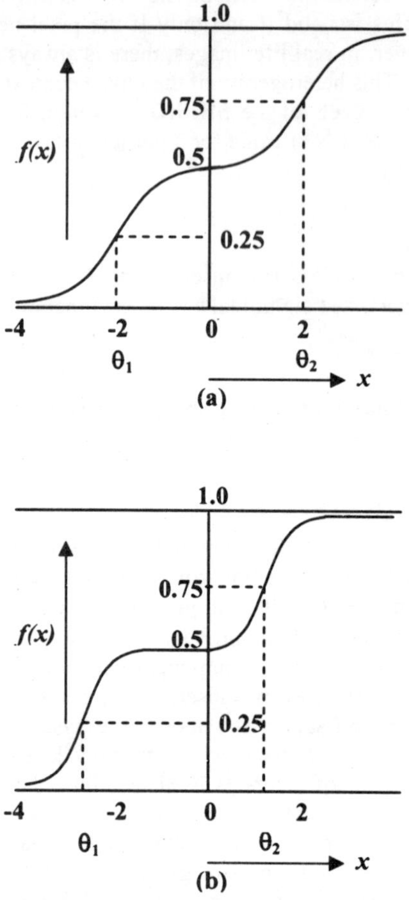

Figure 1: MUSIG with (a) fixed point thresholding (b) with CONSENT

4.2. Object Extraction by a parallel SONN architecture

The input color image is fed to the source layer of the parallel SONN architecture. The source layer extracts the three primary color components (Red, Green, Blue) from the image scene and propagates them to the three component networks. The three component SONNs extract spatially compact graded color regions from the color components by means of self-organization. Each of the component networks use a MUSIG with CONSENT values derived from the pixel hostility indices. At the output layers of these component networks, the linear indices of fuzziness of the different subnormal fuzzy subsets of color intensity levels are used as the system error. This error is back propagated to adjust the interconnection weights. The outputs are then fed back to the input layers for further processing. This is continued until the outputs at the output layers of the component networks are segregated into a number of different graded component color objects depending on the number of levels of the MUSIG activation function. Finally, the component network outputs are propagated to the sink layer for fusion of the different graded color components.

5. Results

An application of the proposed CONSENT based MUSIG activation function for graded color object extraction is demonstrated using a noisy synthetic image (Figure 2a, with different levels of Gaussian noise with standard deviation, $\sigma = 8$ and 10) and a real life cube image (Figure 3a). The noisy synthetic images are shown in Figure 2b and 2c. The extracted synthetic images are shown in Figure 2d and 2e using a three level MUSIG with CONSENT values. The extracted real life cube images using a three and a four level MUSIG with CONSENT values are shown in Figure 3b and 3c respectively.

5.1. Evaluation of image quality of extracted objects

The standard correlation coefficients (ρ) between the original images and the extracted images are used to evaluate the quality of the extracted objects. The correlation coefficients (ρ) for the two images are shown in Tables 1 and 2.

Noise level (σ)	Correlation Coefficient (ρ)
8	0.894557
10	0.872018

Table 1: Standard Correlation coefficient (ρ) for the synthetic image with 3 levels of MUSIG

Number of classes (K)	Correlation Coefficient (ρ)
3	0.975053
4	0.972223

Table 2: Standard Correlation coefficient (ρ) for the real life cubes image with different classes

6. Discussion and Conclusions

A parallel SONN architecture comprising three component SONNs for extraction of graded color objects, taking into account the heterogeneity of color information, is presented in this article. The image heterogeneity is accounted for by computing pixel level hostilities and incorporating them into deciding the threshold values of the characteristic transfer function to be used by the component networks. Such an architecture is capable of extracting different grades of color from a noisy/noisefree color image scene. However, the performance of the network degrades at higher noise levels. The authors are currently engaged in incorporating necessary modifications in the network architecture for increasing the noise immunity.

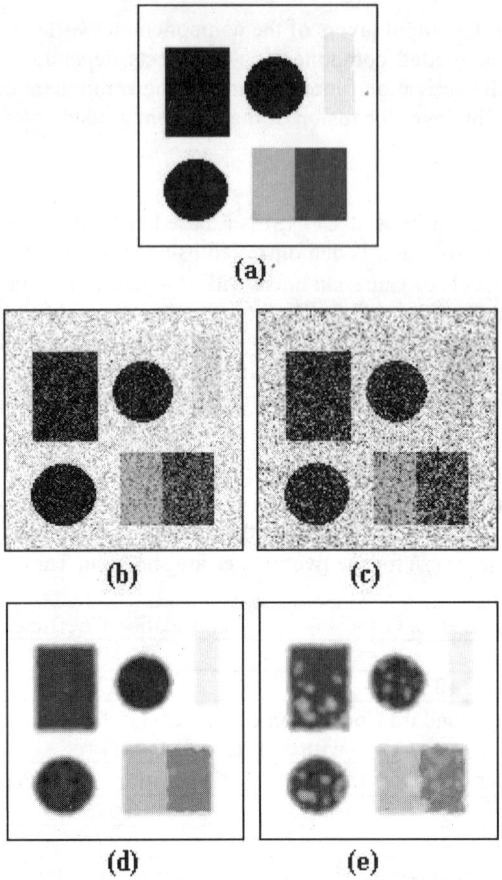

Figure 2(a) Original synthetic image
(b) (c) Noisy images (d) (e) Extracted images

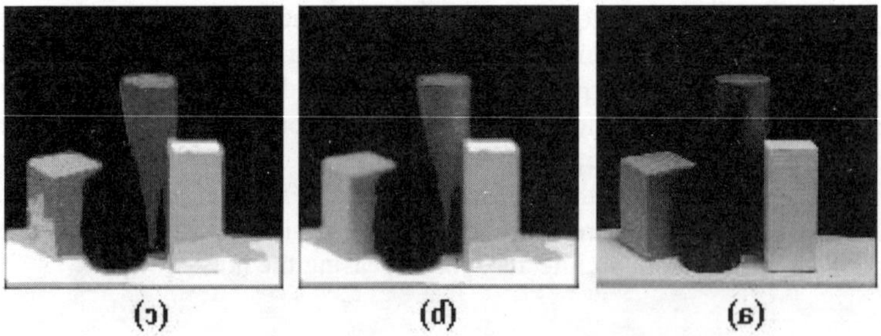

Figure 3(a) Original cubes image
(b) (c) Extracted images for 3 and 4 classes

References

1. J. Hertz, A. Krogh & R.G. Palmer (1991), *Introduction to the theory of neural computation*, Addison-Wesley.
2. M.N. Nasrabadi & W. Li (1991), Object recognition by a Hopfield neural network, *IEEE Trans. on Systems, Man and Cybernetics*, 21 (6): 1523-1535.
3. A. Ghosh, N.R. Pal & S.K. Pal (1993), Self-organization for object extraction using multilayer neural network and fuzziness measures, *IEEE Trans. on Fuzzy Systems*, 1 (1) 54-68.
4. A. Ghosh & A. Sen (2002), Self-organizing neural network for multi-level image segmentation, in *Soft Computing Approach to Pattern Recognition and Image Processing*, World Scientific.
5. S. Bhattacharyya, P. Dutta & U. Maulik (2003), Multi-scale object extraction using self organizing neural network with a multi-level sigmoidal activation function, *Proc. Fifth Int. Conf. on Advances in Pattern Recognition*: 435-438.
6. S. Bhattacharyya & P. Dutta (2004), A Parallel Self-Organizing Neural Network Architecture For Extraction of Graded Color Objects, *Proc. Intl. Conf. on Recent Trends and New Directions of Research in Cybernetics & Systems Theory*.
7. L.A. Zadeh (1965), Fuzzy Sets, *Inform and Control*, 8: 338-353.
8. T.J. Ross & T. Ross (1995), *Fuzzy Logic With Engineering Applications*, McGraw Hill College Div.

Fuzzy Logic and its Application in Technology and Management
Editors: Debjani Chakraborty, S. Nanda and D. Dutta Majumder

An Approach to Design of Fuzzy Rule-Base Using Dempster-Shafer's Evidential Reasoning for Tracking Application

Dwijesh Dutta Majumder[1], Arijit Laha, Sitansu Kumar Das Samit Dutta, Indranil Acharyya, Amitava Chanda and Suman Sarkar

Institute of Cybernetics Systems & Information Technology (ICSIT) Kolkata
[1]Email: ddmdr@hotmail.com

1. Introduction

Fuzzy approaches for solving tracking problem are one of the popular ways, as tracking of any object is closely related to human intelligence. Application of fuzzy mathematics in this type of artificial intelligence related field gives more realistic and reliable tracking result. The problem seems to be more challenging when information related to tracking comes from different sources. Here comes the question of uncertainty management. The problem is to identify which of the data association is correct-while simultaneously estimating the target states.

There exist different approaches to handle uncertainty. Some of them are Probability mass method, Dempster–Shafer (D-S) theory of combination of evidence. Among all of those popular approaches, Dempster–Shafer theory is proved as very efficient in the cases where three-dimensional space related problems are taken care of, such as CT/MR image registration. As our approach is tracking of an object in space, we are using D-S theory for our problem.

2. The Design

2.1. A brief over view of the scheme

In our scheme, the tracking module consists of three fuzzy rule-bases and one D-S combiner. The tusk of combiner is aggregation of the information supplied by fuzzy rule-bases. The real time data for tracking from three independent sources are fed to the three separate fuzzy rule-bases and the results of fuzzy rule-bases are combined by the Dempster-Shafer combiner to generate the final decision.

2.2. Fuzzy Rule-Base

A fuzzy *IF-THEN* rule can be defined as below:
IF v1 is A1 AND v2 is B1 AND...... THEN K
Where *v1,v2...* are the measured values of variables of interest.
A1,A2... fuzzy sets defined on the domain of the variable *v1,v2*.
K is a elements of decision set.
In the case of our tracking problem, there are two crisp variables *V1 and V2*.

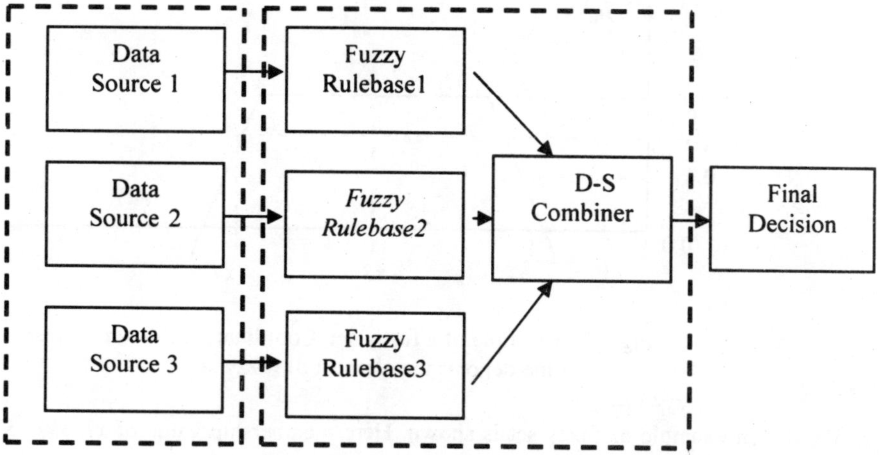

Fig 2.1. Main Building Blocks of tracking software

For each variable there exist three fuzzy set *F1, F2, F3* and in decision set there are three elements, which are D1, D2 and D3. An example of generation of fuzzy rules from the crisp rules is given below:

IF V1 margin is F1 AND V2 is F1 THEN D1.
IF V1 margin is F1 AND V2 is F2 THEN D1.
IF V1 margin is F1 AND V2 is F3 THEN D1.
IF V1 margin is F2 AND V2 is F1 THEN D2.
IF V1 margin is F2 AND V2 is F2 THEN D2.
IF V1 margin is F2 AND V2 is F3 THEN D3.
IF V1 margin is F3 AND V2 is F1 THEN D3.
IF V1 margin is F3 AND V2 is F2 THEN D3.
IF V1 margin is F3 AND V2 is F3 THEN D3.

This can be expressed in a tabular form:

V1 ↓ / V2 →	*F1*	*F2*	*F3*
F1	*D1*	*D1*	*D1*
F2	*D2*	*D2*	*D3*
F3	*D3*	*D3*	*D3*

The rules are supplied by the domain experts.

2.3. Fuzzy Set

A fuzzy set A defined over domain of a variable x is characterized by a membership function $\mu_a(x)$.

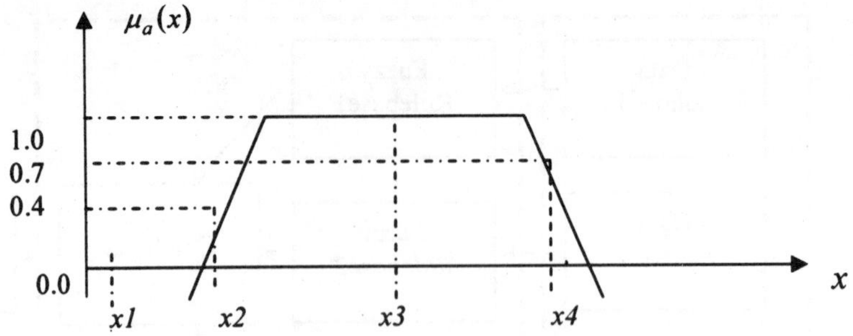

Fig 2.2. Example of a fuzzy set. Coordinates of green colored Line denotes the element of fuzzy set

In Fig 2.3 an example of fuzzy set is shown. Here membership value of *x1, x2, x3* and *x4* are as follows:

$\mu_a(x1)$=0.0, $\mu_a(x2)$=0.4, $\mu_a(x3)$=1.0 and $\mu_a(x4)$=0.7.

In our software normalized *V1* and V2 are taken along *X*. and for each variable three fuzzy sets are assumed for taking decision. For each variables the sets are taken as *D1*, *D2* and *D3*. The concept is illustrated in Fig 2.4.

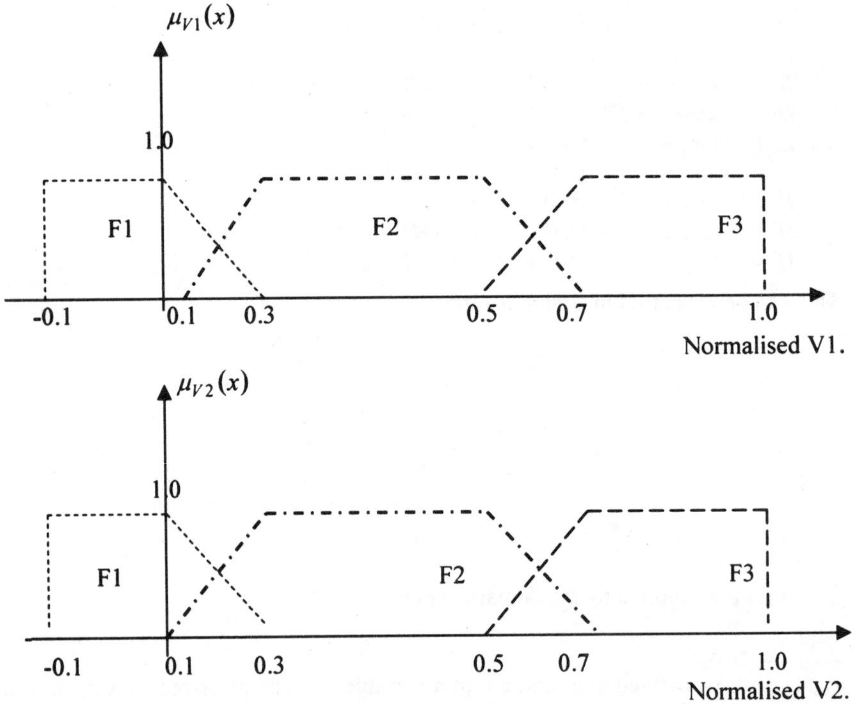

Fig 2.3. Fuzzy sets taken for tracing

Given the values *v1,v2* etc of the variables the firing strength of a rule can be computed as:

$$\alpha_{ij} = \min\{\mu_{aj}(V_{ij})\}, i = 1..k, j = 1..l$$

Where k is the number of set and l is the number of variable.

Formation of sets and rules can be made in two ways. One way is data-mining process and the other option requires expert's opinion. Data-mining process is based on large amount of experimental data. As for the time being enough data is not available, formation of sets and rules are subjected to domain knowledge-base supplied by experts.

2.4. D-S Combiner

Combiner is used for aggregating the outputs generated by the Fuzzy rule-base for three independent sensors. Each output of Fuzzy rule-base is in the form of possibilistic vector. The goal of combiner is to aggregate the information contained in possibilistic vectors using the Dempster–Shafer theory of combination of evidence.

2.5. A Brief Discussion on Dempster-Shafer's Theory of Combination of Evidence

Let θ be a universal set called the *frame of discernment*. In other words θ is a set of all possible hypotheses. Now a *belief structure (BS)* or *body of evidence (BOE)* is characterized by a *basic probability assignment (BPA)* function m:

$$m: P(\theta) \to [0,1] \text{ whenever } m(\phi) = 0 \text{ and } \sum_{A \subseteq \theta} m(A) = 1$$

$m(A)$ Signifies the degree of evidence supporting the claim that the true hypothesis belongs to the set A and given the value of $m(A)$ no more specific comment can be made about the special subset of A. $m(\theta)$ is interpreted as the degree of ignorance. All the sets $A \in P(\theta)$ for which $m(A) > 0$ is called focal element of belief structure m. Given a belief structure, two fuzzy measures Bel (belief measure) and Pl (plausibility measure) can be defined as follows.

Bel: Given a hypothesis, *Bel*(A) is a quantitative measure of belief that the true hypothesis is contained in A and can be formulated as follows:

$$Bel(A) = \sum_{B \subseteq A} m(B)$$

Pl: Given a hypothesis, *Pl(A)* is a quantitative measure of belief that the true hypothesis is not contained in the complimentary set of *A* and can be formulated as follows:

$$Pl(A) = \sum_{B \cap A \neq \phi} m(B) \forall A \subset \theta$$

Now this formalism can be applied to such situation that there are multiple experts who can provide their independent evidences each in the form of a separate belief structure. Now given this multiple set of evidences one is faced with the tusk of aggregating the evidences. This can be accomplished by Dempster-Shafer's rule of combination. Under this rule, evidence obtained in the same context from two distinct sources and expressed by two *BPA*s m^1 and m^2 on some power set $P(\theta)$ can be combined to obtain a joint *BPA* $m^{1,2}$ as:

$$m^{1,2}(A) = \begin{cases} \dfrac{\sum_{B \cap C = A} m^1(B)m^2(C)}{1-K} & \text{if } A \neq \phi \\ 0 & \text{if } A = \phi \end{cases}$$

Where

$$K = \sum_{B \cap C = \phi} m^1(B)m^2(C)$$

A common notation by can express the expression given bellow:

$$m^{1,2} = m^1 \oplus m^2.$$

2.7. A Few Characteristics of this Rule

The rule is commutative and associative. To be combinable two BOEs must be compatible, i.e, their focal elements must not be totally any overlapping. Combination of evidence from k sources is computed as:

$$m = m^1 \oplus m^2 \oplus ... \oplus m^k = \oplus_{i=1}^{k} m^i = \left(...\left(\left(m^1 \oplus m^2\right) \oplus m^3\right) \oplus ... m^k\right)$$

If m^1 is vacuous (i.e., $m^1(\theta) = 1$ and $m^1(A) = 0 \forall A \subset \theta$), then m^1 and m^2 are combinable and $m^1 \oplus m^2 = m^2$.

If m^1 is Bayesian (i.e., $m^1(A) = 0 \forall A$ such that $|A| > 1$) and m^1 and m^2 are combinable, $m^1 \oplus m^2$ is also Bayesian.

2.8. Final Classification Decision

The evidence theory accommodates the concept of belief into the subset of frame of discernment more than one element. Often we use them for more efficient information accumulation. At the decision making stage we would like to find out confidence against singleton element which depends on doubleton or tripleton elements. This can be calculated in three different procedures. These are Pignistic probability by belief measurement or by plausibility measurement. To this end we propose to use Pignistic probability method. Pignisti probability of a singleton element can be formulated as follows:

$$P^{\theta}(\Theta) = \sum_{A \subseteq \theta, \Theta \in A} \frac{m(A)}{|A|}$$

2.9. An example (Courtsy: Klir and Yuan)

An old painting is discovered. Style resembles Raphael's. It may be one by Raphael (R), or one of his discipline (D) or may be a counterfeit (C). Two experts are consulted. The result is as follows:

FEs	Expert 1		Expert 2		Combined	
	m_1	Bel_1	m_2	Bel_2	$m_{1,2}$	$Bel_{1,2}$
R	0.05	0.05	0.15	0.15	0.21	0.21
D	0.0	0.0	0.0	0.0	0.01	0.01
C	0.05	0.05	0.05	0.05	0.09	0.09
R∪D	0.15	0.2	0.05	0.2	0.12	0.34
R∪C	0.1	0.2	0.2	0.4	0.2	0.5
D∪C	0.05	0.1	0.05	0.1	0.06	0.16
R∪D∪C	0.6	1.0	0.5	1.0	0.31	1.0

We calculate the normalizing factor 1-K first. The value of K is calculated by the following expression:

$$K = m_1(R).m_2(D) + m_1(R).m_2(C) + m_1(R).m_2(D \cup C) + m_1(D).m_2(R) + m_1(D).m_2(C) + m_1(D).m_2(R \cup C)$$
$$+ m_1(C).m_2(R) + m_1(C).m_2(D) + m_1(C).m_2(R \cup D) + m_1(R \cup D).m_2(C) + m_1(R \cup C).m_2(D) + m_1(D \cup C).m_2(R)$$
$$= 0.03$$

$$m_{1,2}(R) = [m_1(R).m_2(R) + m_1(R).m_2(R \cup D) + m_1(R).m_2(R \cup C) + m_1(R).m_2(R \cup D \cup C) + m_1(R \cup D).m_2(R)$$
$$+ m_1(R \cup D).m_2(R \cup C) + m_1(R \cup C).m_2(R) + m_1(R \cup C).m_2(R \cup D) + m_1(R \cup D \cup C).m_2(R)]/0.97$$
$$= 0.21$$

$$m_{1,2}(D) = [m_1(D).m_2(D) + m_1(D).m_2(R \cup D) + m_1(D).m_2(D \cup C) + m_1(D).m_2(R \cup D \cup C) + m_1(R \cup D).m_2(D)$$
$$+ m_1(R \cup D).m_2(D \cup C) + m_1(D \cup C).m_2(D) + m_1(D \cup C).m_2(R \cup D) + m_1(R \cup D \cup C).m_2(D)]/0.97$$
$$= 0.01$$

$$m_{1,2}(C) = [m_1(C).m_2(C) + m_1(C).m_2(R \cup C) + m_1(C).m_2(D \cup C) + m_1(C).m_2(R \cup D \cup C) + m_1(R \cup C).m_2(C)$$
$$+ m_1(D \cup C).m_2(R \cup C) + m_1(R \cup C).m_2(C) + m_1(R \cup C).m_2(D \cup C) + m_1(R \cup D \cup C).m_2(C)]/0.97$$
$$= 0.09$$

$$m_{1,2}(R \cup D) = [m_1(R \cup D).m_2(R \cup D) + m_1(R \cup D).m_2(R \cup D \cup C) + m_1(R \cup D \cup C).m_2(R \cup D)]/0.97$$
$$= 0.12$$

$$m_{1,2}(R \cup C) = [m_1(R \cup C).m_2(R \cup C) + m_1(R \cup C).m_2(R \cup D \cup C) + m_1(R \cup D \cup C).m_2(R \cup C)]/0.97$$
$$= 0.2$$

$$m_{1,2}(R \cup D \cup C) = [m_1(R \cup D \cup C).m_2(R \cup D \cup C)]/0.97$$
$$= 0.31$$

Support against final classification will be:

$$P(R) = m(R) + m(R \cup D)/2 + m(R \cup C)/2 + m(R \cup D \cup C)/3$$
$$= 0.21 + 0.12/2 + 0.2/2 + 0.31/3 = 0.47$$
$$P(D) = m(D) + m(R \cup D)/2 + m(D \cup C)/2 + m(R \cup D \cup C)/3$$
$$= 0.01 + 0.12/2 + 0.06/2 + 0.31/3 = 0.2$$
$$P(C) = m(C) + m(R \cup C)/2 + m(D \cup C)/2 + m(R \cup D \cup C)/3$$
$$= 0.09 + 0.2/2 + 0.06/2 + 0.31/3 = 0.32$$

As $P(R) > P(C) > P(D)$, final conclusion could be drawn that painting was done by Raphael.

2.10. Application Of Dempster-Shafer's Theory Of Evidence In Our Software

In our application domain, each data source can be thought as a source of information that is being processed and judged by an expert to produce a belief structure. Each expert consists of the fuzzy rule-base that produces the possibilistic vector that represents the confidence of the rule-base for each alternative decision. This possibilistic vector is further processed by an expert .to generate a belief structure. The nature of the belief structure i.e. the possible focal element and how to assign the BPA into each of them may be designed to incorporate the constraint on the application. There could be several alternatives. Currently we are studying some of them and the final choice can be made based on experimental study and feedback from domain knowledge experts.

As previously discussed the role of fuzzy rule-base is to assign confidence (a fraction number between 0 to 1) against each decision alternatives.

$$f : X \to \alpha,\quad X = \{\textit{Decision alternatives: D1, D2, D3}\}, \alpha \in [0,1]$$

Possibilistic vector can be defined as:

$$\vec{\alpha} = [\alpha_{i=1}, \alpha_{i=2}, \alpha_{i=3}]^T$$

Where $i = 1,2$ and 3 is the number of classes. The frame of discernment is the set of classes $C = \{C_1, C_2, C_3\}$. Schemes for information aggregation for classification depend on how BOEs are defined, which focal elements are used and how BPAa are assigned to the focal elements of each BOE. In our case there are three possibilistic vectors produced by three fuzzy rule-bases from which three BOEs are prepared. The BPA of ith (where $i = 1,2$ and 3) BOE is computed as:

Let q = argmax (α_k^i) ,where $k = 1,2$ and 3

if (q>0.5) then,

$$m^i(\{C_q\}) = \alpha_q^i;$$

$$m^i(\theta) = 1 - \alpha_q^i$$

$$m^i(A) = 0 \forall A \in P(\theta) \setminus \{\theta, \{C_q\}\}$$

else if (q<0.5 && q>0.3) then

if r = argmax $(\alpha_k^i)k \neq q$ then

$$m^i(\{C_q, C_r\}) = (\alpha_q^i + \alpha_r^i)/2$$

$$m^i(\theta) = 1 - m^i(\{C_q, C_r\})$$

$$m^i(A) = 0 \forall A \in P(\theta) \setminus \{\theta, \{C_q, C_r\}\}$$

else if (q<0.3) then

$$m^i(\theta) = 1$$

$$m^i(A) = 0 \forall A \in P(\theta) \setminus \{\theta\}.$$

After computation of three BOEs the evidences are combined into single resultant BOEs the rules as follows:

$$m^{G} = \oplus_{i=1}^{3} m^i = ((m^1 \oplus m^2) \oplus m^3)$$

When two BOEs are combined there may be different situations, some of them are considered here:

Case 1: Let BPAs for which the focal element is common singleton element then such two BPAs combined as:

If $m^i(\{C_q\}) \neq 0$

$$m^i(\theta) \neq 0$$

$$m^i(A) = 0 \forall A \in P(\theta) \setminus \{\theta, \{C_q\}\}.$$

and $m^j(\{C_q\}) \neq 0$

$$m^j(\theta) \neq 0$$

$$m^j(A) = 0 \forall A \in P(\theta) \setminus \{\theta, \{C_q\}\}.$$

then $m^{ij}(\{C_q\}) = 1 - m^i(\theta) m^j(\theta)$

$$m^{ij}(\theta) = m^i(\theta) m^j(\theta)$$

$$m^{ij}(A) = 0 \forall A \in P(\theta) \setminus \{\theta, \{C_q\}\}.$$

***Case 2*:** Let BPAs for which the focal element is different singleton element then such two BPAs combined as :

If $m^i(\{C_q\}) \neq 0$

$$m^i(\theta) \neq 0$$

$$m^i(A) = 0 \forall A \in P(\theta) \setminus \{\theta, \{C_q\}\}.$$

and $m^j(\{C_r\}) \neq 0$

$$m^j(\theta) \neq 0$$

$$m^j(A) = 0 \forall A \in P(\theta) \setminus \{\theta, \{C_r\}\}.$$

then $m^{ij}(\{C_q\}) = \left(m^j(\theta) - m^i(\theta) m^j(\theta)\right) / (1 - K)$

$$m^{ij}(\{C_r\}) = \left(m^i(\theta) - m^i(\theta) m^j(\theta)\right) / (1 - K)$$

$$m^{ij}(\theta) = m^i(\theta) m^j(\theta)$$

$$m^{ij}(A) = 0 \forall A \in P(\theta) \setminus \{\theta, \{C_q\}, \{C_r\}\}.$$

$$K = \left(1 - m^i(\theta)\right)\left(1 - m^j(\theta)\right)$$

Case 3: Let BPAs for one of which the focal element is singleton element and for other focal element is doubleton element having non-zero singleton element as a subset such two BPAs combined as:

If $m^i(\{C_q\}) \neq 0$

$$m^i(\theta) \neq 0$$

$$m^i(A) = 0 \forall A \in P(\theta) \setminus \{\theta, \{C_q\}\}.$$

and $m^j(\{C_q, C_r\}) \neq 0$

$$m^j(\theta) \neq 0$$

$$m^j(A)=0 \forall A \in P(\theta) \setminus \{\theta, \{C_q, C_r\}\}.$$

then $m^{ij}(\{C_q\})=1-m^i(\theta)$

$$m^{ij}(\{C_q, C_r\})=m^i(\theta)-m^i(\theta)m^j(\theta)$$

$$m^{ij}(\theta)=m^i(\theta)m^j(\theta)$$

$$m^{ij}(A)=0 \forall A \in P(\theta) \setminus \{\theta, \{C_q\}, \{C_q, C_r\}\}.$$

Case 4: Let BPAs for one of which the focal element is singleton element and for other focal element is doubleton element having not any non-zero singleton element as a subset such two BPAs combined as:

If $m^i(\{C_q\}) \neq 0$

$$m^i(\theta) \neq 0$$

$$m^i(A)=0 \forall A \in P(\theta) \setminus \{\theta, \{C_q\}\}.$$

and $m^j(\{C_r, C_s\}) \neq 0$

$$m^j(\theta) \neq 0$$

$$m^j(A)=0 \forall A \in P(\theta) \setminus \{\theta, \{C_r, C_s\}\}.$$

then $m^{ij}(\{C_q\})=(m^j(\theta)-m^i(\theta)m^j(\theta))/(1-K)$

$$m^{ij}(\{C_r, C_s\})=(m^i(\theta)-m^i(\theta)m^j(\theta))/(1-K)$$

$$m^{ij}(\theta)=m^i(\theta)m^j(\theta)$$

$$m^{ij}(A)=0 \forall A \in P(\theta) \setminus \{\theta, \{C_q\}, \{C_r, C_s\}\}.$$

$$K=(1-m^i(\theta))(1-m^j(\theta))$$

Case 5: Let BPAs for which the focal element is common doubleton element then such two BPAs combined as :

If $m^i(\{C_q, C_r\}) \neq 0$

$$m^i(\theta) \neq 0$$

$$m^i(A)=0 \forall A \in P(\theta) \setminus \{\theta, \{C_q, C_r\}\}.$$

and $m^j(\{C_q, C_r\}) \neq 0$

$$m^j(\theta) \neq 0$$

$$m^j(A)=0 \forall A \in P(\theta) \setminus \{\theta, \{C_q, C_r\}\}.$$

then $m^{ij}(\{C_q, C_r\})=1-m^i(\theta)m^j(\theta)$

$$m^{ij}(\theta)=m^i(\theta)m^j(\theta)$$

$$m^{ij}(A)=0 \forall A \in P(\theta) \setminus \{\theta, \{C_q, C_r\}\}.$$

Case 6: Let BPAs for which the focal element is different doubleton element then such two BPAs combined as:

If $m^j(\{C_q, C_r\}) \neq 0$

$m^j(\theta) \neq 0$

$m^j(A) = 0 \forall A \in P(\theta) \setminus \{\theta, \{C_q, C_r\}\}$

and $m^j(\{C_r, C_s\}) \neq 0$

$m^j(\theta) \neq 0$

$m^j(A) = 0 \forall A \in P(\theta) \setminus \{\theta, \{C_r, C_s\}\}$.

then $m^{ij}(\{C_q, C_r\}) = m^j(\theta) - m^i(\theta) m^j(\theta)$

$m^{ij}(\{C_r, C_s\}) = m^i(\theta) - m^i(\theta) m^j(\theta)$

$m^{ij}(\theta) = m^i(\theta) m^j(\theta)$

$m^{ij}(A) = 0 \forall A \in P(\theta) \setminus \{\theta, \{C_q\}, \{C_r, C_s\}\}$

Finally decision rule can be computed from Pignistic probability. The Pignistic probability of class C_k:

$$P^C(C_k) = m^G(\{C_k\}) + \frac{\sum_{i=1, l \neq k}^{3} m^G(\{C_k, C_l\})}{2} + \frac{m^G(\{C_k, C_l, C_m\})_{k \neq l \neq m}}{3}$$

Missile's state will be decided for the state for which $P^C(C_k) \geq P^C(G) \forall C_1 \in C$.

3. Conclusion

Our software will do the job for a domain expertise. Introduction of fuzzy mathematics makes the software more realistic in this scenario. There is another aspect of fuzzyfication. We are going to develop a training algorithm with the help of data-mining process so, that the formation of fuzzy-sets can be made more specific. It will make the program autonomous in the sense of knowledge discovery.

Fuzzy Logic and its Application in Technology and Management
Editors: Debjani Chakraborty, S. Nanda and D. Dutta Majumder

Uncertainty Management of A Powered Projectile: A Fuzzy Differential Inclusion Approach

Amitabha Chanda, Sitangsu Das, Samit Dutta, Suman Sarkar
Indranil Acharyya and Dwijesh Dutta Majumder[1]

Institute of Cybernetics Systems & Information Technology (ICSIT) Kolkata
[1]Email: ddmdr@hotmail.com

1. Introduction

Let us suppose that a projectile of mass M, carrying an initial fuel load of mass m_0. We also consider that thrust given by the burning of fuel is constant and so also the burning rate of fuel. The projectile faces air resistance D during its flight. Thrust is constant, but on the other hand mass of the projectile plus fuel decreases. So acceleration imparted by the constant thrust is continuously increasing. It may be noted in this context that thrust must be adequate for imparting acceleration greater than the acceleration due to gravity. Other wise the projectile will be unable to leave the launch pad. Initially, velocity is zero, but as it goes up it increases.

Thrust in this part of motion then is allocated over (i) acceleration of the projectile (ii) air drag and (iii) weight of the projectile plus fuel. Therefore, a balance equation can be written as below.

$$\frac{d(MV)}{dt} = F - D - Mg \tag{1}$$

$$M = M_0 + m,$$

$$D = 0.5\, \rho\, C_D\, AV^2,$$

Where,

M – the instantaneous mass of projectile plus fuel.

M_0 – the mass of the projectile proper.

m – the instantaneous mass of fuel.

D – air drag.

A – area of the projectile normal to its direction of motion.

V – the speed of the projectile.

ρ – density of air.

C_D - coefficient of drag.

The projectile makes an angle θ with the vertical. Its movement may be resolved into two directions, one downrange and other vertical. Components of thrust along vertical and downrange are respectively ($F\cos\theta$) and ($F\sin\theta$). It may be noted that in the equations here i the index of the inertial term will take two values for downrange and vertical and there will be summation over i. The equation for vertical direction is given as

$$\frac{\partial V_y}{\partial t} = -\frac{V_y \dot{m}}{(M_0 + m)} + \frac{F\cos(\theta - \alpha) - 0.5\rho C_D A\left(V_x^{\,2} + V_y^{\,2}\right)\cos\theta}{(M_0 + m)} - g \qquad (2)$$

The equation for downrange direction is given as.

$$\frac{\partial V_x}{\partial t} = -\frac{V_x \dot{m}}{(M_0 + m)} + \frac{F\sin(\theta - \alpha) - 0.5\rho C_D A\left(V_x^{\,2} + V_y^{\,2}\right)\sin\theta}{(M_0 + m)} \qquad (3)$$

α is the angle between thrust and axis of the projectile, which may in many cases taken to be equal to zero. From these two equations it is possible to simulate the movement of the powered projectile. On integrating these two equations two components of velocity can be obtained where from further integration will give the position of the projectile in downrange and vertical coordinates. In the following equations (θ-α) has been replaced by θ only.

Here in this part of the motion the angle θ that the trajectory makes with vertical is again a function of time. Moreover the rate of change of this angle may be different for different intervals. This complicates the numerical method. Over and above two equations have to be interwoven in each step since both Vx and Vy appear in both the equations. This complicates the issue further.

After sometime the vertical component of acceleration imparted by the thrust, less resistance given by atmosphere falls short of the acceleration due to gravity. Hence no positive vertical acceleration is there in this part to force the projectile upward. On the contrary the vertical motion of the projectile undergoes deceleration even it moves in powered.. We will write the components of acceleration as below.

$$\dot{V}_x = \frac{(F - D)\sin\theta}{(M_0 + m)} \qquad (4)$$

$$\dot{V}_y = \frac{(F - D)\cos\theta}{(M_0 - m)} - g \qquad (5)$$

This phase lasts as long as the fuel is not fully consumed. Then the projectile takes simple parabolic phase, if, of course we ignore the air resistance

2. Uncertainty Management in Projectile Flight

It may be observed that while developing dynamical equations we have assumed acceleration to be dependent on a few selected variables like instantaneous location coordinates, velocity components, air resistance etc. In practice, however, the motion of the projectile depends on various other factors. These factors may be divided into two broad categories, internal and external.

Internal factors are the factors that affect the health of the projectile, e.g. trouble in engine, damage of its fins, wings etc. Its firing engine may have fluctuation in its operation. The body of the projectile may go unexpected way because of unforeseen defects in its structure. External factors are factors not linked with the structure or mechanism of the projectile. It may be due to uncertain behavior of the atmosphere. There may be locations in the atmosphere where generation of turbulence, thermal inhomogeneties etc. may take place. These structures may affect the movement of the projectile.

Hence we get by solving the equations the numerous paths the projectile is likely to traverse due to uncertain effect of the internal and external factors, discussed above. Only one path is, however, the calculable path and is the deterministic path of the projectile. This path is the path if the effect of unknown factors, stated above, is nil.

Therefore we should try to incorporate some type of uncertainty management method by which it will be possible to calculate some of the possible paths that the projectile may take up. Here we would follow the method of fuzzy dynamical system, where fuzzy differential inclusions are solved.

3. Fuzzy Dynamical System & Differential Inclusion

Let us write in short the dynamical equation governing the flight of a projectile.

$$\frac{dv}{dt} = f(v.t) \quad (6)$$

Under certain boundary condition and a given set of parameters the equation will give a deterministic solution of velocity. But in case of uncertainty created by randomness in the projectile health and environment it is likely that the number of solutions is more than one. In that case the solutions of the dynamical equation will form a set. So we may rewrite the equation as below.

$$\frac{dv}{dt} \in F(v.t) \quad (7)$$

Here $F(v,t)$ is a set-valued function. And our problem becomes a problem of differential inclusion. It is obvious that the chance of one of the solution from this set to occur is different from that of the other. We would, hence, be interested in attaching a measure to each of the member of the solution set. For that we would in the present case use the fuzzy concept. The deterministic solution is known to us. So we would like to use it as the pivot and would try to quantify the closeness of the other solution to this deterministic solution. Our fuzzy notion that we would like to use is 'closeness to determinism'.

4. Description of Uncertainty

Thrust has three components F_1, F_2, F_3 in three directions respectively downrange, vertical and cross range. In normal health and environment condition $F_3 = 0$. But in real situation where there is large number of unknown factors to disturb the health vis-à-vis the movement of the projectile the components of thrust and velocity as well cannot be deterministic. Cross range velocity is zero in a deterministic situation. In practice the projectile is likely to deviate from the direction of its launch i.e. downrange and also likely to land at a different place from where it was destined to land. That is for each set of initial condition a particular path will be the solution of the dynamical equations. In such uncertain cases redistribution of energy is a likely phenomenon. But in what way this redistribution will take place is a matter of conjecture only. Nothing can be said with certainty.

For the deterministic case we are considering the movement of the projectile in a 2-dimensional plane, downrange and vertical. Our equations are 2-dimensional equations. Velocity is broken down into two components. Now due to the uncertainty factors the projectile may have some movements in the crosswise direction. Hence a 2-dimensional case will be superimposed into a 3-dimensional case. At this stage we would like to make an assumption. Energy remains a constant whether the projectile remains in the deterministic path or an uncertain path. It means that at any point of time total energy remains unaffected by the uncertainty forces. But at any point of time the partitioning of total energy is affected by uncertainty factors. In the ideal case of determinism contribution to total energy comes from two components of velocity. But in the real situation loaded with uncertainty velocity may have a non zero cross range component and hence contribution to total energy may come from all the three components of velocity. The following model may explain this.

$$V_1^2 + V_2^2 + V_3^2 = U_1^2 + U_2^2 = 2E \tag{8}$$

Here U's and V's respectively stand for deterministic and non-deterministic situations. $2E$ is total energy for unit mass. 1 and 2 stand for downrange and vertical directions whereas 3 stand for cross-range direction.
The relation between U's and V's can be modeled as below.

$$V_1^2 = (1-\mu)U_1^2, \quad V_2^2 = (1-\mu)U_2^2, \quad V_3^2 = \mu\left(U_1^2 + U_2^2\right) \tag{9}$$

Here μ is the degree of uncertainty or fuzziness of the flight. We may observe that by this model energy is conserved.

Properties of μ

Source of fuzziness of the trajectory i.e. its degree of uncertainty μ is completely unknown. Hence it may not be correct to model μ as a dependent variable, since we do not know on which factors μ depends. Here we have left the task of determining the instantaneous value of μ with random numbers. Two digit number has been randomly selected and then divided by 200 to get the value of μ rationale behind selecting 200 is to limit the degree of uncertainty to 0.5. Let us now go back to equation 32 to find the value distance traversed in three directions.

$$x = \int_h^{h+1} \sqrt{(1-\mu)}.U_1, \quad y = \int_h^{h+1} \sqrt{(1-\mu)}.U_2, \quad z = \int_h^{h+1} \sqrt{\mu.\left(U_1^2 + U_2^2\right)} \tag{10}$$

Integration has been carried out between two time steps h and $(h+1)$. But still a problem remains unsolved. In all the three integrations square roots has been taken in the integrands. We have ignored the possible negative sign, which may force the projectile to move in negative side of the coordinates. We here ignore random movement in the x and y directions, which mean the projectile is assumed to move forward and upward even under the influence of the uncertainty factor. But for the z direction we consider negative values of the integrand. We make an assumption that the random number remains within a narrow band around 50 we take positive value, otherwise the value is negative. Let us take this band to be (45-55). In this case we have

$$\begin{aligned} z &= +\int_h^{h+1} \sqrt{\mu.\left(U_1^2 + U_2^2\right)} \quad .for\ random\ number\ in\ (45, 55) \\ z &= -\int_h^{h+1} \sqrt{\mu.\left(U_1^2 + U_2^2\right)} \quad .for\ other\ random\ number \end{aligned} \tag{11}$$

We have already discussed the equations governing the motion of the projectile and also the methodology of selecting fuzzy measure and hence to solve fuzzy differential inclusion. Solving these sets of equation it is possible to get the trajectory of the projectile with some degree of uncertainty.

Fuzzy Logic and its Application in Technology and Management
Editors: Debjani Chakraborty, S. Nanda and D. Dutta Majumder

Fuzzy Optimization for Energy Allocation in Shops and Establishments of Rural Areas

Chinmoy Jana

Energy Management Stream, Department of Public Systems Management
Indian Institute of Social Welfare and Business Management (IISWBM)
Management House, College Square West, Kolkata - 700 073
Email: chinmoyjana@yahoo.com

Abstract: Today, mankind is faced with grave 'energy crisis', which, if unresolved, will certainly have dire consequences for civilization in not too distant a future. This means that all avenues of both procuring new supplies and conserving existing ones should be explored. Proper allocation of available sources should be planned not just by instantaneous availability. There is an increasing demand of energy with a fixed, rather decreasing supply. So investigation has to be made as to how best can they be balanced without jeopardizing the overall requirement.

The present investigation attempts to take a close look at the micro level energy scene in shops and establishments in rural areas and undertakes an in-depth study in fuzzy environment. The common sources are taken for supply and available state of art technology is used to satisfy the demand. This study seeks to evolve a technique for energy conservation by optimizing its uses by cost minimization of direct energy, minimization of use of non-local energy sources and minimization of total energy use, i.e., maximization of efficiency. This structured model is applied for testing in the situation of a backward Community Development Block, Narayangarh, which is situated in the state of West Bengal in India. The Multi Objective Fuzzy Linear Programming Approach is applied for arriving at a set of recommendations for the Community Development Block, and draws policy outlines for arresting any environmental deterioration and managing energy use in a self reliant manner.

Keywords: Rural Energy, Micro Level Planning, Shops and Establishments, Fuzzy Optimization.

1. Introduction

In rural energy micro level planning shops and establishment are normally overlooked due to lack of information and tendency to traditional resource allocation strategy. The absence of realistic planning approach, which can reflect the need of the society, is also a major reason. The shops are unorganized sector, which consumes a remarkable part of the total energy use and pollutes the environment without any calculation or thinking on future. It can broadly be termed as an Energy Planning exercise aiming at understanding logically and scientifically the rural energy interaction pattern in a CD Block and hence correcting its various existing imbalances.

The quality and quantity of energy consumption varies spatially and depends on factors like socio-economic status, energy availability, technology acceptability etc. A successful micro-level plan relies upon the key planning tools such as system analysis, Operations research, statistics, socio economic evaluation and information system. The present investigation is an attempt to apply these planning tools at a Block level by identifying a set of control parameters, which determine the alternative policy option.

Several researchers have constructed Linear Programming model with single objective function and most cases that is minimization of total cost (Sinha (1991)). Since single objective does not reflect the reality, multi criteria decision making problem is rapidly growing (Ramanathan (1993)). To handle the data which are more vague than randomness, formulation of a fuzzy decision making model has been found to be suitable (Narshima (1980), Jana (2000)). Jana (2004) has been used FLP for obtaining solution in multi criteria decision making system. Chedid (1999) first used multi objective fuzzy linear programming technique to energy planning.

2. Analysis of Situation

A large section in rural sector consists of a variety of shops making commercial transactions, different establishments like offices, schools, health centers, clubs, libraries, recreation units etc. The establishments like offices, schools, health centers etc and the shops use energy mainly for lighting and some of these like school hostels, hospitals, hotels, sweet shops etc use direct energy for coking. The tasks performed by these units are not same and the time schedule and duration of completion for those tasks vary widely. For an example the yearly lighting schedule in a hospital and school are given in table-1 that is obtained from the pattern of the area. Similar schedules are also calculated for hostels, hospitals etc.

Table-1: Lighting Schedule for Schools and Offices

Time Schedule	Schools		Offices	
	week days	weekend and holidays	week days	weekend and holidays
Midnight to 9:00 am	5 %	5 %	10 %	10 %
9:00 am to 6:00 pm	100 %	5 %	100 %	10 %
6:00 pm to Midnight	5 %	5 %	10 %	10 %

For obtaining optimal alternative solution, some of the currently less used devices like SPV system have been considered for attaining efficiency on one hand and better use of renewable sources on the other.

Fuzzy logic is an international way to represent vagueness in everyday life. It is important because of the fact that most modes of human reasoning, especially common sense reasoning one approximate in nature. Fuzzy interpretations of data structures are a natural and plausible way to formulate and solve the various problems. This theory has the characteristic that it is able to represent the concepts of vagueness that natural language uses for pointing out qualitative variables. In the present study, the linguistic values and the importance scale were depicted as triangular fuzzy numbers as (a_1, a_2, a_3). Here $a_1 \leq a_2 \leq a_3$, a_2 (membership grade = 1) is the most possible value of the linguistic term, and a_1 and a_3 are the

Table-2: Source-Device-Task -User Combinations Considered in This Sector

(*i*) Source	(*j*) Device	(*k*) Task	(*l*) User
(1) Coal	(6) Traditional chullah	(2) Cooking	(2) Hotel
			(3) Hostel
			(4) Sweet Shop
(2) LPG	(7) Traditional oven	(2) Cooking	(2) Hotel
			(4) Sweet Shop
(3) Kerosene	(8) Stove		(5) Tea Stall
	(1) Lamp	(1) Lighting	(1) All Units
	(2) Hurricane		
(4) SPV	(5) CFL		
(5) Firewood	(4) Chullah	(2) Cooking	(2) Hotel
			(3) Hostel
			(4) Sweet Shop
			(5) Tea stall
(6) Electricity	(3) Bulb	(1) Lighting	(1) All Units
	(4) Tube		
	(5) CFL		
(7) Diesel	(9) Generator set	(1) Lighting	(0) All Units

lower and upper bounds (membership grade = 0) used to reflect the fuzziness of the term. The objective functions are also considered fuzzy because they are conflicting in nature. When decision maker needs to minimize total cost, the total energy use may increase. When total energy use is minimized, the cost may increase. So the decision maker should find an aspiration level of each objective function and try to optimize all these together within that aspiration level.

3. Mathematical Structure

The structural equations and expressions depicting the model are as follows:

Obj 1: Minimisation of Total Cost of Direct Energy (O_1)

$$= Min \sum_i \sum_j \sum_k \sum_l C_{ij} x_{ijkl}$$

Obj 2 : Minimisation of Use of Non-local Sources of Energy (O_2)

$$= Min \sum_i \sum_j \sum_k \sum_l R_i x_{ijkl} \quad \forall i = 1, 2, 3, 6 \text{ and } 7$$

Obj 3: Maximisation of Overall Efficiency, i.e.,
Minimisation of Total Energy Use (O_3)

$$= Min \sum_i \sum_j \sum_k \sum_l R_i x_{ijkl} \quad \forall i$$

where x_{ijkl} is the decision variable indicating quantity of source i needed through device j to perform k^{th} task for l^{th} user.
R_i is the conversion factor to MJ for the source i.
C_{ij} is the total cost per unit of ith source through j^{th} device. And given by

$$C_{ij} = (DC)_{ij} + (FC)_i + (SC)_{ij}$$

Of which $(DC)_{ij}$ is the Device Cost per unit fuel, $(FC)_i$ is the unit Fuel Cost and $(SC)_{ij}$ is the Initial Investment Cost per unit Fuel.

Taking rate of interest at r percent per year the initial investment per unit fuel is equivalent to:

$$\sum_{i=1}^{m}\sum_{j=1}^{n}\frac{\left[r\left(1+\frac{1}{(r+1)^{n_j}-1}\right)+m_j\right](INV)_{ij}}{Q_{ij}}$$

Where, n_j : Lifetime of jth System

m_j : Maintenance of jth System per year.

$(INV)_{ij}$: Total Investment Cost for j^{th} System.

Q_{ij} : Quantity of ith source per year to run through j^{th} device.

subject to

$$\sum_{i}\sum_{j} x_{ijkl} \quad \forall k, l \tag{1}$$

where, D_{kl} are the demand for k^{th} task of lth user

$$\sum_{j}\sum_{k}\sum_{l} x_{ijkl} \leq A_i \quad \forall i \tag{2}$$

where, A_i are the availability of ith source in the system.

$$\sum_{i}\sum_{j} x_{ijkl} \geq d^{we} \quad \text{for } i \neq 6, k = 1 \text{ and } l = 1 \tag{3}$$

where, d^{we} is the demand of lighting points where electricity is not available.

$$\sum_{j} x_{ijkl} \leq A_{ik} \qquad \forall \text{possible } i, k \tag{4}$$

where, A_{ik} is availability of ith source for k^{th} task.

$$\left(\frac{1}{\overline{q_{11}}},\frac{1}{q_{11}},\frac{1}{\underline{q_{11}}}\right)x_{1111} \oplus \left(\frac{1}{\overline{q_{12}}},\frac{1}{q_{12}},\frac{1}{\underline{q_{12}}}\right)x_{1211} \oplus \left(\frac{1}{\overline{q_{23}}},\frac{1}{q_{23}},\frac{1}{\underline{q_{23}}}\right)x_{2311}$$

$$\oplus \left(\frac{1}{\overline{q_{24}}},\frac{1}{q_{24}},\frac{1}{\underline{q_{24}}}\right)x_{2411} \oplus \left(\frac{1}{\overline{q_{25}}},\frac{1}{q_{25}},\frac{1}{\underline{q_{25}}}\right)x_{2511} \oplus \left(\frac{1}{\overline{q_{35}}},\frac{1}{q_{35}},\frac{1}{\underline{q_{35}}}\right)x_{3511}$$

$$\oplus \left(\frac{1}{\overline{q_{46}}},\frac{1}{q_{46}},\frac{1}{\underline{q_{46}}}\right)x_{4611} \,\tilde{\geq}\, (\underline{u}, u, \overline{u}) \tag{a}$$

Where, $\frac{1}{\underline{\underline{q}}_{ij}}$: Quantity of source i through device j for 5 working hours.

$\frac{1}{q}$: Quantity of source i through device j for 4 working hours.

$\frac{1}{\underline{q}_{ij}}$: Quantity of source i through device j for 3 working hours.

u : Number of units required in the Block.

$\underline{u}$: Left tolerance for u

$\bar{u}$: Right tolerance for u

The fuzzy constraint (a) becomes as crisp constraints

$$\sum_i \sum_j \frac{1}{\underline{\underline{q}}_{ij}} x_{ij} \geq \underline{u} \tag{5}$$

$$\sum_i \sum_j \frac{1}{q_{ij}} x_{ij} \geq \bar{u} \tag{6}$$

$$\sum_i \sum_j \frac{1}{\underline{q}_{ij}} x_{ij} \leq u \tag{7}$$

4. Solution Technique

The problem of targeting the demand supply balance in an optimized fashion has been solved first by applying LP technique with individual objective function, namely, minimization of total cost of direct energy, minimization of use of non-local sources of energy and Minimization of total energy use and then taking these objectives together in a MOFLP frame work for obtaining a result that could provide maximum level of compromise among the said objectives.

The Objectives O_1, O_2 and O_3 have been solved individually subject to the set of constraints (1) - (7) and the non-negativity of all the decision variables. Let x^1, x^2, x^3 represent the set of solutions for the objectives O_1, O_2 and O_3 respectively.

Then we get,

$U_i = x^i$

$L_i = \text{Max}\,(O_i(x^1), O_i(x^2), O_i(x^3))$ for i=1,2,3

where U_i and L_i indicate the most and the least acceptable values of ith objective. Suppose μ_i is the membership function for the ith objective which is defined by

$$\mu_i = \begin{cases} 0 & \text{when } O_i \geq L_i \\ (L_i - O_i)/(L_i - U_i) & \text{when } U_i \leq O_i \leq L_i \\ 1 & \text{when } O_i \leq U_i \end{cases}$$

Now here a dummy variable λ has been introduced such that $\lambda \leq \mu_i$, $\forall$ t and $\lambda \leq \mu$

Therefore, the final Fuzzy Linear Programming (FLP) Problem becomes

Maximise λ

Subject to constraints

$O_i + (L_i - U_i)\,\lambda \le L_i,$

$\lambda \ge 0$

and all other constraints already mentioned with non-negativity of all x_{ij}
Then the Simplex method has been applied to solve the above FLP task.

5. A Case Study

This structured model is applied for testing the case of Narayangarh Block of West Midnapore district, West Bengal in India. Micro-level energy planning has to ignore many location specific characteristics and thus many of the details are lost which invariably have a great bearing on the local problems.In this study, supply and demand is considered to be steady for a fixed interval of time.

The data are collected through questionnaire surveys and employed fuzzy modeling to analyze and integrate the information. An extensive field survey was conducted to collect relevant information and analysis had been done for measuring the available sources, assessing present energy consumption and estimation the minimum and peak demand in the system. Linguistic terms have been found intuitively easy to use in expressing the subjectiveness and impression of the decision makers' assessment. Therefore interviewees were asked to give their subjective assessments using predetermined linguistic terms.

The model is employed to compute the outputs from knowledge of the inputs and technical coefficients for the year 2003 referred to as the base year in this study. The model is used for computing the output for the block as a whole and the results are compared for model validation.

5.1. Information

So far higher education facilities are concerned, the block has one degree college and an Basic Training (BT) college. It has 6 higher secondary school, 23 high school and 10 junior high schools. Besides there are 226 primary schools spread over the villages of the Block. The BT College and 16 schools has hostel facility and only 7 schools with hostels are electrified. There are one block primary health center, 3 subsidiary health center, 15 primary health centers and 8 other clinical facilities run by Govt. For veterinary services the block has 1 veterinary hospital, 2 govt. dispensaries and a few artificial insemination centers. The block has 17 branches of nationalized bank, 4 branches of central cooperative bank and one branch of land development bank. In the entire Block there are 116 cooperative societies of which some are non-functioning at present. The Block has several administrative and agricultural offices and various line departments to cater to needs of the people. These include Block Development Office, two Police Station, Agricultural Development Office, Block Livestock Office, Office of the Public Health Engineering Department, sub registrar office, settlement office etc. The block has 23 post offices located in 16 GPs.

Altogether 180 villages has been given electric connections through 173 transformers of which 164are of 25 KVA and 9 of 63 KVA of capacity controlled by

2 substations at Belda and Narayangarh. Four health centers, college, 7 school hostels, 5 cinema halls are electrified along with only 1175 commercial units. (Present consumption pattern at shops and establishments are given in table 4.)

Table-3 : Conversion Factor of the Energy Sources with Unit Price

Sl. No.	Name of Source	Conversion factor (MJ)	Price (Rs.)
5	Firewood (ton)	19.7	0.75 per kg
8	LPG ('000 lt)	44	18.6 per kg
9	Kerosene ('000lt)	41.3	10 per lt
10	Electricity (MWh)	11.93	1.5 per KWh
11	SPV (MWh)	11.93	00
12	Diesel ('000lt)	56.31	17 per lt
16	Coal (ton)	25	.5 per kg

Table- 4: Energy Allocation and value of the objective functions for Shops and Establishments Sector

Task	Sources	Present Consumption	Obj-1	Obj-2	Obj-3	Fuzzy
Lighting	Kerosene ('000lt)	50.5	47.1	29.5	29.5	29.5
	Electricity (MWH)	980.65	204.2	17.6	204.2	167.75
	SPV (MWh)	35.0	--	197.4	10.7	47.21
	Diesel ('000lt)	28.8	--	--		
Cooking	Firewood (ton)	1356.0	1713	1713	183.6	1000.5
	LPG(ton)	8.4			421.5	361.7
	Kerosene ('000lt)	39.4			144.4	
Cost ('000 Rs)			4801.57	11868.9	8252.9	8083.4
Non-local energy use (MJ)			4381.3	1429.27	18369.7	9295.86
Total energy use (MJ)			38050.7	37453.1	22101.8	29507.9

5.2. Results and Discussions

Different parameters related to energy conversion factor, unit fuel cost by sources, availability, device efficiency, its life span, fixed and maintenance cost of devices etc have been derived from the available literature sources, local market condition and experimental observations.

The result shows that kerosene is till economically most viable source for lighting while electricity is also used wherever available. Use of diesel for lighting in shops & establishments has been found to be inappropriate. SPV systems have the potential to replace diesel-fueled generator set, which are commonly used for lighting purposes in the shops, establishments, market areas etc. After replacing diesel fueled generator set by SPV, carbon emission will be reduced. In the present situation of the global environment, the certificate of the reduction of carbon emission may be traded to get back the initial investment cost.

CFL may be used in place of traditional tube lights bulbs & this 775 Mwh will reduce electrical energy. In shops & establishments, firewood used for cooking will have to be reduced by 356 tons for stopping illegal extraction of firewood. This will however be supplemented by LPG.

5.3 Policy

A long term solution of this problem may be extensive energy plantation in wasteland deforested areas & vacant public used areas, 200 ha, Barren and waste land (0.4 %), area not available for cultivation (3036.2 ha, i.e. 6.17%). The awareness programme and proper marketing are needed to popularise SPV lighting systems.

6. Conclusion

This is a part of the work where all rural sectors are considered with the inter relation to each other. For more appropriate decision, this sector should be interacted with other sectors like agriculture, domestic, transport etc.

Rampant extraction and unrestricted use of firewood should be discouraged and the long-term demand for this source needs to be outlined in tune with the target of multipurpose energy plantation on suitable lands. The claim for more efficient devices replacing the lesser ones is universal. This claim will have to be realized since it would have significant impact on the economy and the society through saving energy, reducing unit costs, protecting environment and removing drudgery of the users. This energy study provides ample scope for further research in the area, which could not be covered under the present work due to various limitations. In this regard, one should not overlook the possibility of framing non-linear relationship between the dependent and independent variables while developing the structure of the equations and expressions. It has been proven that an optimal trajectory towards a steady state exists and the social planner should follow it.

Acknowledgement

The author is thankful to Department of Science and Technology (DST), Govt. of India for their financial support under SERC FAST Track scheme for the work.

References

1. R. Chedid, T. Mezha & C. Jarronche (1999), A fuzzy programming approach to energy resource allocation, *International Journal of Energy Research*, 23: 303-317.
2. Chinmoy Jana & R.N. Chattopadhyay (2001), Planning for rural energy allocation in the cooking sector – A multiple objective fuzzy linear programming approach. *Indian Journal of Regional Science*, XXXIII (2): 74-82.

3. Chinmoy Jana & R.N. Chattopadhyay (2004), Block level energy planning for domestic lighting – a multiobjective fuzzy linear programming approach, *Energy*, 29 (11): 1819 – 1829.
4. R. Ramanathan & L.S. Ganesh (1993), A Multi Objective Programming Approach to Energy Resourçe Allocation Problems, *International Journal of Energy Research*, 17: 105-119.
5. H. Romelfanger (1989), Interactive decision making in Fuzzy Linear Optimization Problem, *European Journal of Operation Research*, 41: 210-217.
6. J. Romik & J. Rimanek (1985), Inequality relation between fuzzy numbers and its use in fuzzy optimization, *Fuzzy sets and Systems*, 16: 123-138.
7. C.S. Sinha & T.C. Kandpal (1991), Optimum mix of technologies in rural India – the lighting sector, *International Journal of Energy Research*, 15: 653-665.
8. H.J. Zimmerman (1978), Fuzzy Programming and Linear Programming with several objective functions, *Fuzzy sets and systems*, 1: 45-55.

Fuzzy Logic and its Application in Technology and Management
Editors: Debjani Chakraborty, S. Nanda and D. Dutta Majumder

Selection of Maximum Viewer Programme Slots in Television Channels in Fuzzy Environment

Anita Saha, Madhumangal Pal and Tapan K. Pal

Department of Applied Mathematics with Oceanology and Computer Programming
Vidyasagar University, Midnapore - 721 102
e-mail:mmpal@vidyasagar.ac.in

Abstract: In this paper, it is shown that all programmes of all television channels can be modelled as a fuzzy interval graph. The weights of the vertices are taken as interval numbers. The programme slots are taken as the vertices of the graph and if the time duration of two programme slots have non-empty intersection then the corresponding vertices are considered to be connected by an edge. The number of viewers of a programme is taken as the weight of the vertex. We assume that a company wishes to advertise for his product through television channels to broadcast the information about their product to the maximum number of viewers. The constraints of the problem are that any two consecutive programmes are mutually exclusive. The problem is to select some programme slots, which are mutually exclusive with respect to the duration of broadcasting time such that the total number of viewers of the selected programme slots is maximum. This problem is solved by solving the 1-colouring problem on weighted interval graphs. The proposed method takes O(n) time, where n is the number of all programmes of all channels.

Keywords: Modelling, colouring, Interval numbers, Interval graphs.

1. Introduction

In recent years most of the world wide mass is being influenced by advertisement. The daily requirements of man are now being governed by the attractive advertisements. So, advertise companies (ACs) (the companies those are interested to advertise for their products) are also spending a large amount of money for the purpose. The main objectives of the ACs are to catch maximum viewers so that the sell of their product becomes maximum.

There are more than hundred of channels like BBC World, STAR, ESPN, ANIMAL PLANET, ZEE, MGM, AXN etc, running numerous programmes for 24 hours. Among all these programmes there are some which are very popular. The programmes like `Guiness World Record Prime Time' at 9:00 a.m. shown at AXN, `Charlie Chaplin' at 9:00 a.m. on ZEE ENGLISH, `Mission Everest' at 10:00 a.m. on National Geographic channel and many other are found very much popular. Now the problem arises how and where an AC will relay his advertisement to make it viewed to large mass of population. Sometimes, depending on the popularity of the programmes viewers get divided. So, there arises a problem in the selection of the channel.

Each programme slot of a particular channel can thus be represented as an interval on 0 to 24 hours time interval. All the programme slots of all channels can be represented as a collection of intervals on the line segment [0,24]. Each interval has an weight which is equal to the number of viewers watching the corresponding programme. It is well known that the most difficult task is to find out the number of viewers of a programme. But, one acceptable method to find the number of viewers is fuzzy technique. Here, the number of viewers of a programme is taken as an interval number instead of crisp number. The set of programmes schedule form a weighted interval graph G whose weights are interval numbers. An interval graph is a graph whose vertices can be mapped into unique intervals on the real line such that two vertices in the graph are adjacent if and only if their corresponding intervals intersect. An interval graph is called weighted if its vertices have weights. It will be shown that the maximum number of viewers of some disjoint programme slots is equal to the weight of the 1-colourable subgraph of the interval graph G. Thus, this problem can be modelled as an interval graph.

Using greedy heuristic technique, Olariu [1] has solved the optimal colouring problem for interval graphs in $O(n+m)$ time, where n and m are the number of vertices and edges respectively. Pal & Bhattacharjee [2] have designed a parallel algorithm to solve optimal colouring problem on interval graphs which takes $O(n/P+\log n)$ time and P processors. But, no work is available in literature to solve the maximum weight 1-colourable problem on interval graphs when the weights are interval numbers.

In this paper, an algorithm is designed to solve maximum weight 1-colouring problem on interval graphs whose weights of the vertices are interval numbers. The proposed algorithm takes $O(n)$ time, where n is the number of vertices.

2. Modelling of the Problem

At first we assume that an AC wishes to run their advertisement from 0 to 24 hours in one session, that is, in a single duration the advertisement is to be shown in only one channel. The formal definition of our problem is stated below:

Problem P1. *Suppose a company or any organization is interested to run their advertisement through television channels. The constraint is that only one programme slot is selected in any instant. The objective of the problem is to select some programmes slots such that the sum of the number of viewers of the selected programmes is maximum.*

In this problem, a television programme slot is represented as an interval on real line of length 24 hours. Each programme slot i.e., each interval is regarded as a vertex of the interval graph G. That is, all the programme slots of all television channels constitute the set of vertices V of the interval graph G. If there exist intersection of timings in between two programme slots then there is an edge between the vertices corresponding to these programme slots. If finishing time of a programme is the starting time of another programme then we assume that these programmes are non-intersecting. It may be noted that any two programmes in a particular channel are non-intersecting. The number of viewers of each programme is taken as interval number. Thus the weight of the vertices of the interval graph are taken as interval numbers. Finally, all the programme slots of all television channels are modelled as a crisp interval graph, but, interval numbers are the weights of the vertices.

A *colouring* of a graph is an assignment of colours to its vertices so that no two adjacent vertices have the same colour. The vertices of one particular colour form a *colour class*. Any two vertices of a colour class are not adjacent. The maximum weight colouring problem is to find a subset S of V such that no two vertices of S are adjacent and sum of the weights of the vertices of S is maximum. This problem is also known as maximum weight 1-colouring problem.

Thus, the problem P1 is equivalent to the 1-colouring problem on interval graph. The problem P1 can be solved by solving the following problem P2 on the interval graph.

Problem P2. *Find a subgraph $H(G)$ of an interval graph G which can be coloured using only one colour such that $\sum w(u)$ for all $u \in H(G)$ is maximum among all other such subgraphs, where $w(u)$ is the weight of the vertex u. This problem is known as maximum weight 1-colouring problem.*

Here the weight of each vertex is an interval number. So, a study is required on interval arithmetic and comparison of intervals. In the following section, an overview of interval numbers is presented.

3. Interval Numbers

The interval analysis is a well-established area of classical mathematics. In this section, an overview of them is introduced.

Let $A=[a_L, a_R]=\{a: a_L \leq a \leq a_R, R$ is a set of real numbers$\}$, where a_L and a_R are the lower and upper limit of the interval number A. A real number a may also be regarded as a special (degenerated) interval $[a, a]$. Interval A is alternatively represented as $A=<mean(A), width(A)>$, where $mean(A)$ and $width(A)$ are the mid-point and half-width (or width) of interval A, i.e.,

$mean(A)=(a_L+a_R)/2$ and $width(A)=(a_R-a_L)/2$.

The three basic arithmetic operations on closed intervals are defined as follows:

$[a, b] \oplus [d,e]=[a+d, b+e]$

$[a, b] \ominus [d,e]=[a-d, b-e]$

$[a, b] \ominus_R [d,e]=[a-d, b-e]$.

3.1. Comparison between interval numbers

Comparison between two interval numbers is very important in interval arithmetic. Different order relation to compare two intervals are proposed by several authors. Out of these ranking approaches, which seems to be more significant is introduced in [3].

Let $M=[a_L, a_U]$ and $N=[b_L, b_U]$ be two intervals. Then mean and half width (which is simply termed as 'width') of the interval M are denoted by $mean(M)$ and $width(M)$ and are defined as $mean(M)=(a_L+a_U)/2$ and $width(M)=(a_U-a_L)/2$. The intervals M and N can be represented alternatively in mean-width notation as, $M=<mean(M), width(M)>$ and $N=<mean(N), width(N)>$. For $mean(M) \leq mean(N)$, the acceptability index of the premise $M<N$ is defined by

$$A(M<N)=\frac{mean(N)-mean(M)}{width(M)+width(N)},$$

and this is the value judgment by which M is inferior to N or precisely M is less than N.

For the different position of mean and width of the intervals M and N the value of acceptability index is as follows:

(i) $A(M<N) \geq 1$ when $mean(M)<mean(N)$ and $a_U \leq b_L$,

(ii) $0<A(M<N)<1$ when $mean(M)<mean(N)$ and $a_U>b_L$,

(ii) $A(M<N)=0$ when $mean(M)=mean(N)$.

(i) In this case M is preferred over N i.e., M is considered inferior to N with acceptability index greater than or equal to 1 then the decision maker (DM) accepts it with full satisfaction.

(ii) If $0<A(M<N)<1$ then M is preferred over N or precisely $M<N$ with different grades of satisfaction lying between 0 and 1.

(iii) If $A(M<N)=0$ then obviously $mean(M)=mean(N)$. In addition, if $width(M)=width(N)$ then M and N are the same interval and hence there is no question of comparison. But, if $width(M) \neq width(N)$ then the intervals M and N are non-inferior to each other, i.e., acceptability index becomes insignificance. In this case, DM has to negotiate with the widths of M and N. Let M and N be two cost intervals and the problem is to choose the minimum cost interval. If the DM is optimistic then he/she will prefer the interval with maximum width along with the risk of more uncertainty giving less importance. Again if the DM is pessimistic then he/she will pay more attention on the more uncertainty and will choose the interval with minimum width. The case will be converse when M and N are profit intervals.

4. Representation of Intervals as Interval Graph

Let $I=\{I_1, I_2, \ldots, I_n\}$ be the set of all programme slots of all television channels available in a geographical area. Let $G=(V,E)$, $V=\{1, 2, \ldots, n\}$ be the interval graph corresponding to the set of intervals of I. The vertex $j \in V$ of G corresponds to the interval I_j. Each interval I_j has two end values the starting or left end value (a_j) and ending or right end value (b_j) corresponding to the starting and ending times of the jth television programme i.e., $I_j=[a_j, b_j]$, $j=1, 2, \ldots, n$. It may be noted that $0 \leq a_j<b_j \leq 24$, as a_j and b_j represent the time instances in 24 hours clock. The weight of an interval I_j i.e., of vertex j is denoted as $w(j)=[w_L(j), w_R(j)]$, where $w_L(j)$ and $w_R(j)$ respectively represent the lower and upper limit of the weight i.e., the minimum and maximum number of viewers of the programme j. As mentioned earlier there is an edge between the vertices j and k iff $I_j \cap I_k \neq \varphi$.

5. Data Structure and Preliminaries

Without any loss of generality we assume that the intervals in I are indexed by increasing finishing time (right end points) i.e., $b_1 < b_2 <\ldots< b_n$. The duration of programmes, the intervals on real line and vertices of the interval graph are same think.

All the programme slots are stored into an array e_j, $j=1, 2, \ldots, 2n$. Each element e_j has three fields, viz., $e_j.var$, $e_j.type$ and $e_j.wt$ representing respectively the name of the vertex, type of endpoint (left or right) and weight of the vertex $e_j.var$. That is, if e_i is the left (right) endpoint corresponding to the interval I_j then $e_i.var=j$, $e_i.type=a$ (b) and $e_i.wt=w(j)$.

The following lemma is well known.

Lemma 1. If the vertices $u, v, w \in V$ are such that $u<v<w$ in the ordering and u is adjacent to w, then v is also adjacent to w.

Let $S(k)$ be a colour class whose largest vertex is k. Let the interval number $W(k)$ be the weight of the colour class $S(k)$. Several colour classes may be available containing the largest vertex k. But, the intension of our problem is to find a colour class containing k with maximum weight. Let $S_{max}(k)$ be the colour class whose weight (W_{max}) is maximum among all colour class whose largest vertex is k. Then $\max\{W_{max}(k): k \in V\}$ is the weight of the maximum colour class i.e., the number of viewers of the selected programme slots.

Lemma 2. Let $S(k)$ be a colour class whose largest vertex is k. If $b_k < a_j$ for any $j \in V$ then $S(k) \cup \{j\}$ ($=S(j)$) is also a colour class with weight $W(k) \oplus w(j)$.

The following lemma gives the weight of the colour class $S(j)$.

Lemma 3. The maximum weight of the colour class $S_{max}(j)$ is given by

$$W_{max}(j)=w(j) \oplus \{\vee W_{max}(i): i \in V \text{ and } b_i<a_j\} \text{ for any } j \in V.$$

The dynamic programming method is used to solve the problem P2. The problem is solved into two phases. In a first phase, the weight of the maximum weight colour class is determined and in the second phase, the intervals of the maximum weight colour class are identified.

Phase I. In this phase, each end point e_i, starting from e_1, is scanned sequentially. It is easy to observe that at each right end point there is an end of a colour class and at each left end point a colour class begins. Depending on this observation two different operations are done at each end point e_i.

Let *max_wt* be a variable which store the weight of a maximum weight colour class when a right end point is scanned and the variable LV (last vertex) store the vertex where this *max_wt* is attended. Initially, let *max_wt* and LV be both 0. For each end point e_i, $i = 1, 2, \ldots, n$ the following operations are performed.

If $e_i.type = a$ i.e., e_i is a left end point of an interval I_j then $W(j) = max\_wt \oplus w(j)$.
If $e_i.type = b$ i.e., e_i is a right end point of an interval I_j then compare *max_wt* with $W(j)$.
If $max\_wt \vee W(j) = W(j)$ then set $max\_wt = W(j)$ and $LV = j$.

At the end of the Phase I, the value stored to the variable *max_wt* is the weight of the maximum weight colour class. Also, *max_wt* is equal to $W(LV)$ and let it be W^*. Let S^* be the colour class whose weight is W^*. This fact is proved in the Lemma 4. As mentioned earlier, the Phase I gives the weight W^* of maximum weight colour class and Phase II gives the corresponding intervals. Phase II of the algorithm is described below.

Phase II. To identified the intervals whose sum of weights is equal to W^*, a backward searching and checking is done from $W(LV)$ to $W(1)$. Obviously, the vertex stored at LV at the end of Phase I is a member of S^*. Thus initially, $S^*=\{LV\}$. Now, the weight $w(LV)$ is subtracted from *max_wt* i.e., the operation $max\_wt = max\_wt \ominus_R w(LV)$ is performed. Then we search for an interval I_i whose right end point is less than the left end point of I_{LV} and $W(i) = max\_wt$. In this case, *i* becomes a member of S^*, i.e., S^* can be computed as $S^*=S^* \cup \{i\}$. This process is repeated until the value of *max_wt* becomes [0,0].

Lemma 4. The value of *max_wt* at the end of Phase I is the weight of maximum weight colour class S^*, that is, $W^* = max\_wt$.

6. The Algorithm and its Complexity

The formal algorithm to compute a maximum weight colour class S^* of an interval graph is presented below.

Algorithm MWCC

Input: A set of intervals $I=\{I_1, I_2, \ldots, I_n\}$ on a real line and the weight $w(j)$ of the interval I_j, $j=1, 2, \ldots, n$. These intervals are stored into the sorted array $e=\{e_1, e_2, \ldots, e_{2n}\}$.

Output: The set S^* and the weight W^*.

Step 1: Phase I

```
for i=1 to 2n do
    if e_i.type=a of the interval I_j then
      W(j)=max_wt ⊕ w(j);
    endif;
    if e_i.type=b of the interval I_j then
      Compute A(max_wt<W(j))=L (say)
      if (L>0) or (L=0 and width(max_wt)<width(W(j))) then
        max_wt=W(j);
        LV=j;
      endif;
    endif;
endfor;
```

Step 2: Phase II

```
S*=S* ∪ { LV }
max_wt=max_wt ⊖R w(LV);
for j=LV-1 to 1 do
    if W(j)=max_wt and b_j < a_LV then
      S*=S* ∪ {j};
      max_wt=max_wt ⊖R w(j);
      LV=j;
    endif;
endfor;
```

end MWCC.

Lemma 5. The running time of the algorithm MWCC is $O(n)$.

Theorem 1. The maximum weight colour class of an interval graph when the weights of the vertices are taken as interval numbers can be computed using $O(n)$ time, where *n* is the total number of vertices of the graph.

If n is the total number of programmes of all channels of a particular day in a geographical area then the optimal programme slots can be selected using the algorithm MWCC. Thus the time complexity to solve the problem P1 is stated below.

Theorem 2. The optimal programme slots among n programmes of all channels can be selected using $O(n)$ time, when the number of viewers are taken as interval numbers.

References

1. S. Olariu (1991), An optimal greedy heuristic to color interval graphs, *Information Processing Letters*, 37: 21-25.
2. M. Pal & G.P. Bhattacharjee (1996), An optimal parallel algorithm to color an interval graph, *Parallel Processing Letters*, 6: 439-449.
3. A. Sengupta & T.K. Pal (2000), Theory and methodology on comparing interval numbers, *European Journal of Operational Research*, 127: 28 - 43.

Fuzzy Logic and its Application in Technology and Management
Editors: Debjani Chakraborty, S. Nanda and D. Dutta Majumder

Application of Fuzzy Logic in the Context of Start-up Evaluations

Shuvabrata Nandi[1] *and Vihang Patel*[2]

[1]Department Electronics & Communication Engineering, I.I.T. Kharagpur
[2]Department of Mathematics, I.I.T. Kharagpur

Abstract: In this paper we have discussed the problems usually faced by a typical startup company in the small and medium sized enterprises segment in the Technology Sector, and how we can gauge the relative success or failure of that startup, by the application of Fuzzy Logic Principles. In this paper we discuss about a typical generic startup, where the parameters involved in evaluating its performance is more or less fixed and their effect can be effectively summarized by using expert knowledge bases, which also forms the part of the Fuzzy Rule base that we are using for evaluation. For example we take into account the relative effects of *projects at hand, funding, technological know-how, leadership quality of the management, the Human Resources of the Company, the track record of the parent company (if any) etc.* All of these factors are then taken as the parameters on which the degree of success of the startup depends, and then these variables are fuzzified using the Rule base we mentioned earlier, developed from key persons and previous data.

Although the nature of the problem is quite well known, our approach to the solution of it is unique in a way because this is one of the first instances of soft-computing tools used in evaluating company startups. Since the parameters involved in this decision making are themselves not very rigidly defined and they are more or less linguistic in the sense, it is very natural that one should get a reasonably high degree of accuracy in the prediction of success or failure of that company by applying fuzzy concepts. Since the fuzziness is built into the problem only, it will give us better results than the conventional technique.

Keywords: Fuzzy Logic, Decision Making, Entrepreneurship, Real Options, Project Management, Technology Start-ups, Profitability and Investment Decisions.

1. Introduction

High Tech Industries are different from ordinary industries, and high technology firms have, in general, shorter span of product life cycle, heavier R&D investments, higher uncertainties, and more intense competition for new product share. Consequently small and medium sized Technology-based enterprises (SMTE-s) in high tech industries face both the opportunity of a rapid growth as well as a high risk factor. Opportunities result from increased worldwide demands for high-tech products and services whereas the risks are generated from the features inherent in Technology companies as mentioned earlier only. And in case of technology start-ups the problem becomes even more complicated when one has to judge the value of a start-up from an investor point of view. We have to take into account all the key success factors associated with the start-up, which can be either general variables across all sectors or specific variables related to that particular industry in which the company is going to operate. So we felt the need of a framework by which one can evaluate the value of a technology startup, albeit from an investor viewpoint, taking into account the new rules of the new economy where the traditional rules of project management does not suffice to give satisfactory answers. Since the key factors involved are mostly linguistic variables or factors that can't be explicitly expressed in a numerical framework, the need for the incorporation of Fuzzy Logic came into picture. And the use of a Fuzzy Rule base gives us a unique advantage in terms of coming out with an linguistic answer, and in this case neither a binary output nor a numerical figure claiming the index of success is of much use to the end-investor. So in this work we took a slightly different approach of evaluation of the success. Our method of work is given below:

Finding the key factors of success → Gauge their relative weights using linguistic variables using an expert rule base → Find out Fuzzy Numbers for each of the variables (which can include fuzzy weighted average to get the value itself) → Use another Fuzzy Rule Base to get the criteria of success → the final verdict, which is a fuzzy variable itself.

In the following section we will take a look at the various factors affecting the success of a start-up in detail and their relative weightings. And these weights can be found out either by using a rule base (industry specific or general, whatever may be the case) or by answering a few questions in a linguistic form, which is shown in that section. The next section defines the variables used in our model to actually generate the rule base for evaluation purpose. And then we look into the feasibility of the start-up using real options theory and calculate the SFG (???) for the startup to give the answer to the most important question namely the cash flow in the new company formed. And then we finally present the complete model and conclude.

2. Success Factors Of Technology Start-Ups

We look into the factors affecting the success factors of Technology Startups in a categorized fashion. First we find out the broad areas of focus and then look deep in that particular area to gauge the relative importance of that factor on the overall effect. The key areas of a technology start-up are given below:

1. Leader and Team (or the quality of Human Resources available)
2. Innovative Technology or New Business Model
3. Marketing Strategy
4. Funding Strategy
5. Operational Efficiency.

Now we look into all of these factors separately and in depth.

A. *Leader and Team:*

a. CEO: optimistic, resilient, inspiring, visionary

b. COO: resilient, patient, meticulously attentive to details

c. CTO: smart, competitive, experienced

A successful start-up needs at least two of them. And the rest of the human resources also need to have past collaborative experiences, have the same vision of the company, flexible in job and hard-working.

B. Innovative Technology:

To assess the degree of innovation which we call here '*Innovator Position*' [1], we need to answer a set of questions given below:

a. Why is the model/ technology innovative and how much innovative is it?

b. Is it possible to protect the technology through IPR?

c. Can we assess the life-cycle of the technology?

d. What all are the competitive technologies and what is their chance of success over us?

e. Do we have a fair/ unfair advantage?

C. Market Strategy:

The market strategy is another important factor in the evaluation of a start-up. And for determining the factor, answers to the following questions are necessary:

a. OEM/ODM or brand name?

b. Market Segment? Mass or Niche Market?

c. What is the market awareness of the firm or the technology?

D. Funding Strategy:

This is the single-most important factor in determining the economic success of the start-up. And this we discuss at large in the coming sections. First we apply real options to test the feasibility of the start-up and then calculate fuzzy SFG of the firm to determine the growth and profitability which is very important. Nonetheless the following funds should be ready before one should start thinking of building his own company, and these phases are outlined below:

a. Incubation Phase → Seed Funding, R&D focus and Business Plan

b. Start-up Phase → 1st Round Funding, Business Development, Product Ready

c. Expansion Phase → 2nd Round Funding, Marketing, Alliances

E. Operational Efficiency:

The operational efficiency is the reason for the success or failure of the start-up in a way because if the company has all the features demanded by the above factors and yet low on operational efficiency, it will burn out. The efficiency of operation is basically the efficient integration of *Command, Control and Communication.*

3. Fuzzy Theory Basics And Application To Our Problem

The fuzzy theory of sets was proposed by Zadeh [2] in 1965. Let U be the universe of discourse, $U = \{(u_1, f_A(u_1)), (u_2, f_A(u_2))\ldots (u_n, f_A(u_n))$ and let A be a fuzzy set of U. Then the fuzzy set A can be represented as

$$A = \{(u_1, f_A(u_1)), (u_2, f_A(u_2))\ldots (u_n, f_A(u_n))\} \quad (1)$$

Where f_A, f_A: $U \rightarrow [0, 1]$, is the membership function of the fuzzy set A; $f_A(u_i)$ indicates the degree of membership of u_i in A. If the universe of discourse U is an infinite set, then the fuzzy set A can be expressed as follows:

$$A = \sum_{i=1}^{n} f_A(u_i)/u_i, \quad u_i \in U \quad (2)$$

If the universe of discourse is an infinite set, then the fuzzy set A can be expressed as follows:

$$A = \int_U f_A(u_i)/u_i \,, \qquad u_i \in U \tag{3}$$

Membership functions can be of various shapes depending on choice and usage like sigmoid, Gaussian, trapezoidal, triangular etc, see figure 1.

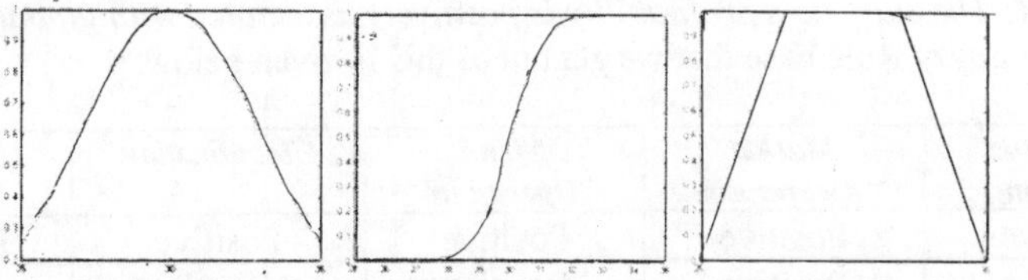

Figure 1: (a) Gaussian (b) sigmoid (c) trapezoidal membership functions

In our case we are going to use Sigmoid membership functions or triangular membership functions and we need to rate a particular variable say 'degree of innovativeness' in a scale of 'low', 'medium' and 'high' which will then be mapped to (0,1) and the membership value calculated. Thus the human nature of speaking in terms of linguistic variables is taken into account by use of Fuzzy Rule base.

4. Development Of Fuzzy Rule Base

We pick up a few variables from the wide list of factors affecting start-up profitability to design our fuzzy rule base. And the key variables are as given below:

- Innovator Position:

Innovator strategy offers considerable benefits to the SMTE start-up. First, when a new product emerges in the market, sales volume is normally small and large production capacity is not needed. Second, employees of a start-up are more innovative than that of the big established players [3] due to their higher inertia. So the first rule is:

RULE 1: The start-up's profitability is positively associated with its innovator position or the degree of 'innovativeness'.

- Market Awareness:

Innovator Strategy is usually accomplished with wide market awareness [5]. Market awareness is more important in High-tech industries as the new products on offer can be dramatically different from the existing products and not at all familiar to potential customers [3]. Moreover market awareness builds brand-power and distributor confidence. So the next rule is:

RULE 2: The start-up's profitability is positively associated with its market awareness.

- Niche Operation:

Technology start-ups should restrict their operations in certain market niches and concentrate their limited resources there. Such focus strategy relaxes size constrains and enables the co. to maintain their technological leadership [6]. First, the co can carve out niches with very few or no competitors and charge relatively high prices [7]. Second, focus orientation helps the co. to avoid extending beyond its core capabilities [8]. Therefore our next rule is:

RULE 3: The start-up's profitability is positively associated with its niche operation.

- Globalization:

The SMTE start-up will benefit from globalization. First, the co. has to leverage its R&D costs across a greater volume of products. Second, globalization increases economies of scale. Third, globalization generates extra profits to sustain R&D growth. So the next rule is:

RULE 4: The start-up's profitability is positively associated with globalization.
Hence the Fuzzy Rule base that we get out of this is given below:

Innovator Position	***Market Awareness***	***Niche Operation***	***Globalization***	***Decision***
Positive	Positive	Positive	Positive	Excellent
Positive	Positive	Positive	Negative	Good
Negative	Positive	Positive	Positive	Poor
Negative	Negative	Negative	Negative	Worst

Thus we can develop a huge rule base of Decisions regarding the success of the start-up and we need to include many more fuzzy variables as well as the linguistic classes (instead of just using positive or negative). That would improve the results even more.

5. Self-Financeable Growth & Funding Strategy

Startups typical take a go with a fixed seeding capital and then as required, plunge into the first round funding, second round funding and so on. Striking a balance between consumption of cash and generating it is one of the key challenges a startup or an SME faces. Failure in doing so and even a thriving company can soon find itself out of business [9].

A Framework to manage the growth is given by Churchill and Mullins [9].
The above framework takes into account three critical factors:

1. A company's operating cash cycle - the amount of time the company's money is tied up in inventory and other current assets before the company is paid for the goods and services it produces.
2. The amount of cash needed to finance each dollar of sales, including working capital and operating expenses.
3. The amount of cash generated by each dollar of sales.

These three factors determines the *self-financeable growth(SFG)* - the growth at which a company can sustain its growth through the revenues it generates without going hat in hand to financiers.

The usefulness of this framework goes beyond the calculation of sustainable growth rate. It gives us an insights into how efficiently the operations are running, how profit margins affect their ability to fuel faster growth, which of their product lines and customer segments holds growth potential and what kinds of businesses might be attractive investment targets.

Calculation of the SFG (Annual SFG rate):
Annual SFG rate = Cash generated per Sales Rupees / Amount of Cash Tied up per sales Rupees.

The three levers of growth are a) Speeding up Cash flow, b) Reducing cost, c) Raising prices. These three implicitly have direct impact (positive or negative) on the

SFG. By pulling the three levers of growth; we can get two growth factors:

1. Leveraged SFG (the positive growth direction).
2. Restricted SFG (the negative or pessimist growth direction).

6. Feasibility Analysis Using Real Options

The very first decision an investor or for that matter even the entrepreneur has to take is how viable the project is. The decision or decisions which he has to take can be broken down into several steps and each of these step offers an option to the investor or the entrepreneur as to whether to go further or not. A technique very often used by the corporate managers lately is that of valuing decisions using *real options.* Here, the underlying assumption is that the value of the growth of a company is a function of the underlying portfolio of the growth projects at hand. Just like the financial options, these 'real' options are right to invest but no obligation whatsoever. But an inherent problem lies with the tried and tested methodology of applying real options. Unlike financial options, the applicability of these is much more complex as the variables under scrutiny are far more and are sometimes abstract. Hence just by applying says, Black-Scholes-Merton model for valuing options has found to be inefficient and sometimes ineffective. A better model, based on *binomial lattice* has been suggested by Copeland and Tufano (2004) [10].
The model works as follows:

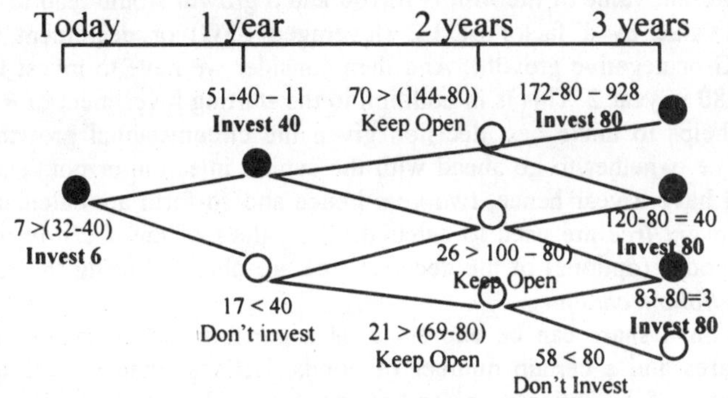

Fig. 01: Decision Tree for Real Options

Firstly, a *decision tree* (Fig 01) is created. This tree illustrates the dynamics of the project mapping how its value changes over time. The nodes of the decision tree are essentially *options* the value of which helps the investor/entrepreneur in deciding what the best option to pick is. For instance, after putting in the seeding capital, is it the right decision to chip in further round funding or not.

A growth cycle of a startup is lined up with these phases of funding and a decision of how much or whether or not to fund is a critical question faced by an investor. Now the decision tree could for investing in the company as a whole or a part / project of the company. In our case, we need to evaluate say whether out very first step of investing the company is right. So for that we calculate the value of the company as it would be if it existed today and then its value in the future. This involves creating another type of tree called the *event tree* (Fig 02). It shows how the

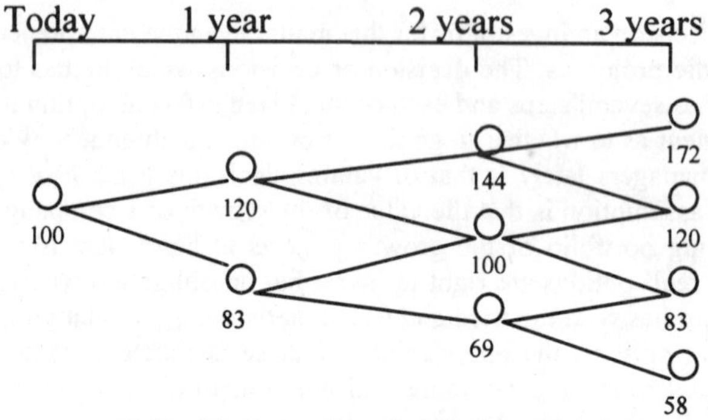

Fig. 02: Event Tree for Real Options

company's likely values are determined by fluctuations in the spread or deviation in the possible growth values. Now these can be the leveraged SFG and restricted SFG. Now say the present value of the firm is Rs100 and a growth would lead to increase in the firm's value by a factor of 1.2 (leveraged SFG) or decrement by 0.83 (restricted SFG or negative growth). And then consider we have to invest Rs 40 in year 1 and Rs 80 in year 2. This is in-addition to the starting investment of Rs 6. The decision tree helps to make best decision given the circumstantial growth at that point in time, i.e. whether to go ahead with the capital injection or not. The values the firm could have a year hence, two year hence and so-forth are calculated. The values of the *event tree* are used to calculate the values of the *decision tree*. The values of the nodes (options) of the decision tree are obtained using the following *replicating portfolio technique*.

Any option on a share can be expressed as a portfolio consisting of a certain number of shares and a certain number of bonds. Let's assume we are trying to estimate the value of the option at the end of year two under a scenario in which the company has a current value of Rs 144 (the highest-value end-of-year-two scenario). An option that is kept alive will have two possible pay-offs: Rs 93 or Rs 40. This means a certain proportion of company value (the equivalent of equity) less a certain amount of bonds with interest (we'll assume a risk-free rate of 8%) will also produce either Rs 93 or Rs 40 after a year, depending on whether the company is worth Rs172 or Rs 120 at the end of the year. Mathematically, we could express this as two formulae: m (172)-(1+.08) (B) =93 and m (120)-(1+.08) (B) =40, where m is the proportion of the company value and B is the number of par-value one-rupee bonds. Since we have two equations hence the two unknowns (m and B) can be solved. In this case, the replicating portfolio for an option in the top node in year two calls for the company worth Rs144 and minus74 par-value one-rupee bonds. Thus the option is worth Rs. 70.

7. Conclusion

The above model is tries to elucidate how an investor or an entrepreneur can take various factors like Innovativeness of the entity, product scope, marketing awareness and internationalization into consideration and make a decision as to whether to go about it or not. The growth trees helps one take the stance on the second critical factor of next round funding and to some extend the amount of funding too. The inherent linguist nature of the key variable in the above model comes from the fact that in real world scenario, decisions as complicated as these cannot be made using rigid mathematical model. Evidently there is much greater scope of work including empirical testing of the model and calibration of the model variable but the current model in abstract has taken the most important factor into consideration.

References

1. Gongming Qian & Lee li (2003), Profitability of Small and Medium Sized Enterprises in High-Tech industries: The case of Biotechnology Industry; *Start. Mgmt. J.*, 24: 881-887.
2. L.A. Zadeh (1965), Fuzzy sets, *Information and Control,* 8: 338-353.
3. Moriarty RT, Konsik TJ (1989), *High Tech Marketing: Concepts, continuity and change;* Sloan Management Review 30(4): 7-17
4. L.A. Zadeh (1978), Fuzzy sets as a basis for a theory of probability, *Fuzzy sets and Systems*, 1: 3-28.
5. M. Lambkin & G. Day (1989), Evolutionary Processes in competitive markets: beyond the product life cycle; *Journal of Marketing*, 53(4): 4-20.
6. T.O. Kohn (1997), Small Firms as international players, *Small Business Economics* 9: 45-51.
7. A. Fiegenbaum, S. Hart & D. Schendel (1996), *Strategic Reference point theory.* Start. Mgmt. J. 17(2): 219-235.
8. Covin et al (1990) *Content and Performance of growth-seeking strategies.* Journal of Business Venturing 5: 391-412.
9. N. Churchill & J. Mullins (2001), *How Fast Can Your Company Afford To Grow* Harvard Business Review
10. Tom Copeland & Peter Tufano (2004March). *A Real-World Way to Manage Real Options*, Harvard Business Review.
11. Richard Makadok (2003), *Doing the Right Thing and Knowing the Right Thing to Do: Why the Whole is Greater Than Sum of the Parts.* Strategic Management Journal, 24: 1043-1055.

Fuzzy Logic and its Application in Technology and Management
Editors: Debjani Chakraborty, S. Nanda and D. Dutta Majumder

Fuzzy VaR: A New Approach for VaR Evaluation and Application to the Portfolio Optimization Problem

Amarendra Kumar[1], *Abhrashikhar Banerjee*[2] *and Debjani Chakraborty*[3]

[1]Georgia Institute of Technology, Atlanta, USA
[2]University of Chicago, Chicago, Illinois, USA
[3]Department of Mathematics, Indian Institute of Technology Kharagpur
Email: -[1]amrendrakumar@gatech.edu [2]banerji@uchicago.edu, [3]debjani@maths.iitkgp.ernet.in

Abstract: This paper describes the financial modeling of an optimized portfolio with risk measured in terms of Value at Risk. Value at Risk is now in widespread use to measure the risk involved with any assets. The definition of Value at Risk (VaR) labels it being a measure of uncertainty in itself. In this paper we have developed the methodology to determine the fuzzy behavior of VaR by introducing the concept of *fuzzy VaR* and displaying its subsequent use in the optimization of portfolio for the maximum return or minimum risk by involving parameters that have fuzzy behavior.

Keywords: Probability, Fuzzy sets, Genetic Algorithms, Risk Management, Value-at-Risk.

1. Introduction

Value at Risk is an important tool for the measuring the risk involved in any kind of assets traded in the financial market. The investor setting up a portfolio always has a degree of risk in his mind, exposure to which he is comfortable with, the measurement of which was more or less initiated by Markowitz in 1952. This approach required an accurate forecast of the rate of return. The modern era of risk measurement for foreign exchange positions began in 1973. That year saw both the collapse of the Bretton Woods system of fixed exchange rates and the publication of the Black-Scholes option pricing formula. The collapse of the Bretton Woods system and the rapid transition to a system of more or less freely floating exchange rates among many of the major trading countries provided the impetus for the measurement and management of foreign exchange risk, while the ideas underlying the Black-Scholes formula provided the conceptual framework and basic tools for risk measurement and management.

This approach has been more or less the convention in the financial world until JP Morgan's Risk Metrics in 1994. Several methods have been proposed and implemented to varying degrees in the intermediate time to forecast the adequate rate of returns such as the regression based method proposed by Elton & Gruber [7]. A portfolio rebalancing problem is mostly solved by reducing the problem to an optimization problem and considering it either as a minimization of the total exposed risk or maximization of the expected rate of return on the portfolio [5]. Several other methods have been proposed incorporating the non-ideal market conditions [9, 21, and 22]. All these approaches use variance of the asset price as the measure of risk involved. If the risk is measured in a way which can answers the question "How bad can things get?" [13], then the abovementioned Value at Risk (VaR) method is the measure for evaluating risk. Consequently, several methods have been proposed to evaluate the optimal portfolio with risk measured in terms of VaR. The VaR has been much maligned since its' inception due to many reasons [2] leading to initial apprehension about its' acceptability as a risk forecasting tool, but in the recent times much of those apprehensions have been put to rest as increasingly more traders worldwide are switching over from the conventional mean-variance analysis.

While exposing a portfolio to a slightly greater risk fetches more returns, the probability of losses also increases. It just takes a little more exposure to risk in a particular asset in the portfolio to bring in huge losses across the returns of the portfolio as a whole. This argument is the strongest purveyor in our logic that the risk which an investor is exposed to in any form is fuzzy and thereby the motivation for the mathematical and logical foundation of fuzzy VaR.

Considering the VaR as the measure of risk and also taking into account the fuzzy nature involved in it, we introduce the concept of fuzzy VaR. The risk free return is the return that we get for an investment of extremely low risk, but a risk by its very nature of its meaning is about uncertainty. A fuzzy approach keeps the simplicity of VaR intact but provides more freedom to make a choice over a range of possible portfolios with the return associated with it. The fuzzy VaR approach is more importunate to the fuzzy nature of the risk taking ability of the investors [3] in the sense that one could define a fuzzy model of the risk profile of the investor and compare it with the risk profile of the investment. This would be the scenario of ideal application of the ideas presented in the paper.

The next section starts off with the formal definition and concept of the VaR [12]. Section 3 follows with defining the fuzzy VaR leading to section 4, where we present the model of finding optimal portfolio with constraints on fuzzy VaR. In section 5, a methodology of solving the above optimization problem efficiently is discussed with the modified genetic algorithm followed by our concluding remarks in section 6.

2. Value at Risk

Value at risk is the measure of risk in the worst case scenario. It indicates how big a loss could be in the worst-case scenario. Let x be a portfolio and $P(x)$ denotes value of portfolio. For almost all practice purpose we can assume that $P(x)$ follows the normal distribution [12]. However as with Campbell et al [3], whose work was later commented upon by Huang in 2003 [12] as having some errors; we have tried to avoid that shortcoming while keeping the idea of VaR simple and alive.

Let $P(x) \sim N(\mu_X, \sigma_X)$, and $\overline{P}(x) = \mu_X$ is the mean value of return for the different distributions shown in Fig.1, of the portfolio X and $P_\theta^*(x)$ is the quartile of $P(x)$ corresponding to the specific portfolio X and confidence interval θ. In other words, the value of the portfolio x exceeds or equal $P_\theta^*(x)$ with probability θ. Using this notation VaR can be defined as,

$$v = VaR(x,\theta) = R_0(x) - P_\theta^*(x) \tag{1}$$

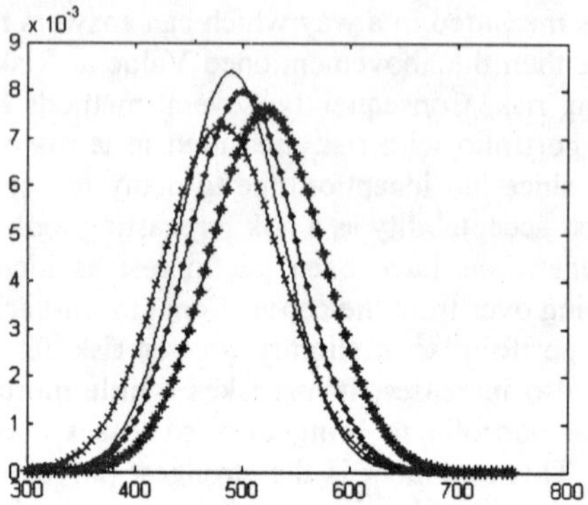

Fig. 1: Value of return for different distribution

where, R_0 denote the benchmark of the portfolio return, which have been considered here as a subject parameter depending upon the kind of assets and behavior of return.

3. Fuzzy VaR

We have introduced the concept of fuzzy VaR in this section. A VaR by its very definition measures the fuzziness involved in the risk [13]. A VaR takes two parameters to measure the risk, namely the confidence interval (CI) denoted by $(1-\theta)$ and the expected maximum loss possible in that case. Now a close look at the nature of these two parameters, reveal its fuzzy nature attributed by the fuzzy decision making habits of the investors. While assigning the membership function to this fuzzy parameter, the priorities of the investors is kept in mind and a membership value of one is assigned to a value at which the investor is very comfortable. Mathematically *fuzzy VaR* is defined as follows,

Let for fixed θ_0, the value of the VaR is v_0 for a portfolio X. A fuzzy VaR $V_{\circ f}$ can now be defined as a fuzzy set A, with membership function (MF) given as follows,

$$\bar{A} : x \to [0,1], x \in X \tag{2}$$

Where, membership function $\bar{A}$ is defined as,

$$\bar{A}(v) = f_{v_0,\mu,\sigma}(v) \tag{3}$$

Where, function f takes v_0, μ and σ as the parameters and v is the variable, such that $f_{v,\mu,\sigma}(v_\circ) = 1$. Membership function is subjective and can be changed to suit ones desires of choosing the shape of its curve.

We have taken MF as the Gaussian membership function.

$$\overline{A}(v) = e^{-\frac{(v-v_0)^2}{2(k\sigma)^2}} \tag{4}$$

where, $k > 1$ and the value of k is subjective.

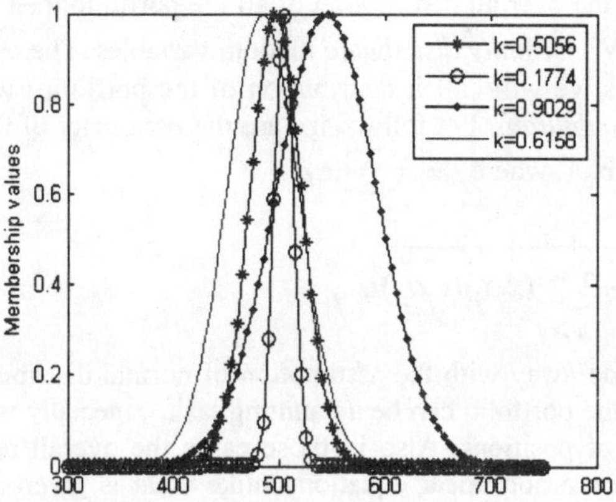

Fig.2. Plot of different Gaussian membership functions

A next level of fuzziness can also be introduced here by taking the Θ as the fuzzy parameter. The membership function for Θ can be determined either theoretically or by taking the observation as to at which value of Θ is likely to be appropriate for the amount of risk that the investors are willing to take. For that value of Θ the membership value can be assigned as one and thereon, for other values of Θ the order of preference can be scaled to different values of membership function. The general trend of operation for investors is between a confidence interval of 95-99%, i.e. a Θ ranging from 1-5%. Any value lying in this range can be assigned a membership value of 1 and remaining value of Θ could be an assigned an appropriate membership value depending upon the psyche (in this case, the risk taking ability) of the investor.

Let us now consider the physical interpretation of the abovementioned concept; the notion of *fuzzy VaR* is so that it measures varying amounts of capital that can be lost for a particular confidence interval across a varying degree of confidence intervals over the course of the investment horizon.

4. Portfolio Rebalancing Problem Formulation

Our portfolio consists of finite number of assets $i = 1,2,.....,N$. Each assets can be of any kind of financial assets like stocks, bonds and derivatives. We have assumed here that distribution of each kind of assets is known and moving one step ahead we have also assumed that each of the assets have normal distribution. Although this formulation can work well for all parametric and nonparametric distribution we have assumed normal distribution to present out idea in a simple and easy way. Our portfolio x is characterized by positions in different assets.

$$x = (x_1, x_2, ...x_N)$$

Each of the assets has the following parameters, (μ_i, σ_i) Where $i = 1,2,\ldots\ldots, N$.
Each of μ_i represents the expected return on i^{th} assets, however one can also consider $R_{o,i}$ as the benchmark of the ith asset [2].
If we will consider the over all distribution of all the portfolio then it will turn out to be summation of N *normally distributed* random variables. The overall distribution (from now onwards we will call it distribution of the portfolio) will eventually be another *normal distribution.* Let followings are the parameter of the distribution of the portfolio: (μ_0, σ_0), where $\mu_0 = \sum x_i \mu_i$ (4a)

and

$$\sigma_0 = \sqrt{\Sigma(x_i\sigma_i)^2 + 2\sum_{i>j}(x_i\sigma_i)(x_j\sigma_j)\rho_{i,j}} \quad (4b)$$

In practice, if one do away with the assumption of normal distribution then finding the distribution of the portfolio can be a daunting task, especially when the portfolio contains hundreds of positions. Also in those cases the overall return may not be liner but may follow a non-linear equation unlike what is given in (4a). Financial literature contains discussion of various parametric and simulation approaches to this problem [1, 10, 11, 16].

***Definition*:** A *fuzzy Vector* is a vector that's each element is a *fuzzy number*. A fuzzy vector X is denoted here by $\hat{X}_f$.
With all the notation defined and explained the optimization problem can be formulated as:

$$\max \quad \hat{X}'_f \hat{\mu}$$

Subject to, (5)

$$\sum_N \hat{X}_f = 1 \quad \text{and,} \quad P_\theta - R_0 \in (V_{\theta_f})_f$$

Where $\hat{X}_f = (X_{1f}, X_{2f} \cdots X_{Nf}), P_0 = R(x), P_\theta = P_\theta^*(x)$ is the expected price of the portfolio corresponds to θ and $V_{f,\theta}$ is the *fuzzy VaR* specified by the owner.

The above problem seems hard to solve but the underlying difficulty lies due to the introduction of fuzzy variables and parameters. This can be handled. Once the above problem is solved, it gives solution as the *fuzzy Vector* $\hat{Xo}_f = (Xo_{1f}, Xo_{2f}, \ldots\ldots Xo_{Nf})$. With each of the N *fuzzy numbers* will have there MF defined alike will have σ_i values corresponds to each *i*, (2) and (3)

5. Modified GA Required for Solution of the Fuzzy Optimization Problem

Let X_f be the fuzzy vector corresponds to a fuzzy portfolio. $\hat{\mu}$ is a mean vector such that, , where, μ_i is the mean value of the return for the ith asset.

The above problem is solved by considering a fuzzy number as a set of many crisp values with their appropriate membership function. First the fuzzy problem is converted into a crisp valued optimization problem and then finds out the solution for every possible permutation. We can take a particular instance of the possible permutation to find out the optimum portfolio. For this we can convert the above fuzzy optimization problem into a crisp valued optimization problem as follows,

$$\max \quad \hat{X}_i' \mu_i$$

Subject to,

$$\sum_N \hat{X}_i = 1 \tag{6}$$

$$\text{and,} \qquad P_\theta - R_0 < V_\theta$$

Here $\hat{X}_i$ is a crisp values vector and V_θ is the VaR corresponding to a particular instance of fuzzy θ.

The initial population is generated randomly which should satisfy the following:

$$\sum_N \hat{X}_i = 1 \quad \text{and} \quad P_\theta - R_0 < V_\theta \tag{7}$$

the term P_θ corresponds to the distribution of portfolio obtained, defined in (4a) and (4b), for the crossover population the fitness value (fval) is obtained by,

$$fval = \sum \hat{X}_i' \mu_i \tag{8}$$

Corresponds to the fitness value obtained, population for cross over and mutation can be used as is done in any standard genetic algorithm.

6. Conclusion

We developed here a framework of defining Fuzzy VaR and using the VaR as a constraint for finding optimal portfolio. The fuzzy constraints here produce a fuzzy portfolio with a corresponding fuzzy expected return. There are mainly two advantages for using this concept. Firstly, from the result of the fuzzy optimization problem formulated in the preceding section, the investor can choose a configuration of portfolio which corresponds to an interval of risk which depends on the psyche of the investor. Secondly, it obviates the need for basing the investment judgments on the conventional models of evaluating VaR which deduce the risk from a limited number of scenarios in the tail distribution. However, one glaring disadvantage of this process of computing fuzzy portfolios is the computational complexity. But we feel that this can be reduced by the use of parallel algorithms or grid computing. We recommend using the MATLAB environment for all the computation.

References

1. C.O. Alexander & C.T. Leigh (1997), On the covariance matrices used in value at risk models, *Journal of Derivatives* 4: 50–62.
2. G.J. Alexander & A.M. Baptista (2002), Economic implications of using a mean-VaR model for portfolio selection: A Comparison with mean-variance analysis, *Journal of Economic Dynamics & Control*, 26: 1159-1193.

3. R. Campbell, R. Huisman & K. Koedijk (2001), Optimal portfolio selection in a value at risk framework, *Journal of Banking and Finance*, 25: 1789–1804.
4. M.A.H. Dempster (2002), *Risk Management: Value at Risk and Beyond*, Cambridge University Press, London.
5. X.T. Deng, Z.F. Li & S.Y. Wang (2005), A minimax portfolio selection strategy with equilibrium, *European Journal of Operation Research*, 166: 278-292.
6. D. Duffie & J. Pan (1997), An overview of value at risk, *Journal of Derivatives*, 4: 7-49.
7. E.J. Elton & M.J. Gruber (1995), *Modern Portfolio Theory and Investment Analysis*, Wiley, New York.
8. C.D. Feinstein & M.N. Thapa (1993), Notes: A reformulation of a mean-absolute deviation portfolio optimization model, *Management Science*, 39: 1552–1553.
9. A.A. Gaivoronsk, S. Krylov & N.V.D. Wijst (2005), Optimal portfolio selection and dynamic benchmark tracking, *European Journal of Operational Research*, 163: 115-131.
10. H. Glyn (1998), Simulating value-at-risk, *Risk* 60–63.
11. T. Ho, M. Chen & F. Eng (1996), VAR analytics: Portfolio structure, key rate convexities, and VAR betas, *The Journal of Portfolio Management*, 89–98.
12. H.H. Huang (2005), Comment on Optimal portfolio selection in a Value-at-risk framework, *Journal of Banking and Finance*.
13. J.C. Hull (2003), *Options, Futures and Other Derivatives*, 5th. Edition, Prentice Hall International, London.
14. H. Konno & A. Wijayanayake (2001), Portfolio optimization problem under concave transaction costs and minimal transactions unit constraints, *Mathematical Programming*, 89: 233–250.
15. M.K. Luhandjula (1996), Fuzziness and randomness in an optimization framework, *Fuzzy Sets and Systems*, 77: 291–297.
16. Manoj K. Singh (1997), Value at risk using principal components analysis, *The Journal of Portfolio Management*, 24: 101–112.
17. H.M. Markowitz (1991), Foundations of portfolio theory, *Journal of Finance*, 46 (2): 469–477.
18. H.M. Markowitz (1959), *Portfolio Selection: Efficient Diversification of Investments*, Wiley, New York.
19. J. Philippe (1996), Risk2: Measuring the risk in value at risk, *Financial Analysts Journal*, November/December: 47–56.
20. W.F. Sharpe (1970), *Portfolio Theory and Capital Market*, McGraw-Hill, New York.
21. M. Taboga (2002), Portfolio selection with two-stage preferences, *Finance Research Letters*, 2: 152-164.
22. H.G. Xue, C.X. Xu & Z.X. Feng (2005), Mean-variance portfolio optimal problem under concave transaction cost, *Applied Mathematics and Computation*.
23. H.J. Zimmermann (1996), *Fuzzy Set Theory and its Application*, Kluwer Academic Publishers, Boston.
24. Z. Zmeskal, (2005), Value at Risk methodology under soft condition approach (fuzzy-stochastic approach), *European Journal of Operation research*, 161 (2): 337-347.

Part 5

Fuzzy Data Analysis in Medical Technology

Fuzzy Logic and its Application in Technology and Management
Editors: Debjani Chakraborty, S. Nanda and D. Dutta Majumder

Fuzzy Correlation Study to Assess the Association Between Clinico-Epidemiological Features and Progression of Oral Submucous Fibrosis – A Precancerous Condition

Ranjan Rashmi Paul[1], *Keya Chaudhuri*[2]
Debjani Chakraborty[3], *Chandan Chakraborty*[3], *Mousumi Pal*[4]
Hirendranath Seal[4] *and Jyotirmoy Chatterjee*[5]

[1]Department of Oral & Maxillofacial Pathology, R Ahmed Dental College, Kolkata
[2]Human Genetics & Genomics Group, Indian Institute of Chemical Biology, Kolkata
[3]Depatment of Mathematics, Indian Institute of Technology Kharagpur
[4]Institute of Interdisciplinary Scientific Research (IISR), Kolkata
[5]Calcutta Medical College, Kolkata

Abstract: Oral submucous fibrosis (OSF), a progressive precancerous condition of oral cavity and oropharynx, having high degree of malignant potentiality. It occurs mainly among the people of Indian subcontinent with sporadic existence in other parts of the world including UK and USA. The etiopathology of this precancerous condition is very poorly understood. Researchers reported varied clinico-epidemiological attributes of OSF and inability to open the mouth (trismus) as the pivotal clinical feature of this pathosis in respect to its progression. Till date no such study is available to correlate the clinico-epidemiological variables and the progression of this disease process. Accordingly, besides classical correlation studies, fuzzy correlation techniques are applied to evaluate the clinico-epidemiological variables since some attributes are in fact qualitative and their joint influence on trismus. Some of the clinico-epidemiological feature variables are fuzzy in nature, which have the power of capturing large spectrum of information. This paper aims to study the important feature variables that have more influence on trismus using fuzzy correlation technique.

Keywords: Oral Submucous Fibrosis (OSF); Trismus; Clinico-epidemiology; Pre-cancer; Fuzzy number; Classical correlation; Fuzzy correlation

1. Introduction

Oral submucous fibrosis (OSF) is an insidious, chronic, progressive, precancerous condition of the oral cavity and oropharynx having a high degree of malignant potentiality [1, 2]. A significant number of OSF cases (7.6%) transform into squamous cell carcinoma [3]. The disease occurs almost exclusively among the people of the Indian subcontinent but sporadic cases have been observed in other parts of the world including UK and USA. Recently it has been reported that this precancerous condition is becoming epidemic in India [4]. Across all socio-economic status of India, the incidence rate of this disease exhibits inter-regional variations being highest in the southern region of the country [3]. The etiopathology of this precancerous condition is very poorly understood. Irritants like arecanut and its derivatives including pan-masala and chillies as well as imbalance of trace metals [5] and malnutrition are said to play a very crucial role in the development of OSF [3, 6-8]. In this context recent observation regarding association of arecanut use with an increase in the prevalence of OSF in the lower age group of Indian population is a matter of deep concern [4]. Clinically the disease is manifested with varied clinical attributes like discomfort, burning sensation, pain and dryness of mouth, excessive salivation, ulceration, inability to open mouth (trismus), blanched opaque appearance of oral mucosa, vesiculation, depapillation of tongue, hypomobility of tongue, mucosal pigmentation etc. Amongst all these clinicians give pivotal importance to degree of trismus with the progression of the disease. Our preliminary work suggests that amongst different epidemiological variables age, economic status and oral habit of the patients have possible significant contribution to the disease process. But it is very difficult to resolve the actual role played by the different linguistic epidemiological variables in the genesis and progression of this pathosis. In fact, some of the variables assess linguistic values to capture more information. Accordingly, to have a better idea in this regard, fuzzy correlation coefficients have been employed in this study to evaluate the correlation between progression of the disease and varied clinico-epidemiological variables where some of them are practically linguistic.

On the basis of clinician's visualization and perception, the OSF related clinico-epidemiological variables may be subdivided into two categories: (i) clinical and (ii) epidemiological. In this study, trismus, a clinical variable, is directly associated with the progression of the disease. Therefore, an investigatation is required to check whether there exists any relationship among the variables with trismus and to identify those having more influence on trismus. The classical correlation and regression techniques are powerful tools to analyze the interrelationship among the crisp variables and enable us to obtain the joint effect of all the crisp variables in association with trismus. Experts feel comfortable to express their opinion in terms of natural language that involves fuzziness. Therefore 'trismus' is considered as a fuzzy variable that takes the linguistic values to capture large spectrum of information. The medical experts define the linguistic terms for trismus variable here as follows, Severe = $(0, 1, 1.5)_{LR}$; Moderate = $(1.5, 2, 2.5)_{LR}$; Mild = $(2.5, 3, 3.5)_{LR}$ in the sacle of [0, 3.5]. Actually, crisp number is a special case of fuzzy number. Therefore, in this paper we represent a fuzzy extension of classical correlation in order to investigate the relationship among the fuzzy variables.

The paper is organized as follows: Section 2 describes the clinico-epidemiological variables considered in the medical experiment. The classical technique of correlation analysis and linear regression analysis is studied in Section 3. In Sub-section 3.1, the correlation coefficients are calculated (using SPSS [16]) to study the extent of influence of *pain, burning sensation* and *ulceration* on trismus. And also stepwise regression modeling technique is used to achieve a regression model some (not all) of the predictors have high predictive power in Sub-section 3.2. Section 4 describes the fuzzy correlation coefficients in subsequent Subsections 4.1, 4.2, and 4.3 for fuzzy variables whose values are linguistic terms (quantized in fuzzy logic framework). Results and discussions are drawn in Section 5.

2. Materials and Methods

In this study clinico-epidemiological variables of 852 clinico-pathologically confirmed OSF cases were included. The clinico-epidemiological variables were age (20-80 years), religion (Hindu / Muslim), sex (male/ female), habitat (urban/rural), economic status (low-economically dependent to Rs 4000/- pm, middle-Rs 3000 to Rs 7000/- pm, high- Rs 6000/- to Rs 15,000/- pm), food habit (Veg/Nonveg), oral habits (yes/no); discomfort (yes/no), pain (yes/no), burning sensation (yes/no) and ulceration (yes/no) of oral mucosa; inability to open the mouth [trismus- mild (inter-incisal distance 2.5cm $\leq$ IID < 3.5cm, moderate 1.5 cm$\leq$ IID < 2.5 cm, severe IID <1.5cm, considering IID$\geq$3.5 cm as normal); sites of lesion in oral mucosa (buccal-B, buccal palatal-BP, buccal palatal labial- BPL, buccal palatal labial tongue-BPLT, buccal palatal labial tongue and faucial pillars-BPLTF).

Initially, an attempt is made to study the correlations among the crisp variables though trismus is fuzzy using classical correlation coefficients. Also some predictors have been investigated those have more influence on trismus.

3. Classical correlation and regression models

3.1. Correlation

The correlation techniques viz., simple correlation, partial correlation and multiple correlation coefficients [7] play important role in analyzing the interrelationship and strength of associationship of variables. Simple correlation coefficient is used to find the relationship between two random variables. If there are more than two variables, partial correlation measures the relationship between any two variables where others are partialed out. In multivariate regression analysis, multiple correlation measures the strength of predictive power of all predictors in association with the dependent variable. In studying the bivariate correlations it is observed that "*sites of lesion*" variable is highly correlated (0.725) with "*trismus*". *Pain, burning sensation* and *ulceration* has also certain influence on *trismus*, but others are not much related. Negative correlation indicates that trismus increases / decreases according to certain feature decreases / increases. Though there exists diverse relationships, stepwise regression modeling is done to achieve a final regression model with the predictors having more predictive capability.

3.2. Refinement of Models

The fitted model seems to be quite good. But some of the predictors i.e. *age, sex* and *oral habit* should not be considered in the model since their p-values are insignificant. On the other hand, the predictors *burning, site of lesion, ulceration and pain* are very significant to predict *trismus* whereas *discomfort, economic status, food habit, habitat and religion* predictors are quite significant. All the predictors except *age, sex* and *oral habit* are highly associated (0.739) with trismus. Therefore the model is fitted based on only the predictors *burning, site of lesion, ulceration, pain, discomfort, economic status, food habit, habitat and religion.* And therefore the revised F-value (112.330) is higher than the previous one (84.014), which implies that the model has been improved. Hence the fitted model is written as follows:

$$Trismus = 0.722 - 0.197 \times Burning + 0.53 \times Site\_lesion + 0.131 \times Ulceration - 0.06 \times Pain - 0.296 \times Discomfort - 0.241 \times Economic\_status - 0.474 \times Food\_habit - 0.0312 \times Habitat + 0.047 \times Religion. \quad (1)$$

Following the equation 2.1, the predictors' *economic status, food habit, habitat and religion* fall in clinical framework according to medical experts' those of which has lesser importance on *trismus*. Therefore an attempt is made to fit a model based on only the epidemiological predictors (which has high predictive power) and hence the fitted model is

$$Trismus=0.655-0.199\times Burning+0.533\times Site\_lesion+0.136\times Ulceration- .581\times Pain-0.276\times Discomfort \quad (2)$$

The model is good fitted since its revised F-value is high (201.003). Also it can be remarkably observed that the p-values for *discomfort and pain* are quit high with respect to others. Therefore a decision could be again made that *burning sensation, ulceration and site of lesion* features have more impact as compared to *discomfort and pain.* Hence an improvement can be done in the model based on only these three important predictors and hence the optimal regression model is as follows

$$Trismus=0.532-0.198\times Burning+0.537\times Site\_lesion+0.120\times Ulceration \quad (3)$$

There is significant improvement in F-value (331.809) and leads to obtain the best-fitted model. A decision can be made that the disease "*trismus*' is due to mainly based on the discrimination of *burning sensation, ulceration and site of lesion. Trismus* is considered as a crisp variable through out the analysis, but practically it is a fuzzy variable whose values are based on medical expert's perception, experience and knowledge. Therefore a modification to the classical correlation is essential to find out the fuzzy correlation among fuzzy variables.

4. Fuzzy Correlation Coefficients

Let us consider two fuzzy variables say $\widetilde{X}_i=(X_i^C-X_i^L, X_i^C, X_i^U-X_i^C)_{LR}$ and $\widetilde{X}_j=(X_j^C-X_j^L, X_j^C, X_j^U-X_j^C)_{LR}$. The multiplication of two LR-type fuzzy variables generating another fuzzy variable their expectation is well defined in [2]. In view of this, the concept of fuzzy covariance between $\widetilde{X}_i$ and $\widetilde{X}_j$ is developed in [1] that is basically a fuzzy extension of the classical covariance. Therefore,

$$Cov(\widetilde{X}_i,\widetilde{X}_j) = E(\widetilde{X}_i \circ \widetilde{X}_j) - E(\widetilde{X}_i)\mathrm{E}(\widetilde{X}_j) = \widetilde{\sigma}_{ij} = (\sigma_{ij}^L, \sigma_{ij}^C, \sigma_{ij}^U)_{LR} \quad (4)$$

where

$$\sigma_{ij}^L = 2Cov(X_i^C, X_j^C) - Cov(X_i^C, X_j^L) - Cov(X_i^L, X_j^C);$$

$$\sigma_{ij}^C = E(X_i^C X_j^C) - E(X_i^C)E(X_j^C) = Cov(X_i^C, X_j^C) \text{ and}$$

$$\sigma_{ij}^U = Cov(X_i^C, X_j^U) + Cov(X_j^C, X_i^U) - 2Cov(X_i^C, X_j^C).$$

In general, the covariance matrix for fuzzy '*p*' variables can be written as

$$\Sigma = ((\widetilde{\sigma}_{ij}))_{p\times p} = \begin{bmatrix} (\sigma_{11}^L, \sigma_{11}^C, \sigma_{11}^U)_{LR} & .. & .. & (\sigma_{1p}^L, \sigma_{1p}^C, \sigma_{1p}^U)_{LR} \\ .. & .. & .. & .. \\ .. & .. & .. & .. \\ (\sigma_{p1}^L, \sigma_{p1}^C, \sigma_{p1}^U)_{LR} & .. & .. & (\sigma_{pp}^L, \sigma_{pp}^C, \sigma_{pp}^U)_{LR} \end{bmatrix}.$$

4.1. Simple Fuzzy Corrrelation

A simple fuzzy correlation index for any two fuzzy variables $\tilde{X}_i$ and $\tilde{X}_j$ is given by as $\tilde{\rho}_{ij} = (\rho_{ij}^L, \rho_{ij}^C, \rho_{ij}^U)_{LR}$ where

$$\tilde{\rho}_{ij} = \frac{\tilde{\sigma}_{ij}}{\sqrt{\tilde{\sigma}_{ii}} \otimes \sqrt{\tilde{\sigma}_{ij}}}. \qquad (5)$$

The multiplication and division of fuzzy numbers are well defined [2].

4.2. Partial Fuzzy Correlation

If there are more than two fuzzy variables then partial correlation is studied to investigate the correlation between any two where other's effect is partialed out. In general, for p fuzzy variables $\tilde{X}_1, \tilde{X}_2, \ldots, \tilde{X}_i, \ldots, \tilde{X}_j, \ldots, \tilde{X}_p$ a partial fuzzy correlation index between $\tilde{X}_i$ and $\tilde{X}_j$ where others are partialed out can be defined as

$$\tilde{\rho}_{ij\bullet 12..(i-1)(i+1)..(j-1)(j+1)..p}$$

$$= (\rho^L_{ij\bullet 12..(i-1)(i+1)..(j-1)(j+1)..p}, \rho^C_{ij\bullet 12..(i-1)(i+1)..(j-1)(j+1)..p}, \rho^U_{ij\bullet 12..(i-1)(i+1)..(j-1)(j+1)..p})_{LR}$$

$$\text{where } \tilde{\rho}_{ij\bullet 12..(i-1)(i+1)..(j-1)(j+1)..p} = \frac{(-1)^{i+j}\tilde{\sigma}_{ij\bullet 12..(i-1)(i+1)..(j-1)(j+1)..p}}{\tilde{\sigma}_{i\bullet 12..(i-1)(i+1)..(j-1)(j+1)..p} \otimes \tilde{\sigma}_{j\bullet 12..(i-1)(i+1)..(j-1)(j+1)..p}} \qquad (6)$$

Here $\tilde{\sigma}_{ij\bullet 12..(i-1)(i+1)..(j-1)(j+1)..p}$, $\tilde{\sigma}_{ii\bullet 12..(i-1)(i+1)..(j-1)(j+1)..p}$ and

$\tilde{\sigma}_{jj\bullet 12..(i-1)(i+1)..(j-1)(j+1)..p}$ are the cofactors of $\tilde{\sigma}_{ij}$, $\tilde{\sigma}_{ii}$ and $\tilde{\sigma}_{jj}$ in Σ.

4.3. Multiple Fuzzy Correlation

The concept of multiple fuzzy correlation is needful to measure the strength of predicting the output ($\tilde{Y}$) in association with a set of fuzzy variables. It is useful in multiple fuzzy regression analysis [3-5]. Here, the multiple fuzzy correlation index can be derived as $\tilde{\rho}_{\tilde{y}\bullet 12\ldots p} = (\rho^L_{\tilde{y}\bullet 12\ldots p}, \rho^C_{\tilde{y}\bullet 12\ldots p}, \rho^U_{\tilde{y}\bullet 12\ldots p})_{LR}$ such that

$$\tilde{\rho}_{\tilde{y}\bullet 12\ldots p} = \sqrt{(0,1,0) - \frac{1}{\tilde{\sigma}_{\tilde{y}\tilde{y}} \otimes \tilde{\sigma}^{\tilde{y}\tilde{y}}}} \qquad (7)$$

with the membership function

$$\mu_{\tilde{\rho}_{\tilde{y}\bullet 12\ldots p}}(r) = \begin{cases} L\left(\dfrac{\rho^C_{\tilde{y}\bullet 12\ldots p} - r}{\rho^C_{\tilde{y}\bullet 12\ldots p} - \rho^L_{\tilde{y}\bullet 12\ldots p}}\right) & \text{for } \rho^L_{\tilde{y}\bullet 12\ldots p} \le r \le \rho^C_{\tilde{y}\bullet 12\ldots p} \\ R\left(\dfrac{r - \rho^C_{\tilde{y}\bullet 12\ldots p}}{\rho^U_{\tilde{y}\bullet 12\ldots p} - \rho^C_{\tilde{y}\bullet 12\ldots p}}\right) & \text{for } \rho^C_{\tilde{y}\bullet 12\ldots p} \le r \le \rho^U_{\tilde{y}\bullet 12\ldots p} \end{cases}$$

$\tilde{\sigma}_{\tilde{y}\tilde{y}} = (\sigma^L_{\tilde{y}\tilde{y}}, \sigma^C_{\tilde{y}\tilde{y}}, \sigma^U_{\tilde{y}\tilde{y}})_{LR}$ and $\tilde{\sigma}^{\tilde{y}\tilde{y}} = (\sigma_L^{\tilde{y}\tilde{y}}, \sigma_C^{\tilde{y}\tilde{y}}, \sigma_U^{\tilde{y}\tilde{y}})_{LR}$ denote the element of the fuzzy covariance matrix Σ and Σ^{-1} corresponding to $(\tilde{y},\tilde{y})^{th}$ position.

5. Results & Discussion

The classical correlations and fuzzy correlations are computed based on the above methodologies. The test of significance of the variables is based on the *p*-values. And F-statistics is considered as model fitting cr4iteria. Using SPSS it can be primarily studied that all the predictors are not important in assessing *trismus*. Therefore, the following table shows various correlations for important ones.

Table 1: The correlations for important predictors and crisp "*trismus*" variable

Simple correlation	*Partial correlation*
$\rho_{Trismus*Burning} = 0.148$	$\rho_{(Trismus*Burning)\bullet Ulceration} = 0.1279$
$\rho_{Trismus*Ulceration} = 0.301$	$\rho_{(Trismus*Ulceration)\bullet Burning} = 0.2919$
$\rho_{Trismus*Site\_loc} = 0.725$	$\rho_{(Trismus*site\_loc)\bullet Burning} = 0.724$
$\rho_{Trismus*Age} = 0.023$	$\rho_{(Trismus*site\_loc)\bullet Ulceration} = 0.6951$
$\rho_{Trismus*Economic\ status} = 0\ .035$	$\rho_{(Trismus*Buring)\bullet Site\_loc} = 0.1396$
$\rho_{Trismus*Oral\ habit} = -0.011$	$\rho_{(Trismus*Ulceration)\bullet Site\_loc} = 0.097$
	$\rho_{(Trismus*Oral\ habit)\bullet Site\_loc} = 0.0043$
	$\rho_{(Trismus*Economic\ status)\bullet Site\_loc} = 0.0589$
	$\rho_{(Trismus*Age)\bullet Site\_loc} = -0.0321$
Multiple Correlation: $\rho_{Trismus\bullet(Burning*Ulceration*Site\_loc)} = 0.735$	

Table 2: Fuzzy correlations for important predictors affecting fuzzy "*trismus*" variable

Simple fuzzy correlation	*Partial fuzzy correlation*
$\tilde{\rho}_{Trismus*Burning} = (0.145,0.145,0.148)_{LR}$	$\tilde{\rho}_{(Trismus*Burning)\bullet Ulceration} = (0.124, 0.127,0.127)_{LR}$
$\tilde{\rho}_{Trismus*Ulceration} = (0.29,0.30,0.30)_{LR}$	$\tilde{\rho}_{(Trismus*Ulceration)\bullet Burning} = (0.2906,0.2919,0.2919)_{LR}$
$\tilde{\rho}_{Trismus*Site\_loc} = (0.704,0.725,0.725)_{LR}$	$\tilde{\rho}_{(Trismus*site\_loc)\bullet Burning} = (0.715,0.724,0.734)_{LR}$
$\tilde{\rho}_{Trismus*Economic\ status} = (0.03,0.04,0.04)_{LR}$	$\tilde{\rho}_{(Trismus*site\_loc)\bullet Ulceration} = (-0.742,-0.695,-0.695)_{LR}$
	$\tilde{\rho}_{(Trismus*Buring)\bullet Site\_loc} = (0.1390,0.1396,0.14)_{LR}$
	$\tilde{\rho}_{(Trismus*Ulceration)\bullet Site\_loc} = (-0.140,-0.097,-0.097)_{LR}$
Multiple Fuzzy Correlation: $\tilde{\rho}_{Trismus\bullet(Burning*Ulceration*Site\_loc)} = (0.720, 0.735, 0.775)_{LR}$	

From both the above tables it can be observed that fuzzy correlations are more general that covers the classical one. Obviously, a decision can be floated that *trismus* is much affected by mainly three factors as *burning sensation, ulceration* and *sites of lesion.* As a future work, a fuzzy regression model can be fitted to predict *trismus* (a fuzzy output variable) based three crisp inputs.

References

1. J.P. Caniff, W. Harvey & M. Harris (1986), Oral submucous fibrosis: its pathogenesis and management, *Br. Dent. J.* 160: 429-433.
2. J.J. Pindborg, P.R. Murti, R.B. Bhonsle, P.C. gupta, D.K. Daftary & F.S. Mehta (1984), Oral submucous fibrosis as a precancerous condition, *Scan J. Dent. Res.* 92: 224-229.
3. S.R. Aziz (1997), Oral submucous fibrosis: an unusual disease, *J N. J Dent Assoc* Spring 68: 17-19.
4. P.C. Gupta, P.N. Sinor, R.B. Bhonsle, V.S. Pawar & H.C. Mehta (1998), Oral submocous fibrosis in India: A new epidemic, *The National Medical Journal of India*, 2: 113-116.
5. R.R. Paul, J. Chatterjee, A.K. Das, M.L. Cervera, M. De la Guardia & K. Chaudhuri (2002), Altered elemental profile as indicator of homeostatic imbalance in pathogenesis of oral submucous fibrosis, *Biol Trace Elem Res.* 87: 45-56.
6. Jayanti, C.S.J. Probert, K.S. Sher & J.F. Maybery (1992), Oral submucous fibrosis- a preventable disease, *Gut* 33: 4-6
7. P.R. Murti, P.C. Gupta, R.B. Bhonsle, D.K. Daftari, F.S. Mehta, J.J. Pinborg (1990), Effect on the incidence of intervention in the arecanut chewing habit, *J. Oral Pathol Med.* 19: 95-100
8. P.R. Murti, R.B. Bhonsle, P.C. Gupta, D.K. Daftari, J.J. Pindborg, F.S.Mehta (1995), Etiology of oral submucous fibrosis with special reference to arecanut chewing, *J. Oral Pathol. Med.* 24: 145-152
9. C. Chakraborty & D. Chakraborty (2004), Interrelationship study among market predictors using fuzzy correlation, *Proc. of National Seminar on Recent Advances in Applied Mathematics*, Vidyasagar University, W.B., India.
10. D. Dubois & H. Prade (1980), Fuzzy Sets and Systems: Theory and Applications, *Academic Press*, NY.
11. M.S. Yang & T.S. Lin (2002), Fuzzy least-squares linear regression analysis for fuzzy input-output data, *Fuzzy Sets and Systems* 126: 389-399
12. P. D'Urso & T. Gastaldi (2000), A least-square approach to fuzzy linear regression analysis, *Computational Statistics and Data Analysis* 34: 427-440.
13. C. Kao & C. L. Chyu (2002), A fuzzy linear regression model with better explanatory power, *Fuzzy Sets and Systems* 126: 401-409.
14. L.A. Zadeh (1965), Fuzzy Sets, *Inform. and Control* 8: 338-353.
15. T.W. Anderson (2003), An Introduction to Multivariate Statistical Analysis, 3rd Ed., Willey Pub., NY.
16. SPSS (Statistical Package for Social Sciences) version 7.5 for Windows

Fuzzy Logic and its Application in Technology and Management
Editors: Debjani Chakraborty, S. Nanda and D. Dutta Majumder

Fuzzy Trend Analysis of Healing Wounds Treated with Honey

Provas Banerjee[1], Debjani Chakraborty[2], Chandan Chakraborty[2], Snigdha Palchowdhury[3] Jyotirmoy Chatterjee[4], Shyamal K Basu[5], A.K. Hui[1], N.C. Chakraborty[1], P.K. Debnath[6] and Keya Chaudhuri[7]

[1]Department of Anatomy, Burdwan Medical College, Burdwan, West Bengal
[2]Department of Mathematics, Indian Institute of Technology, Kharagpur
[3]Department of Geology, Presidency College, 87 College Street, Kolkata
[4]Institute of Interdisciplinary Scientific Research (IISR), Kolkata
[5]Instiitute of Health and Family Welfare, Salt Lake, Kolkata
[6]J B Roy State Ayurvedic Medical College & Hospital, Kolkata
[7]Human Genetics & Genomics Group, Indian Institute of Chemical Biology

Abstract: Besides the use of different antibiotics and chemotherapeutic agents, the role of varied traditional natural substances like honey and varied plant products are also looked into through its recurrent and persistent usages over time. Treatments of wound with honey are known from the records of ancient Egyptians to Hippocrates and Koran. Despite the worldwide use of honey in healing wounds through ages but its scientific evaluation has not been properly reported in literature. Though sporadic reports indicate that a variety of serious and dangerous wound infections encountered in surgical and medical practices that appear less responsive to conventional topical antibiotic therapy, have been cured by application of honey in its purest natural form. Our preliminary work already documented the physico-chemical characteristics of natural honey and we propose that honey is effective in wound healing as topical agent and possibly acts in healing through a multi-factorial and multidimensional mechanism. The aim of the present paper is to analyze the fuzzy trend of various types of wounds (viz. fresh superficial thermal burn wound, infected burn wound, diabetic ulcer and traumatic ulcer) treated with honey. The trend is actually captured by statistical model dealing with fuzzy data.

Keywords: Honey; Burn wound; Diabetic wound; Infected wound; Fuzzy linear regression.

1. Introduction

Honey has been recognized for medicinal properties since antiquity. It is mentioned in the Bible, the Koran and the Torah for healing purposes. In the past century there has been sporadic scientific report for its use in wounds and infection control [1].

Honey is a natural product of bees of the genera of Apis and Meliponinae. The bees collect nectar from flowers and then subjected to enzymatic processing in vivo in both the collecting bee and in a processing bee inside the hive. Various reports indicated that honey contains various inorganic and organic substances, which help in wound healing in many ways. The antimicrobial effects of honey are variably ascribed to its acidic pH, the low hydrogen peroxide content, the osmotic effect and as yet unidentified compounds like inhibines [1-3]. In fact, the deodorizing epithelialysing, nonadhesive, chemical debridement and styptic properties of honey are observed in the treatment of wounds where this natural product is used as a topical agent. Honey also shows the wound healing without contracture. Keeping these facts in mind, this paper tries to develop a quantitative trend analysis technique regarding varied healing wounds under honey through analyzing the area of wound bed over the time of healing by using fuzzy trend analysis system.

Trend analysis is very important mathematical and statistical aspect in various fields of research to make a valid prediction about future based on the past information. As the situation/system becomes more complex day-by-day, the available information is inaccurate and inexact blended with fuzziness [7, 8]. In such imprecise environment, the trend analysis is to be revised, called as fuzzy trend analysis that is done through fuzzy regression model fitting. The paper initially starts with the medical problem. Practically there are various types of wounds amongst which four types are considered here: (i) *fresh superficial thermal wound*; (ii) *infected burn wound*; (iii) *diabetic ulcer and* (iv) *traumatic ulcer*. These four wounds are treated with honey and observed its progress at different days. The medical experiment is performed for various samples and come to an approximate wound area that transforms the original data for every day of observation into fuzzy to capture more information. Fuzzy set theory is the tool that fractures the fuzzy information for decision. Therefore, the main objective of this paper is to capture the inherent trend based on fuzzified wound areas by means of fitting fuzzy regression model and its pictorial representations.

Section 2 describes the fuzzy trend analysis technique for the given data. And the results are discussed and compared in Section 3.

2. Fuzzy Trend Analysis

In analogy to the classical regression analysis the fuzzy regression tool plays very important role in analyzing and predicting the incomplete, imprecise and inexact data, since the system becomes more complex in our day-to-day life. In fitting the fuzzy regression model, the output variable and model parameters are usually assumed to be fuzzy where the input variable may be fuzzy / crisp depending on the situation. In this medical problem four types of wounds have been treated with honey and observed at different days. Therefore, the 'day of observation' is considered as independent crisp variable where the 'area of would' observed at a particular day is dependent fuzzy variable. Assume that the 'area of would' and 'day of observation' represented by $\widetilde{Y}$ and X respectively. Now a simple fuzzy regression model is assumed to capture its fuzzy trend as follows:

$$\widetilde{Y} = \widetilde{a}_0 + \widetilde{a}_1 \times X .$$

The observed data $\{(\tilde{y},x)\}$ collected on $\{(\tilde{Y},X)\}$ are transformed into *LR*-type fuzzy numbers. Here the data form of fuzzy variable and fuzzy parameters are designed as triangular fuzzy numbers, i.e., $\tilde{y}=(m_y,\alpha_y,\beta_y)_{LR}$ and $\tilde{a}=(m_a,\alpha_a,\beta_a)_{LR}$ with membership functions. As fuzzy arithmetic is well defined [4], the model can be explicitly written as follows:

$$\tilde{a}_0+\tilde{a}_1\times X = (m_{a_0},\alpha_{a_0},\beta_{a_0})_{LR}+ (m_{a_1},\alpha_{a_1},\beta_{a_1})_{LR}\times X \cong (m,\alpha,\beta)_{LR}$$

where $m=m_{a_0}+m_{a_1}x$; $\alpha=\alpha_{a_0}+\alpha_{a_1}x$ and $\beta=\beta_{a_0}+\beta_{a_1}x$ for the observed data. The model parameters are estimated on the basis of the random sample of size '*n*', say $\{(\tilde{y}_i,x_i)\}_{i=1,2,\dots,n}$ using the least square method [6]. Hence the model can be reformulated for the ith sample is as

$$\tilde{y}_i \cong (m_i,\alpha_i,\beta_i)_{LR} \text{ i.e., } (m_y,\alpha_{y_i},\beta_{y_i})_{LR} \cong (m_i,\alpha_i,\beta_i)_{LR} \ \forall\ i = 1,2,\dots,n.$$

The estimated parameters say $\hat{\tilde{a}}_0=\left(\hat{m}_{a_0},\hat{\alpha}_{a_0},\hat{\beta}_{a_0}\right)_{LR}$ and $\hat{\tilde{a}}_1=\left(\hat{m}_{a_1},\hat{\alpha}_{a_1},\hat{\beta}_{a_1}\right)_{LR}$ are obtained by minimizing the following objective function:

$$J(\tilde{a}_0,\tilde{a}_1)=\sum_{i=1}^{n}[(m_i-m_{y_i})^2+\{(m_i-\alpha_i)-(m_{y_i}-\alpha_{y_i})\}^2+\{(m_i+\beta_i)-(m_{y_i}+\beta_{y_i})\}^2].$$

The partial derivatives of the objective functions with respect to the associated parameters are equated to zero and hence

$$\begin{pmatrix}\hat{m}_{a_0}\\ \hat{m}_{a_1}\end{pmatrix}=M^{-1}\begin{pmatrix}\sum_{i=1}^{n} m_{y_i}\\ \sum_{i=1}^{n} m_{y_i}x_i\end{pmatrix};$$

$$\begin{pmatrix}\hat{\alpha}_{a_0}\\ \hat{\alpha}_{a_1}\\ \hat{\alpha}_{a_2}\end{pmatrix}=M^{-1}\begin{pmatrix}\sum_{i=1}^{n}\{\hat{m}_{a_0}+\hat{m}_{a_1}x_i-(m_{y_i}-\alpha_{y_i})\}\\ \sum_{i=1}^{n}\{\hat{m}_{a_0}+\hat{m}_{a_1}x_i-(m_{y_i}-\alpha_{y_i})\}x_i\end{pmatrix}$$

$$\begin{pmatrix}\hat{\beta}_{a_0}\\ \hat{\beta}_{a_1}\end{pmatrix}=M^{-1}\begin{pmatrix}\sum_{i=1}^{n}\{(m_{y_i}+\beta_{y_i})-\hat{m}_{a_0}-\hat{m}_{a_1}x_i\}\\ \sum_{i=1}^{n}x_i\{(m_{y_i}+\beta_{y_i})-\hat{m}_{a_0}-\hat{m}_{a_1}x_i\}\end{pmatrix} \text{ where } M=\begin{pmatrix}n & \sum_{i=1}^{n}x_i\\ \sum_{i=1}^{n}x_i & \sum_{i=1}^{n}x_i^2\end{pmatrix}$$

If spreads are negative, the analogies proposed by Yang & Ko [5] are as follows:

(i) If $\alpha_a<0$, $\beta_a<0$, then $\tilde{a}=(m_a,-\beta_a,-\alpha_a)_{LR}$;

(ii) If $\alpha_a<0$, $\beta_a>0$, then $\tilde{a}=\begin{cases}(m_a,0,\beta_a)_{LR} & \text{for } |\beta_a|>|\alpha_a|\\ (m_a,0,-\alpha_a)_{LR} & \text{for } |\beta_a|<|\alpha_a|\end{cases}$;

(iii) If $\alpha_a>0$, $\beta_a<0$, then $\tilde{a}=\begin{cases}(m_a,\alpha_a,0)_{LR} & \text{for } |\alpha_a|>|\beta_a|\\ (m_a,-\beta_a,0)_{LR} & \text{for } |\alpha_a|<|\beta_a|\end{cases}$.

Hence the fitted fuzzy regression model is written as: $\hat{\tilde{Y}} = \hat{\tilde{a}}_0 + \hat{\tilde{a}}_1 \times X$.

3. Results and Discussion

In fact, there are various types of wounds amongst which we consider the following
(i) fresh superficial thermal wound;
(ii) infected burn wound;
(iii) diabetic ulcer and
(iv) traumatic ulcer.

These wounds are treated with honey and observed its progress in terms of area of each of the wounds in different days. In reality, these areas cannot be clearly specified because of the difficulty to differentiate the transition between affected and non-affected area. In view of this, an experiment is performed for various samples and hence the fuzzification process is used to obtain the fuzzified wound areas in table 1.

Table 1: Fuzzification of the experimental data

Day of Obs	*Wound area of fresh_ther*	*Wound area of infec_burn*	*Wound area of dia_ulcer*	*Wound area of tra_ulcer*
0	$(100, 0, 0)_{TFN}$	$(100, 0, 0)_{TFN}$	$(100, 0, 0)_{TFN}$	$(100, 0, 0)_{TFN}$
2	$(94.76, 4.66, 3.83)_{TFN}$	$(97.02, 4.54, 2.46)_{TFN}$	$(96.84, 3.65, 3.14)_{TFN}$	$(97.20, 2.78, 2.25)_{TFN}$
3	$(81.70, 10.37, 16.41)_{TFN}$	$(91.42, 7.14, 6.87)_{TFN}$	$(91.25, 3.86, 6.73)_{TFN}$	$(91.67, 3.24, 4.76)_{TFN}$
7	$(60.20, 11.13, 11.03)_{TFN}$	$(85.03, 9.02, 9.48)_{TFN}$	$(75.13, 14.85, 14.78)_{TFN}$	$(68.87, 15.33, 15.68)_{TFN}$
11	$(43.60, 12.87, 10.99)_{TFN}$	$(69.45, 8.03, 6.94)_{TFN}$	$(64.05, 14.7, 15.14)_{TFN}$	$(70.02, 11.43, 7.99)_{TFN}$
15	$(21.56, 12.65, 9.77)_{TFN}$	$(64.71, 12.49, 15.72)_{TFN}$	$(54.68, 9.74, 10.24)_{TFN}$	$(47.41, 16.88, 13.88)_{TFN}$
22	$(10.45, 5.06, 9.5)_{TFN}$	$(49.73, 9.24, 8.2)_{TFN}$	$(38.04, 16.62, 7.41)_{TFN}$	$(33.38, 8.19, 9.74)_{TFN}$
30	$(3.08, 1.79, 2.95)_{TFN}$	$(17.65, 3.44, 4.26)_{TFN}$	$(24.94, 8.45, 5.48)_{TFN}$	$(30.00, 12.65, 10.19)_{TFN}$
45	$(0, 0, 0)_{TFN}$	$(15.82, 6.64, 9.8)_{TFN}$	$(14.11, 5.1, 6.12)_{TFN}$	$(14.21, 3.98, 3.72)_{TFN}$

In order to get the idea about how honey is effective for wounds, we apply the fuzzy linear regression model where the 'day of observation' is considered as independent crisp variable where the 'wound area' is dependent fuzzy variable. The fitted fuzzy models for the wounds are obtained and discussed as follows.

Case I: Fresh Superficial Thermal Burn Wound

The fitted fuzzy regression model for fresh superficial burn is given as

$$\hat{\tilde{Y}}_{\text{superficial\_burn}} = (81.716, 8.707, 9.409)_{TFN} - (2.371, 0.147, 0.15)_{TFN} \times X_{\text{day}}.$$

In Fig. 1, '—' and '-----' imply respectively the centerline with highest membership value 1 and the borderline with lowest membership value 0. This convention is maintained through out all the graphs.

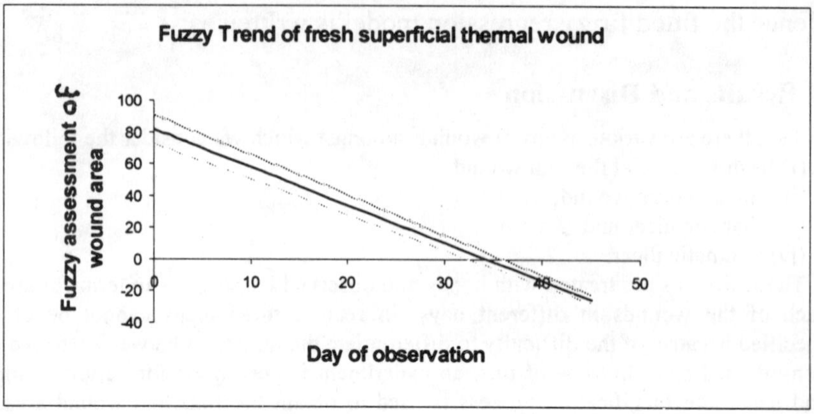

Fig. 1. Fuzzy trend line for fresh superficial thermal wound area

Case II: Infected Burn Wound

The fuzzy model for infected type of burn is fitted as

$$\hat{\tilde{Y}}_{\text{infected\_burn}} = (96.861, 6.171, 5.429)_{TFN} - (2.081, 0.11, 0.037)_{TFN} \times X_{\text{day}}\,.$$

The graphical representation of fuzzy assessment of infected burn area is

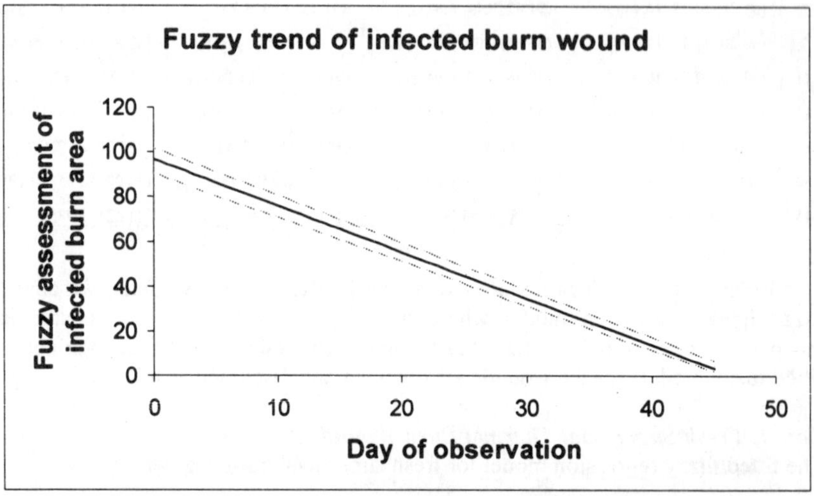

Fig. 2. Fuzzy trend line for infected burn wound area

Case III: Diabetic Ulcer

The regression model for diabetic burn is

$$\hat{\tilde{Y}}_{\text{diabetic\_burn}} = (92.647, 7.508, 7.526)_{TFN} - (2.035, 0.009, 0.07)_{TFN} \times X_{\text{day}}$$

The fuzzy trend line for diabetic ulcer is represented below:

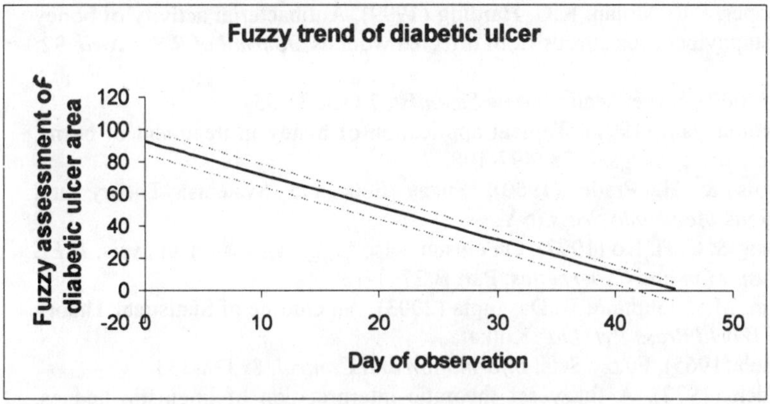

Fig. 3. Trend line for diabetic ulcer wound area

Case IV: Traumatic ulcer

The fitted fuzzy regression model for traumatic wound is as

$$\hat{\tilde{Y}}_{\text{traumatic\_burn}} = (91.424, 7.245, 6.835)_{TFN} - (2.0, 0.049, 0.069)_{TFN} \times X_{\text{day}}$$

The fitted fuzzy regression line for traumatic ulcer is shown as follows:

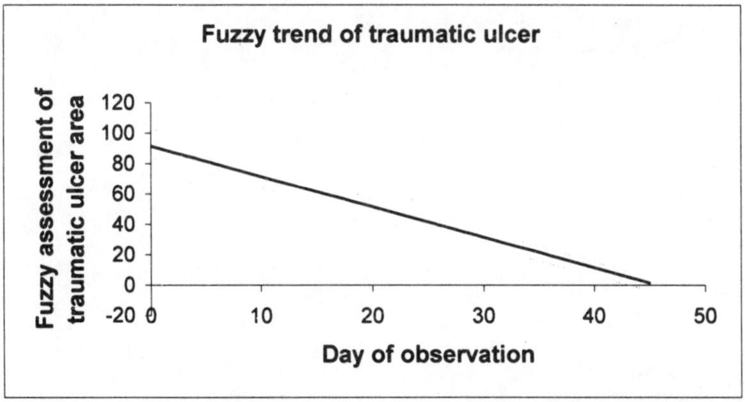

Fig. 4. Fuzzy trend line for the assessment of traumatic ulcer wound area

The colored and dotted lines have been used to indicate the fuzzy trends for four wounds treated with honey. It can be observed from the graphs that there is a decreasing trend for all types wounds treated with honey. Therefore the honey treatment can be prescribed to cure these wounds on preference basis: highly recommended for fresh superficial thermal wound, normal advice for traumatic and diabetic ulcer and lastly lesser importance for infected burn.

Reference

1. R.A. Cooper, P.C. Molan, K.G. Harding (1999), Antibacterial activity of honey against Staphylococcus aurous from infected wounds, *Journal of R Soc Med*, 92: 283-285.
2. A. Lord (2002), Sweet healing, *New Scientist*, 7 Oct: 32-35.
3. N. Subrahmanyam (1991), Topical application of honey in treatment of burns, *British Journal of Surgery*, 78: 497-498.
4. D. Dubois & H. Prade (1980), Fuzzy Sets and Systems: Theory and Applications, *Academic Press,* NY.
5. M. S. Yang & C. H. Ko (1997), On cluster-wise fuzzy regression analysis, *IEEE Trans. Syst. Man and Cybernetics*: Part B 27: 1-13
6. A.M. Gun, M.K. Gupta & B. Dasgupta (2003), An Outline of Statistical Theory – II, *The World Press Pvt. Ltd.*, Kolkata.
7. L.A. Zadeh (1965), Fuzzy Sets, *Information and Control*, 8: 338-353
8. L.A. Zadeh (1972), A fuzzy set theoretic interpretation of linguistic hedges, *Journal of Cybernetics*, 2 (3): 4-34.

Fuzzy Logic and its Application in Technology and Management
Editors: Debjani Chakraborty, S. Nanda and D. Dutta Majumder

Fuzzy Regression Analysis of Tissue Trace Metal Content of Radiation Workers

Debjani Chakraborty[1], *Chandan Chakraborty*[1], *Jyotirmoy Chatterjee*[2], *Shyamal K Basu*[3], *Arabinda K Das*[4] *Snigdha Palchowdhury*[5], *Santanu Chakraborty*[6] *and Keya Chaudhuri*[7]

[1]Department of Mathematics, Indian Institute of Technology Kharagpur
[2]Department of Radiology (Diagnosis), Medical College Hospitals, Kolkata
[3]Institute of Health & Family Welfare, Govt. of West Bengal, Kolkata
[4]Department of Chemistry, University of Burdwan, Burdwan, West Bengal
[5]Department of Geology, Presidency College, 87 College Street, Kolkata
[6]TLD Division, VECC, BARC, Salt Lake City, Kolkata
[7]Human Genetics & Genomics Group, Indian Institute of Chemical Biology, Kolkata

Abstract: Various radiation workers are occupationally exposed to chronic low dose ionizing radiations in addition to natural background radiations. But till date there is no such well-accepted biomarker to resolve actual effect (hazardous or beneficial) of chronic low dose radiation on human subjects, though many schools of thoughts are prevailing in this regard. Present study investigates the zinc status in the peripheral blood and scalp hair of medical radiographers in comparison to age and economy matched normal healthy individuals in fuzzy environment. To capture more information it is always necessary to integrate the experts' knowledge, experience and perception with the experimental information. In view of this, according to medical experts', the overall decisions of zinc content for several period (2-34 years) of occupational radiation exposures are considered as fuzzy. Therefore, a fuzzy regression technique is suggested to capture the trend of tissue zinc content based on time of exposures in fuzzy environment.

Keywords: Chronic low dose x-radiation; Radiographers; Zinc; Atomic Absorption Spectrometry (AAS); Fuzzy logic; Fuzzy regression model; Least squares method.

1. Introduction

Radiation workers are occupationally exposed to low dose ionizing radiations in a chronic way. As per radiation safety standard of Atomic Energy Regulatory Board of India, cumulative effective dose limit for each consecutive block of five years is 100 mSv (1 mSv = 0.1 cGy for X-ray) for individual radiation workers and annual effective dose in a calendar year should not exceed the limit of 30 mSv (Chatterjee et al [4]). On an average an Indian diagnostic medical radiation worker receives 1.46 mSv ionizing radiation (X-ray) per year occupationally in addition to natural background radiation. However, till date researchers differ regarding actual biological effects of chronic low dose radiations on human subjects. Paucity of radiobiological data in this regard is also evident in Kuzin [1], Chatterjee et al [4], Fabrikant [8], Sagan [9], Wolff [14] and Ikushima [15]. In fact, most of the low dose radiobiological predictions suffer from intrinsic uncertainty since these are generally based on linear extrapolation of data from high dose radiation effect studies (Kuzin [1], Sagan [9] and Wolff [14]). Further, the researchers of low dose radiobiological study mostly depend on physical dosimeters like thermoluminescence device (TLD) in radiobiological dosimetry. But, this physical approach is unable to calculate or assess the actual radiation exposure in an extensive mammalian system like human body. In recent times, researchers are trying to overcome this radiobiological dosimetric problem through identifying suitable biomarkers, which can simultaneously assess the radiation dose and its effect at the biosystem levels. Our group has already demonstrated the significant changes in the trace metal contents of different tissues of radiation workers and chronic low dose x-irradiated rats (Chatterjee et al [5-7] and Majumdar et al [13]). The present study has opted to assess the zinc status in blood and hair of medical radiation workers. The trace metal like zinc is not only vital in maintaining biological integrity of mammalian systems but also a potent indicator of the pathology of the system in its excess and its deficiency states in Chatterjee et al [4] and Paul et al [12]. In this context, it also may be noted that the biological data from the effect research are gaining attentions of the mathematicians. Besides nonlinearity they have noticed overlapping, ambiguity and inexactness in biological data (Chatterjee et al [7]). In view of this, fuzzy logic (Zadeh [10]) is recommended here as an appropriate tool. Here radiographers and normal healthy individuals of mixed age groups are investigated to capture the inherent trend of zinc content in their hair and blood over the period of radiation exposures. Practically, it is very difficult for the medical experts to assess the perfect amount of zinc rather some approximated value for certain age group. Rather the experts feel comfortable to express their opinion (Chakraborty [2]) in natural languages aiming at the experimental observations because the factors are really uncertain. In this way the subjectivity of the experts in terms of the cognitive factors like perception, experience, thinking etc gets involved in making a suitable decision in medical science. E.g. suppose an expert is asked to make a general comment on the zinc status for radiographer and normal healthy individuals of an approximate age group. Then s/he naturally answers like that "If the time of exposure is *approximately* 2.25 years then the zinc content in radiographer's (i) hair is *around* 150.3μg/g and (ii) blood is *approximately* 10.3μg/ml" depending upon the experimental observations. Here the predicates "approximately", "more or less" and "around" etc are fuzzy. Thus the two factors viz. zinc content and time of exposure become fuzzy. Now the paper develops a fuzzy regression model based on the fuzzy input and fuzzy output to find the trend of zinc for radiographer and normal individuals. Fuzzy regression analysis has become an important tool for prediction under fuzzy framework. Many researchers (Yang et al [11, 16]) have worked on it.

Section 2 demonstrates the medical study where subsection 2.1 describes the materials and methods under study. And the procedure for the estimation of zinc with atomic absorption spectrometry (AAS) for normal and medical radiographers is studied in subsection 2.2. Also subsection 2.3 enhances the fuzzification process to transform the raw data into the reliable fuzzy data. In section 3, a methodology for fitting nonlinear fuzzy regression model is devised for linguistic variables. Finally the model is fitted with the medical data for normal person and radiographers and results are discussed in section 4.

2. Medical Study and Fuzzification

2.1. Materials and Methods

In this study, scalp hair and peripheral blood from 58 medical radiographers (27- 60 years of age) those of who are occupationally exposed to x-radiation works for 2-34 years and economy matched normal healthy volunteers were collected for estimation of zinc with atomic absorption spectrometry (AAS). From each medical radiographer and healthy volunteer 2 ml of peripheral blood and 50-100 mg (about 4-5 cm long) distal ends of scalp hair were collected.

2.2. Estimation of Zinc with AAS

Zinc content in blood and hair samples was estimated with AAS (Model AA646 Shimadzu, Japan) using following procedure. Each blood sample was slowly wet digested with 15 ml H_2O_2 for 30 min, then 5 ml of conc. HNO_3 was added and slowly digested for 2 h. Finally 3 ml of 70% $HClO_4$ was added and strongly heated until it was converted into white dry mass. It was dissolved in water and volume was made upto 10 ml in a volumetric flask. Hair filaments were first washed in acetone (AR, Merck, Germany) and then with double distilled water. Next each sample was slowly wet digested in 20 ml of conc. HNO_3 and 1 ml of 70% $HClO_4$ was added. Then the solution was heated until it was converted into a white dry mass. It was dissolved in deionized and double distilled water and volume was made upto 10 ml in volumetric flask. In zinc estimation with AAS, 1 ppm working solution was prepared from 5000 ppm stock solution of $ZnCl_2$. Finally zinc was estimated with flame AAS against reagent blank by using standard procedures (Chatterjee et al [7], Majumdar et al [13]) with required instrumental conditions as follows

Table1: Instrumental conditions for AAS

Analyte	Wavelen gth (nm)	Hollow cathode lamp current (mA)	Slit width (nm)	Flame	C_2H_2 flow rate (lit/min)	Burner height (mm)
Zn	213.9	6	0.38	Air- C_2H_2	2.4	4

In hair and blood, zinc is estimated as *µg/g* and *µg/ml* respectively. For each age category, the content of zinc in blood and hair is observed for a particular time of exposure (TOE). The process is repeated a number of times with the reasonable variation of TOE and note the content of zinc corresponding to different TOE. In practice, the amount of zinc in blood and hair is varying quantity that can be estimated by transforming into fuzzy numbers. Therefore the TOE and zinc content are considered here as linguistic variables.

2.3. Fuzzification Method

Definition (*Fuzzy number*): A fuzzy number $\tilde{A} = (m, \alpha, \beta)$ is a convex and normalized fuzzy set whose membership function is of the following form:

$$\mu_{\tilde{A}}(t) = \begin{cases} 1 - \left(\dfrac{m-t}{\alpha}\right) & \text{for } m - \alpha \le t \le m \\ 1 - \left(\dfrac{t-m}{\beta}\right) & \text{for } m \le t \le m + \beta \end{cases}$$

It is convinced that a fuzzy number is more informative than a crisp number though vagueness is inherent there. In view of this, a heuristic fuzzification process is devised here. If there is no outlier, it can be statistically said that the more concentration is at mean of all the observations for that specific age group. Spreads are computed in terms of range. Suppose we have 'p' experimental observations, $x_1, x_2, \ldots, x_p$ of TOE for certain age group. Therefore, a triangular fuzzy number $\tilde{A} = (m, \alpha, \beta)$ where

$$m = \text{more concetration} = \frac{1}{p}\sum_{i=1}^{p} x_i ; \alpha = m - \min_i(x_i) \text{ and } \beta = \max_i(x_i) - m .$$

Hence the linguistic variables TOE and zinc content consider the fuzzy numbers as experimental output. For the sake of simplicity, all the fuzzy numbers are assumed to be triangular in the form through out the paper.

3. Trend Analysis with Fuzzy Data

The fuzzy regression analysis (Yang et al [11, 16]) has been developed to deal with imprecise data. In this paper, the aim is to fit a regression model with the fuzzy input and fuzzy output. It is observed that the present raw data set fits well with the second-degree polynomial ignoring the fuzziness. So when the data truly converted into a fuzzy data through the fuzzification process, a second-degree fuzzy polynomial model, called here parabolic fuzzy regression model is required to fit for trend analysis. In practice, the dependent variable 'zinc content' and the independent variable 'time of exposure (TOE)' become linguistic by integrating with radiation biologists' experience, perception and thinking etc. Therefore an attempt is made here in fitting a fuzzy parabolic model with these linguistic variables. Let us assume that the 'zinc content in hair' and 'time of exposure' denoted by $\tilde{Y}$ and $\tilde{X}$ respectively. Therefore a fuzzy parabolic model for fuzzy input-output to be fitted is as follows

$$\tilde{Y} = \tilde{a}_0 + \tilde{a}_1 \otimes \tilde{X} + \tilde{a}_2 \otimes \tilde{X}^2 \quad (1)$$

where $\tilde{a}_j$ are also fuzzy. '$\otimes$' denote the extended multiplication operator for fuzzy numbers [7]. The observed data $\{(\tilde{y}, \tilde{x})\}$ collected on $\{(\tilde{Y}, \tilde{X})\}$ are transformed into triangular fuzzy numbers. Here the linguistic variables and fuzzy parameters are also expressed by triangular fuzzy numbers, i.e.,

$$\tilde{y} = (m_y, \alpha_y, \beta_y),\ \tilde{x} = (m_x, \alpha_x, \beta_x) \text{ and } \tilde{a}_j = (m_{a_j}, \alpha_{a_j}, \beta_{a_j}).$$

Therefore the R.H.S of model (1) can be re-written as follows:

$$\tilde{a}_0 + \tilde{a}_1 \otimes \tilde{X} + \tilde{a}_2 \otimes \tilde{X}^2 = (m_{a_0}, \alpha_{a_0}, \beta_{a_0}) + (m_{a_1}, \alpha_{a_1}, \beta_{a_1}) \otimes (m_x, \alpha_x, \beta_x)$$
$$+ (m_{a_2}, \alpha_{a_2}, \beta_{a_2}) \otimes (m_x, \alpha_x, \beta_x) \otimes (m_x, \alpha_x, \beta_x) \equiv (m, \alpha, \beta), \text{ say} \quad (2)$$

where $m = m_{a_0} + m_{a_1} m_x + m_{a_2} m_x^2$;

$$\alpha = \alpha_{a_0} + m_{a_1} \alpha_x + m_x \alpha_{a_1} + 2 m_{a_2} m_x \alpha_x + \alpha_{a_2} m_x^2 \text{ and}$$

$$\beta = \beta_{a_0} + m_{a_1} \beta_x + m_x \beta_{a_1} + 2 m_{a_2} m_x \beta_x + \beta_{a_2} m_x^2 .$$

The fuzzy regression parameters are estimated by least squares method based on the 'n' observations $\{(\tilde{y}_i, \tilde{x}_i)\}_{i=1,2,\ldots,n}$. In particular, the relationship between the actual and approximated i^{th} fuzzy data point is

$$\tilde{y}_i \cong (m_i, \alpha_i, \beta_i) \text{ i.e., } (m_{y_i}, \alpha_{y_i}, \beta_{y_i}) \cong (m_i, \alpha_i, \beta_i) \ \forall\, i = 1,2,\ldots,n. \tag{3}$$

Therefore the Euclidean distance between $(m_{y_i}, \alpha_{y_i}, \beta_{y_i})$ and (m_i, α_i, β_i) that leads to form the objective function for estimation of the parameters. Assume that $\hat{\tilde{a}}_0 = (\hat{m}_{a_0}, \hat{\alpha}_{a_0}, \hat{\beta}_{a_0})$, $\hat{\tilde{a}}_1 = (\hat{m}_{a_1}, \hat{\alpha}_{a_1}, \hat{\beta}_{a_1})$ and $\hat{\tilde{a}}_2 = (\hat{m}_{a_2}, \hat{\alpha}_{a_2}, \hat{\beta}_{a_2})$ are estimated by minimizing the objective function

$$J = J_1 + J_2 + J_3 \tag{4}$$

where $J_1(m_{a_0}, m_{a_1}, m_{a_2}) = \sum_{i=1}^{n} (m_i - m_{y_i})^2$;

$$J_2(\alpha_{a_0}, \alpha_{a_1}, \alpha_{a_2}) = \sum_{i=1}^{n} \{(m_i - \alpha_i) - (m_{y_i} - \alpha_{y_i})\}^2 \text{ and}$$

$$J_3(\beta_{a_0}, \beta_{a_1}, \beta_{a_2}) = \sum_{i=1}^{n} \{(m_i + \beta_i) - (m_{y_i} + \beta_{y_i})\}^2 .$$

The normal equations are obtained by taking partial derivatives of eq. (4) equating to 0 as follows

$$\begin{pmatrix} \hat{m}_{a_0} \\ \hat{m}_{a_1} \\ \hat{m}_{a_2} \end{pmatrix} = M^{-1} \begin{pmatrix} \sum_{i=1}^{n} m_{y_i} \\ \sum_{i=1}^{n} m_{y_i} m_{x_i} \\ \sum_{i=1}^{n} m_{y_i} m_{x_i}^2 \end{pmatrix} \text{ where } M = \begin{pmatrix} n & \sum_{i=1}^{n} m_{x_i} & \sum_{i=1}^{n} m_{x_i}^2 \\ \sum_{i=1}^{n} m_{x_i} & \sum_{i=1}^{n} m_{x_i}^2 & \sum_{i=1}^{n} m_{x_i}^3 \\ \sum_{i=1}^{n} m_{x_i}^2 & \sum_{i=1}^{n} m_{x_i}^3 & \sum_{i=1}^{n} m_{x_i}^4 \end{pmatrix} \text{ and}$$

$$\begin{pmatrix} \hat{\alpha}_{a_0} \\ \hat{\alpha}_{a_1} \\ \hat{\alpha}_{a_2} \end{pmatrix} = M^{-1} \begin{pmatrix} \sum_{i=1}^{n} \{\hat{m}_{a_0} + \hat{m}_{a_1} m_{x_i} + \hat{m}_{a_2} m_{x_i}^2 - \hat{m}_{a_1} \alpha_{x_i} - 2\hat{m}_{a_2} m_{x_i} \alpha_{x_i} - (m_{y_i} - \alpha_{y_i})\} \\ \sum_{i=1}^{n} m_{x_i}^2 \{\hat{m}_{a_0} + \hat{m}_{a_1} m_{x_i} + \hat{m}_{a_2} m_{x_i}^2 - \hat{m}_{a_1} \alpha_{x_i} - 2\hat{m}_{a_2} m_{x_i} \alpha_{x_i} - (m_{y_i} - \alpha_{y_i})\} \\ \sum_{i=1}^{n} m_{x_i}^2 \{\hat{m}_{a_0} + \hat{m}_{a_1} m_{x_i} + \hat{m}_{a_2} m_{x_i}^2 - \hat{m}_{a_1} \alpha_{x_i} - 2\hat{m}_{a_2} m_{x_i} \alpha_{x_i} - (m_{y_i} - \alpha_{y_i})\} \end{pmatrix}$$

$$\begin{pmatrix} \hat{\beta}_{a_0} \\ \hat{\beta}_{a_1} \\ \hat{\beta}_{a_2} \end{pmatrix} = M^{-1} \begin{pmatrix} \sum_{i=1}^{n} \{(m_{y_i} + \beta_{y_i}) - \hat{m}_{a_0} - \hat{m}_{a_1} m_{x_i} - \hat{m}_{a_2} m_{x_i}^2 - \hat{m}_{a_1} \beta_{x_i} - 2\hat{m}_{a_2} m_{x_i} \beta_{x_i}\} \\ \sum_{i=1}^{n} m_{x_i} \{(m_{y_i} + \beta_{y_i}) - \hat{m}_{a_0} - \hat{m}_{a_1} m_{x_i} - \hat{m}_{a_2} m_{x_i}^2 - \hat{m}_{a_1} \beta_{x_i} - 2\hat{m}_{a_2} m_{x_i} \beta_{x_i}\} \\ \sum_{i=1}^{n} m_{x_i}^2 \{(m_{y_i} + \beta_{y_i}) - \hat{m}_{a_0} - \hat{m}_{a_1} m_{x_i} - \hat{m}_{a_2} m_{x_i}^2 - \hat{m}_{a_1} \beta_{x_i} - 2\hat{m}_{a_2} m_{x_i} \beta_{x_i}\} \end{pmatrix}$$

The above unconstrained minimization technique may lead to negative spreads. Yang et al. [16] have been proposed the reformulation of fuzzy number when spreads are negative as follows

(i) If $\alpha_a < 0,\ \beta_a < 0$, then $\tilde{a} = (m_a, -\beta_a, -\alpha_a)$;

(ii) If $\alpha_a < 0,\ \beta_a > 0$, then $\tilde{a} = \begin{cases} (m_a, 0, \beta_a) & \text{for } |\beta_a| > |\alpha_a| \\ (m_a, 0, -\alpha_a) & \text{for } |\beta_a| < |\alpha_a| \end{cases}$;

(iii) If $\alpha_a > 0,\ \beta_a < 0$, then $\tilde{a} = \begin{cases} (m_a, \alpha_a, 0) & \text{for } |\alpha_a| > |\beta_a| \\ (m_a, -\beta_a, 0) & \text{for } |\alpha_a| < |\beta_a| \end{cases}$. (5)

Therefore the fitted parabolic fuzzy regression model is written as follows:

$$\hat{\tilde{Y}} = \hat{\tilde{a}}_0 + \hat{\tilde{a}}_1 \otimes \tilde{X} + \hat{\tilde{a}}_2 \otimes \tilde{X}^2 \qquad (6)$$

4. Results

The data has been generated by repeating the experiment a number of times for each of the age groups. Since initially it is difficult to consider a crisp value that never include expert's experience, knowledge and perception. In view of this, the fuzzification process has been applied to cover all the data repeated for each age group. Fuzzy logic is capable of integrating experience, perception and feelings about the status of radiographer's zinc in hair and blood respectively.

Case I: Zinc Content in Radiographer's Blood

Now the transformed data for radiographer's blood are represented as follows

Table 2: Fuzzification of the raw medical data for blood

Time of Exposure $(\tilde{X})$	*Zinc Content* $(\tilde{Y}_{NB})$	*Zinc Content* $(\tilde{Y}_{RB})$
(2.25, 0.25, 0.25)	(11.2, 1.8, 1.4)	(10.3, 0.9, 1)
(5, 0.5, 0.5)	(11.2, 1.01, 1.4)	(6.28, 0.98, 1.2)
(7.5, 0.5, 0.5)	(10.4, 1.3, 1.7)	(6.5, 1.3, 0.9)
(10, 1, 0.5)	(10.7, 1.6, 1.9)	(6.1, 0.8, 0.9)
(15, 1, 1)	(11, 1.2, 1.2)	(6.27, 0.97, 1.2)
(20, 1, 1)	(11.3, 1.4, 1.2)	(6.2, 2, 1)
(24, 1, 1)	(10.7, 1.5, 1.8)	(4.9, 0.8, 1)
(30, 2,2.5)	(8.2, 1.8, 1.3)	(5.24, 1, 1.2)

The fitted model for normal person's blood is as

$$\tilde{y}_{NB} = (10.34, 1.67, 1.52) + (0.16, 0.01, 0.072) \otimes \tilde{x} + (-0.007, 0.003, 0.0007) \otimes \tilde{x}^2 \qquad (7)$$

Hence the fitted model for radiographer's blood is

$$\tilde{y}_{RB} = (9.49, 0.78, 1.113) + (-0.375, 0.082, 0.009) \otimes \tilde{x} + (0.008, 0.0004, 0.003) \otimes \tilde{x}^2 \qquad (8)$$

A comparative assessment between radiographer's and normal person's blood can be inferred from the graph. Here two fuzzy bands are drawn for two fuzzy models with their corresponding left points, centers and right points respectively.

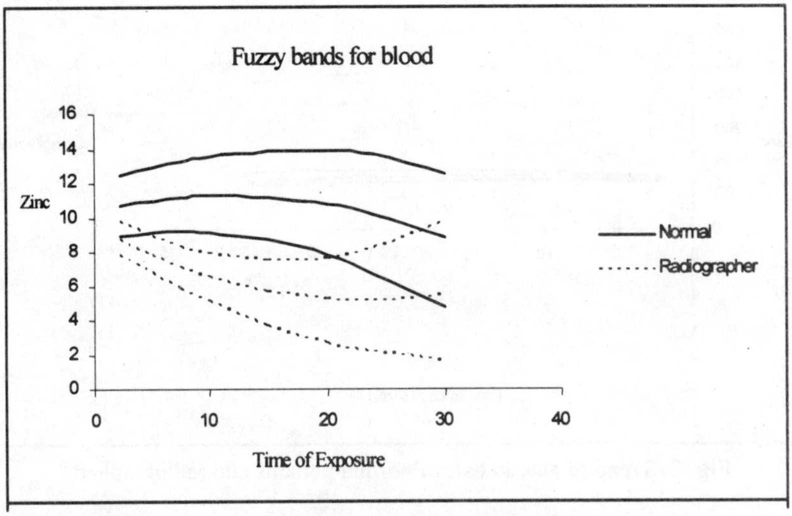

Fig.1. Graph showing reverse fuzzy bands of zinc content for normal person and radiographer's blood

***Case II:** Zinc Content in Radiographer's Hair*

Table 5: Fuzzification of the raw medical data for hair

Time of Exposure $(\widetilde{X})$	Zinc Content $(\widetilde{Y}_{NH})$	Zinc Content $(\widetilde{Y}_{RH})$
(2.25, 0.25, 0.25)	(149.8, 8.5, 11.5)	(150.3,10, 13)
(5, 0.5, 0.5)	(160.9, 11.6, 5.39)	(239.8, 42, 40.7)
(7.5, 0.5, 0.5)	(160.1, 15.7, 5.77)	(276.1, 76.8, 50.1)
(10, 1, 0.5)	(165.5, 10.2, 7.63)	(245.3, 65.9, 95)
(15, 1, 1)	(162.8, 10.2, 7.71)	(256.5, 66, 46.1)
(20, 1, 1)	(166.9, 16.6, 14.4)	(216, 65, 44)
(24, 1, 1)	(150.5, 20.3, 10.4)	(247.5, 62.5, 41.5)
(30, 2,2.5)	(149.3, 12.9, 13)	(116.7, 17.5, 11.9)

The fitted model for normal person's hair is

$$\widetilde{y}_{NH} = (148.35, 7.73, 8.59) + (2.24, 0.5, 0) \otimes \widetilde{x} + (-0.076, 0, 0.024) \otimes \widetilde{x}^2 . \quad (9)$$

Hence the fitted regression model for radiographer's hair is

$$\widetilde{y}_{RH} = (144.2, 0.48, 12.44) + (17.38, 8.23, 5.36) \otimes \widetilde{x} + (-0.604, 0.14, 0.22) \otimes \widetilde{x}^2 . \quad (10)$$

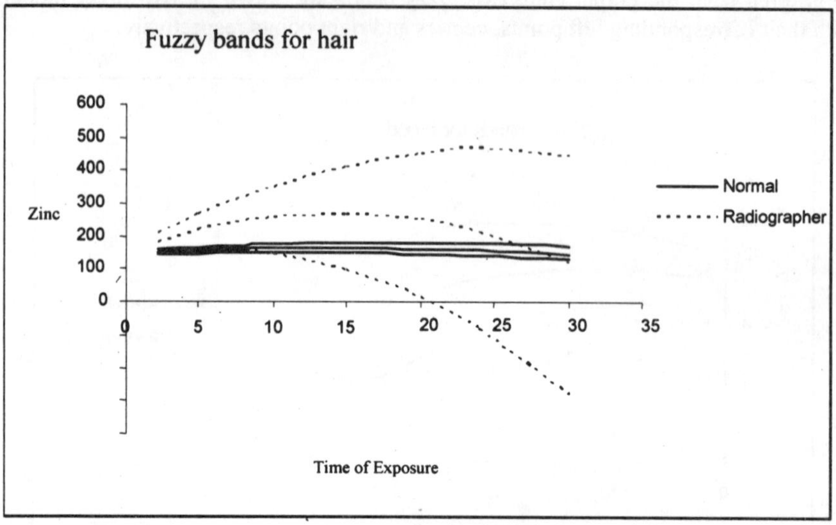

Fig 2. Trend of zinc in hair of normal persons and radiographers

From the above graphs, it is inferred that the zinc content in radiographer's hair behaves in nonlinear fashion i.e., the zinc content increases upto around 25 years age and then steadily decreases with time of exposure. On the other hand the same in case of normal person seems to be linear i.e., the zinc content in normal person's hair doesn't differ much as compared to radiographers.

5. Discussion

The imprecision is characterized within the fuzzy mathematical paradigm. Is the effect of chronic low dose radiation beneficial (hormetic) or hazardous? Researchers differ in this regard but we cannot deny the poorness of the low dose radiobiological data especially with respect to long period of studies. Major application of linear mathematical tools in the analysis of such biological events also failed to resolve the problem. The present study significantly reveals the fuzzy trend of vital bioelement like zinc in blood and hair of medical radiographers. The nature of the fitted fuzzy curve for radiation workers is parabolic in nature. it is observed from Fig1 that radiographer's blood zinc content exhibits an opposite trend to that of normal person. Initially the curve starts decreasing and attends stability. In case of hair, there is much of fluctuation in radiographer's zinc content, which initially shows increasing behaviour followed by decreasing trend. In contrast to the normal, this shows more or less stability of zinc with TOE. No doubt the data set is small but interesting application of fuzzy trend analysis gives meaningful interpretations of the observations. A large data set with detailed information will enable us to achieve more accuracy in interpreting the low-dose radiation affects on human subjects as well as in establishing the zinc as low dose radiobiological biomarker.

References

1. M. Kuzin (1991), The problem of low level radiation and hormesis in radiobiology, *Journal of Radiobiology*, 31: 16-21
2. Debjani Chakraborty (2001), Structural quantization of vagueness in linguistic expert opinions in an evaluation programme, *Fuzzy Sets and System* 119: 171-186
3. D. Dubois & H. Prade (1980), *Fuzzy Sets and Systems* (Theory and Applications), *Academic Press*, NY.
4. J. Chatterjee & K. Chaudhuri (1997), A trace metal (Zinc and Iron) study on low dose x-radiation response in rat skin, *Health Physics*, 73 (2): 362-368
5. J. Chatterjee & K. De (1993), Collagen, zinc and iron contents of rat skin irradiated with chronic low dose X-ray, *Ind. Jnl. of Med. Res.*, 98: 243-247
6. J. Chatterjee & K. De (1994), Low level x-ray exposures on rat skin: Hyperkeratinization and concomitant changes in biometal concentration, *Biol Trace Elem Res.* 46: 203-210
7. J. Chatterjee & B.B. Mukherjee (1994), Trace metal levels of X-ray technicians blood and hair, *Biol Trace Elem Res.* 46: 1-15
8. J. Fabrikant (1987), Adaptation of cell renewal systems under continuous irradiation, *Health Physics* 52: 561-570
9. J. Sagan (1989), On radiation paradigms and hormesis, *Science,* 245: 574-621
10. L.A. Zadeh (1965), Fuzzy Sets, *Information and Control* 8: 338-353
11. M.S. Yang & T. S. Lin (2002), Fuzzy least-squares linear regression analysis for fuzzy input- output data, *Fuzzy Sets and Systems* 126: 389-399
12. R.R. Paul, J. Çahtterjee, A.K. Das, M.C. Luisa, D.L.G. Miguel, K. Chaudhuri (2002), Altered elemental profile as indicator of homeostatic imbalance in pathogenesis of oral submucous fibrosis, *Biol Trace Elem Res.* 87: 45-55.
13. S. Majumdar, J. Chatterjee, K. Chaudhuri (1999), Ultrastructural and trace metal studies on radiographer's hair and nail, *Biol Trace Elem Res* 67: 127-138
14. S. Wolff (1989), Are radiation-induced effects hormetic, *Science*, 245: 571-621
15. T. Ikushima et al. (1996), Radioadaptive response; efficient repair of radiation induced DNA damage in adapted cells, *Mutation Research*, 358: 193-198
16. M.S. Yang & C.H. Ko (1997), On cluster-wise fuzzy regression analysis, *IEEE Transaction on Systems, Man and Cybernetics* B 27: 1-13

Fuzzy Logic and its Application in Technology and Management
Editors: Debjani Chakraborty, S. Nanda and D. Dutta Majumder

Neuro-Fuzzy Rule Generation Enhances Medical Knowledge Base

Suptendra Nath Sarbadhikari

Formerly at the School of Medical Science and Technology
Indian Institute of Technology, Kharagpur
Presently at TIFAC-CORE in Biomedical Technology
Amrita Vishwa Vidyapeetham, Amritapuri, and
Centre for Digital Health, Amrita Institute of Medical Sciences and Research Centre
Kochi 682 026
Email: supten@amrita.edu, supten@gmail.com

Abstract: Medical Knowledge base is dynamic. At the same time, it is also imprecise. Therefore, the inferencing and rule generation ability of neuro-fuzzy computing has a great potential of enriching the medical knowledge base.

Keywords: Medical knowledge base; rule generation and inferencing

1. Introduction

Medical Knowledge base is vast, complex and dynamic. With progressive ease of complex computations, informatics and soft computing tools are helping manage medical databases more efficiently. The inferencing and rule generation ability of neuro-fuzzy computing is actually enriching the medical knowledge base [1].

Reaching a foolproof diagnosis is never an easy job for a clinician. Often, a simple diagnostic procedure or test is overlooked and the disease eludes diagnosis. Clinical reasoning and decision-making are phased. Initially there is a clinical evaluation (history taking and physical examination), followed by precise laboratory investigations. Then integration of clinical findings and test results is done. After that, comparative benefits and risks are weighed among the alternative courses of actions, like drug interactions. Finally, the patient's preferences are taken into account, along with ethical and other considerations like cost of therapy, compliance expectations and, finally, a therapeutic plan is developed. Right from the first step (history taking) to the final one, computers can be of immense help to the clinician. For medical diagnosis, there are scopes for ambiguities in inputs, like, history (patient's description of the diseased condition), physical examinations (especially in cases of uncooperative or less intelligent patients), laboratory-tests (faulty methods or equipment). Moreover, for treatment, there are chances of drug reactions and specific allergies, patients' non-compliance of the therapy due to cost or number of times for taking medication or adverse reactions. Because of advent of new modalities of treatment, almost daily, decision making towards a particular treatment regime to be adopted for each individual patient becomes a complex process. More often, a large amount of information has to be processed, much of which is quantifiable. Intuitive thought processes involve rapid unconscious data processing and combines available information by law of average and therefore has a low intra- and inter-person consistency. So, the clinician of today should move towards analytic decision-making, which albeit typically slow, is conscious, consistent and clearly spells out the basis of decision..

2. Expression of Medical Knowledge Base

Medical knowledge is mostly represented in a qualitative and semi-quantitative (fluid) manner. Therefore, fuzzy logic can express it better. Moreover, numeric and textual information may be interspersed; different symbols can be used with same meaning, redundancy often exists, erroneous/misspelt medical terms are common, data is often sparsely distributed, and, robust preprocessing system is required to extract any kind of knowledge from even medium-sized medical data sets

The data must not only be cleaned of errors and redundancy, but also organized in a fashion that makes sense for the problem.

Connectionist Expert Systems (CES) are ANN (artificial neural network) based ES where the ANN generates inferencing rules *e.g.*, *fuzzy-MLP* where linguistic and natural form of inputs are used. Apart from that, 'rough set theory' may be used for encoding knowledge in the weights better and also GAs (genetic algorithms) may be used to optimize the search solutions better.

The medical applications [2-8] of these soft computing tools include diagnosis, bio-medical signal processing (mainly filtering and prediction), image processing, pattern recognition, knowledge discovery and data mining (KDD), control of processes (*e.g.*, anesthesia control), modeling of biological processes, and evaluation. These may be applied to the main disciplines of medicine, genetics, pharmacology, and psychology. They may also deal with applications in prostheses, or equipment related to health.

3. Neuro-Fuzzy Rule Generation for Diagnosis

3.1. Background

Fuzzy logic and neural systems have very contrasting application requirements. For example, fuzzy systems are appropriate if sufficient expert knowledge about the process is available, while neural systems are useful if sufficient process data are available or measurable. Both approaches build nonlinear systems based on bounded continuous variables. While neural systems are treated in a numeric quantitative manner, fuzzy systems are treated in a symbolic qualitative manner. Fuzzy systems, however, exhibit both symbolic and numeric features. For example, when treated as collections of objects encapsulated by linguistic labels they lend themselves to symbolic processing via rule-based operations, whereas, by referring to the definitions of the linguistic labels, their membership functions are also suitable for numeric processing. Therefore, the integration of neural and fuzzy systems leads to a symbiotic relationship where fuzzy systems purvey a powerful framework for expert knowledge representation, while neural networks provide learning capabilities and exceptional suitability for computationally efficient hardware implementations. The significance of this integration becomes even more apparent by considering their disparities.

In the clinical domain, as already mentioned in the preceding section, data may be missing for various reasons. To cite a few examples, some examinations can be risky for the patient or contraindications can exist, an urgent diagnostic decision may need to be made and some very informative but prolonged test results may have to be excluded from the feature set, or appropriate technical equipment may not be available. In such cases, the network can query the user for additional information only when it is particularly necessary to infer a decision.

3.2 An Example

A trained (fuzzy) multi-layer Perceptron [3] may be used for rule generation in '*If-Then*' form. These rules describe the extent to which a pattern belongs or not to one of the classes in terms of antecedent and consequent clauses. For this, one has to backtrack along maximal weighted paths using the trained net and utilize its input and output activations. After training, the connection weights of an MLP encode among themselves (in a distributed fashion) all the information learned about the input-output mapping. Therefore, any link weight with a large magnitude reflects a strong correlation between the connecting neurons. This property is utilized in evaluating the importance of an input node (*feature*) on an output layer mode (*decision*). For this the path weights from each output node to each input node through the various hidden nodes is computed. The maximum output-input path (through any hidden node) denotes the importance of that input feature for arriving at the corresponding output decision. One or more hidden layers may be considered for evaluating the path weights. An increase in the number of hidden layers leads to an increase in the possible variety of paths being generated through the various hidden nodes in the different layers. A heuristic allows selection of those *currently active* input neurons contributing *the most* to the final conclusion (among those lying along the maximum weighted paths to the output node *j*) as the clauses of the antecedent part of a rule. Hence, it enables the *currently active* test pattern inputs (current evidence) to influence the generated "knowledge base" (connection weights learned during training) in producing a rule to justify the *current* inference. The complete *If* part of the rule is found by *ANDing* the clauses corresponding to each of the features.

3.3. Rule Validation

The rules so generated must be validated by domain experts, *viz.*, clinicians, for accuracy, fidelity (number of times network and rule base output agree), confusion (should be restricted within minimum number of classes), coverage (a rule base with smaller uncovered region *i.e.*, percent test set for which no rules are fired, is better), rule base size (smaller the number of rules, more compact is the rule base), and, certainty (confidence of rules).

3.4. Limitations

A severe drawback in this approach lies in the amount of heuristics involved in writing the grammar. Also, it is almost impossible for an untrained clinician, to inspect the grammar and make the necessary modifications. Knowledge-based approaches purvey a way of increasing the involvement of the expert (doctor) in the design process and also allow the utilization of contextual information to a larger degree. These systems apply a body of knowledge ("knowledge-base") to the input data and subsequently derived facts ("data-base"). Typically, the knowledge base comprises IF<premise>THEN<action> type rules which presumably reflect the human expertise. Another advantage of this type of flexible problem solving approach is that the collection of rules may be modified from time to time.

4. Conclusions and Future Directions

For various clinical disorders pertaining to all the systems and organs of the body, many decision support systems are being built [4-7]. Many of them are incorporating neuro-fuzzy networks for inferencing and rule generation. Albeit some work has already been done in this arena, a lot more is yet to be done, especially for the tropical diseases so common in India [8].

Interested readers may go through the detailed expositions provided in [9] and [10].

References

1. S.N. Sarbadhikari & S.K. Pal (2002), Automated Techniques for Identifying Depression from EEG, In: C.T. Leondes, Ed., *Handbook of Computational Methods in Biomaterials, Biotechnology and Biomedical Systems*, Kluwer Academic Publishers, Vol. 4, Chapter 3: 51-81
2. G.J. Klir & Bo Yuan (1997), Fuzzy Sets and Fuzzy Logic: Theory and Applications, Prentice Hall of India, New Delhi: 443 – 450.
3. S. Mitra, S.N. Sarbadhikari & S.K. Pal (1996), An MLP-based model for identifying qEEG in depression. *Int. J. Biomed. Comput.* 43: 179 – 187.
4. E. Sanchez (1992), Fuzzy logic knowledge systems and artificial neural networks in medicine and biology, in: R.R. Yager & L.A. Zadeh, eds., *An Introduction to Fuzzy Logic Applications in Intelligent Systems*, Kluwer, Dordrecht-Boston, pp 235 – 252.
5. P. Eklund, M. Fogstrm & J. Forsstrm (1992), A generic neuro-fuzzy tool for developing medical decision support, in: P. Eklund ed., *Proceedings of the MEPP'92*, Abo Akademi Press, Abo., pp 11 – 27.
6. R. Brause & F. Friedrich (2000), A Neuro-fuzzy approach as medical diagnostic interface, Proc. ESANN-2000, Elsevier, pp 201 – 206.
7. M. Pfaff, K. Weller, D. Woetzel, R. Guthke, K. Schroeder, G. Stein, R. Pohlmeier & J. Vienken (2000), Prediction of cardiovascular risk in hemodialysis patients by data mining, *Methods Inf Med.* 43: 106 – 113.
8. S.N. Sarbadhikari (2005), The state of Medical Informatics in India: A roadmap for optimal organization, *J. Medical Systems*, 29: 125 – 141.
9. S. Mitra & Y. Hayashi (2000), Neuro–Fuzzy Rule Generation: Survey in Soft Computing Framework, *IEEE Trans. Neural Networks*, 11: 748 – 768.
10. S.K. Pal, Soft computing, machine intelligence and data mining, *http://www.ewh.ieee.org/r10/nsw/subpages/ch_comp/ch_comp_archive/IEEE-DV1.PPT* (Accessed July 2006).

